CAMBRIDGE LIBRARY COLLECTION

Books of enduring scholarly value

Life Sciences

Until the nineteenth century, the various subjects now known as the life sciences were regarded either as arcane studies which had little impact on ordinary daily life, or as a genteel hobby for the leisured classes. The increasing academic rigour and systematisation brought to the study of botany, zoology and other disciplines, and their adoption in university curricula, are reflected in the books reissued in this series.

The Geographical Distribution of Animals

Alfred Russel Wallace (1823–1913) was a British biologist and explorer whose theories of evolution, arrived at independently, caused Darwin to allow their famous joint paper to go forward to the Linnean Society in 1858. Considered the nineteenth century's leading expert on the geographical distribution of animals, Wallace carried out extensive fieldwork in areas as diverse as North and South America, Africa, China, India and Australia to document the habitats, breeding, migration and feeding behaviour of thousands of species around the world, and the influence of environmental conditions on their survival. First published in 1876, this two-volume set presents Wallace's findings, and represents a landmark in the study of zoology, evolutionary biology and palaeontology which remains relevant to scholars in these fields today. Volume 1 focuses on the classification of species, migration processes, factors influencing extinction, and the characteristics of a range of zoological regions worldwide.

Cambridge University Press has long been a pioneer in the reissuing of out-of-print titles from its own backlist, producing digital reprints of books that are still sought after by scholars and students but could not be reprinted economically using traditional technology. The Cambridge Library Collection extends this activity to a wider range of books which are still of importance to researchers and professionals, either for the source material they contain, or as landmarks in the history of their academic discipline.

Drawing from the world-renowned collections in the Cambridge University Library, and guided by the advice of experts in each subject area, Cambridge University Press is using state-of-the-art scanning machines in its own Printing House to capture the content of each book selected for inclusion. The files are processed to give a consistently clear, crisp image, and the books finished to the high quality standard for which the Press is recognised around the world. The latest print-on-demand technology ensures that the books will remain available indefinitely, and that orders for single or multiple copies can quickly be supplied.

The Cambridge Library Collection will bring back to life books of enduring scholarly value (including out-of-copyright works originally issued by other publishers) across a wide range of disciplines in the humanities and social sciences and in science and technology.

The Geographical Distribution of Animals

With a Study of the Relations of Living and Extinct Faunas

VOLUME 1

ALFRED RUSSEL WALLACE

CAMBRIDGE
UNIVERSITY PRESS

CAMBRIDGE UNIVERSITY PRESS

Cambridge, New York, Melbourne, Madrid, Cape Town,
Singapore, São Paolo, Delhi, Tokyo, Mexico City

Published in the United States of America by Cambridge University Press, New York

www.cambridge.org
Information on this title: www.cambridge.org/9781108037846

© in this compilation Cambridge University Press 2011

This edition first published 1876
This digitally printed version 2011

ISBN 978-1-108-03784-6 Paperback

THE GEOGRAPHICAL

DISTRIBUTION OF ANIMALS.

VOL. I.

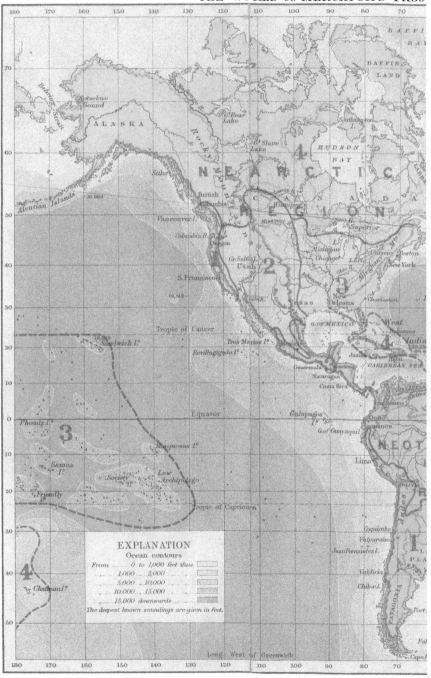

EXPLANATION

Ocean contours

From	0 to 1,000 feet thus	
	1,000 „ 5,000 „	
	5,000 „ 10,000 „	
	10,000 „ 15,000 „	
	15,000 downwards	

The deepest known soundings are given in feet.

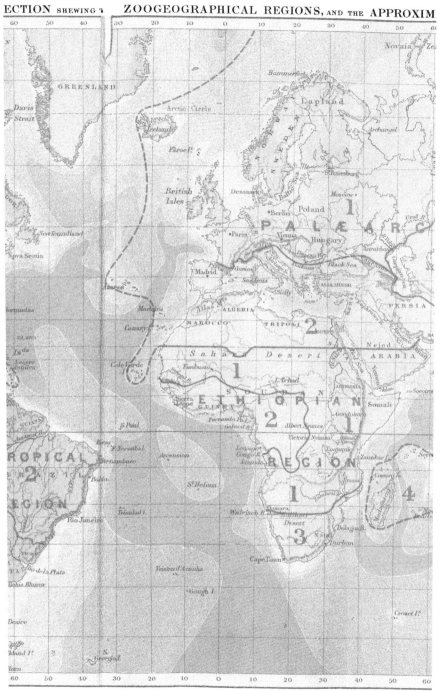

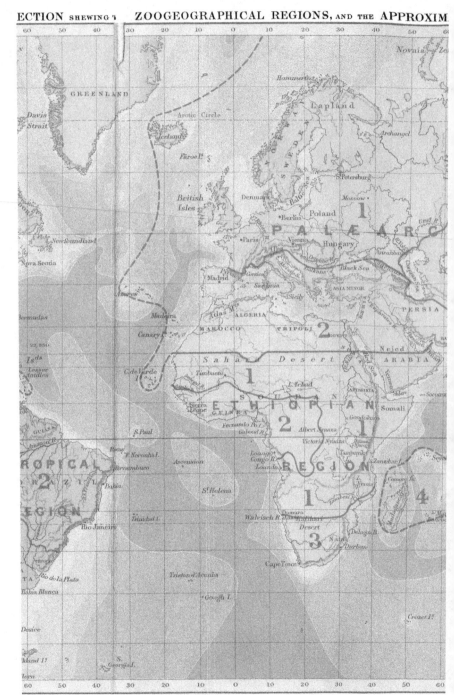

GREENLAND

Davis
Strait

Arctic Circle

Iceland

Faroe Is

British
Isles

Newfoundland

Nova Scotia

Bermudas

22,930

1s ds
Lesser
Antilles

GUIANA

Amazon R.

ROPICAL

2
BRAZIL

REGION

Rio Janeiro

Rio de la Plata
Bahia Blanca

Desire

kland Is

Horn

Azores

Madeira

Canary Is

C. de Verde
Is

S. Paul

Rocas
F. Noronha I.
Pernambuco

Bahia

Trinidad I.

Ascension

St Helena

Tristan d'Acunha

Gough I.

S.
Georgia I.

Hammerfest

Lapland

S. Petersburg

Denmark

Berlin

Paris

Madrid

Atlas Mts

Marocco

Timbuctu

Sierra
Leone

GUINEA

Fernando Po I.
Gaboon R.

Loango
Congo R.
Loando

Saha

Corsica
Sardinia

ALGERIA

TRIPOLI

Desert

SOUDAN

1

L.Tchad

ETHIOPIAN

2

Albert Nyanza

Victoria Nyanza

REGION

Damara
Walvisch R.
Desert

1

3

Cape Town

PALÆARC

1

Poland
Moscow

Vienna
Hungary

Black Sea

ASIA MINOR

Sicily

2

NO
S
W
E
D
E
N

Archangel

Ural R.

Astrakhan

PERSIA

Nejed

ARABIA

ABYSSINIA

Somali

Gondokoro

Tanganyika
Zanzibar

L.Nyassa

Zambesi R.

Delagoa B.

Natal
Durban

Socotra

Comoro Is

4

Croset Is

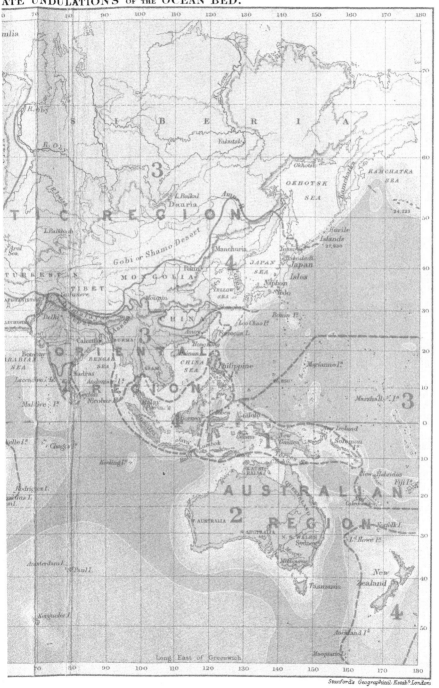

THE GEOGRAPHICAL

DISTRIBUTION OF ANIMALS

WITH A STUDY OF

THE RELATIONS OF LIVING AND EXTINCT FAUNAS

AS ELUCIDATING THE

PAST CHANGES OF THE EARTH'S SURFACE.

BY

ALFRED RUSSEL WALLACE,

AUTHOR OF "THE MALAY ARCHIPELAGO," ETC.

WITH MAPS AND ILLUSTRATIONS.

IN TWO VOLUMES.—VOLUME I.

London:

MACMILLAN AND CO.

1876.

LONDON:
R. CLAY, SONS, AND TAYLOR, PRINTERS,
BREAD STREET HILL.

PREFACE.

The present work is an attempt to collect and summarize the existing information on the Distribution of Land Animals; and to explain the more remarkable and interesting of the facts, by means of established laws of physical and organic change.

The main idea, which is here worked out in some detail for the whole earth, was stated sixteen years ago in the concluding pages of a paper on the "Zoological Geography of the Malay Archipelago," which appeared in the *Journal of Proceedings of the Linnean Society* for 1860 ; and again, in a paper read before the Royal Geographical Society in 1863, it was briefly summarized in the following passage :—

"My object has been to show the important bearing of researches into the natural history of every part of the world, upon the study of its past history. An accurate knowledge of any groups of birds or of insects and of their geographical distribution, may enable us to map out the islands and continents of a former epoch,—the amount of difference that exists between the animals of adjacent districts being closely related to preceding geological changes. By the collection of such minute facts, alone, can we hope to fill up a great gap in the

past history of the earth as revealed by geology, and obtain
some indications of the existence of those ancient lands which
now lie buried beneath the ocean, and have left us nothing but
these living records of their former existence."

The detailed study of several groups of the birds and insects
collected by myself in the East, brought prominently before me
some of the curious problems of Geographical Distribution;
but I should hardly have ventured to treat the whole subject,
had it not been for the kind encouragement of Mr. Darwin and
Professor Newton, who, about six years ago, both suggested that
I should undertake the task. I accordingly set to work; but
soon became discouraged by the great dearth of materials in
many groups, the absence of general systematic works, and the
excessive confusion that pervaded the classification. Neither
was it easy to decide on any satisfactory method of treating
the subject. During the next two years, however, several important catalogues and systematic treatises appeared, which
induced me to resume my work; and during the last three years
it has occupied a large portion of my time.

After much consideration, and some abortive trials, an outline
plan of the book was matured; and as this is, so far as I am
aware, quite novel, it will be well to give a few of the reasons
for adopting it.

Most of the previous writings on Geographical Distribution
appeared to me to be unsatisfactory, because they drew their
conclusions from a more or less extensive *selection* of facts; and
did not clearly separate groups of facts of unequal value, or
those relating to groups of animals of unequal rank. As an
example of what is meant, I may refer to Mr. Andrew Murray's
large and valuable work on the Geographical Distribution of
Mammalia, in which an immense number of coloured maps are

used to illustrate the distribution of various groups of animals. These maps are not confined to groups of any fixed rank, but are devoted to a selection of groups of various grades. Some show the range of single species of a genus—as the lion, the tiger, the puma, and a species of fox ; others are devoted to sections of genera,—as the true wolves ; others to genera,—as the hyænas, and the bears; others to portions of families,—as the flying squirrels, and the oxen with the bisons; others to families, —as the Mustelidæ, and the Hystricidæ; and others to groups of families or to orders,—as the Insectivora, and the opossums with the kangaroos. But in no one grade are all the groups treated alike. Many genera are wholly unnoticed, while several families are only treated in combination with others, or are represented by some of the more important genera.

In making these observations I by no means intend to criticise Mr. Murray's book, but merely to illustrate by an example, the method which has been hitherto employed, and which seems to me not well adapted to enable us to establish the foundations of the science of distribution on a secure basis To do this, uniformity of treatment appeared to me essential, both as a matter of principle, and to avoid all imputation of a partial selection of facts, which may be made to prove anything. I determined, therefore, to take in succession every well-established family of terrestrial vertebrates, and to give an account of the distribution of all its component genera, as far as materials were available. Species, as such, are systematically disregarded, —firstly, because they are so numerous as to be unmanageable ; and, secondly, because they represent the most recent modifications of form, due to a variety of often unknown causes, and are therefore not so clearly connected with geographical changes as are the natural groups of species termed genera ; which may be considered to represent the average and more permanent

distribution of an organic type, and to be more clearly influenced
by the various known or inferred changes in the organic and
physical environment.

This systematic review of the distribution of families and
genera, now forms the last part of my book—Geographical
Zoology; but it was nearly the first written, and the copious.
materials collected for it enabled me to determine the zoo-
geographical divisions of the earth (regions and sub-regions) to
be adopted.   I next drew up tables of the families and genera
found in each region and sub-region ; and this afforded a basis
for the geographical treatment of the subject—Zoological Geography—the most novel, and perhaps the most useful and
generally interesting part of my work. While this was in progress
I found it necessary to make a careful summary of the distribution of extinct Mammalia.   This was a difficult task, owing
to the great uncertainty that prevails as to the affinities of many
of the fossils, and my want of practical acquaintance with
Palæontology; but having carefully examined and combined
the works of the best authors, I have given what I believe is
the first connected sketch of the relation of extinct Mammalia
to the distribution of living groups, and have arrived at some
very interesting and suggestive results.

It will be observed that man is altogether omitted from
the series of the animal kingdom as here given, and some explanation of this omission may perhaps be required.   If the
genus *Homo* had been here treated like all other genera, nothing
more than the bare statement—" universally distributed "—
could have been given;—and this would inevitably have pro-
voked the criticism that it conveyed no information.   If, on the
other hand, I had given an outline of the distribution of the
varieties or *races* of man, I should have departed from the plan
of my work for no sufficient reason.   Anthropology is a science

by itself; and it seems better to omit it altogether from a zoological work, than to treat it in a necessarily superficial manner.

The best method of illustrating a work of this kind was a matter requiring much consideration. To have had a separate coloured or shaded map for each family would have made the work too costly, as the terrestrial vertebrates alone would have required more than three hundred maps. I had also doubts about the value of this mode of illustration, as it seemed rather to attract attention to details than to favour the development of general views. I determined therefore to adopt a plan, suggested in conversation by Professor Newton; and to have one general map, showing the regions and sub-regions, which could be referred to by means of a series of numbers. These references I give in the form of diagrammatic headings to each family; and, when the map has become familiar, these will, I believe, convey at a glance a body of important information.

Taking advantage of the recent extension of our knowledge of the depths of the great oceans, I determined to give upon this map a summary of our knowledge of the contours of the ocean bed, by means of tints of colour increasing in intensity with the depth. Such a map, when it can be made generally accurate, will be of the greatest service in forming an estimate of the more probable changes of sea and land during the Tertiary period; and it is on the effects of such changes that any satis-factory explanation of the facts of distribution must to a great extent depend.

Other important factors in determining the actual distribution of animals are, the zones of altitude above the sea level, and the strongly contrasted character of the surface as regards vege-tation—a primary condition for the support of animal life. I

therefore designed a series of six maps of the regions, drawn on
a uniform scale, on which the belts of altitude are shown by
contour-shading, while the forests, pastures, deserts, and perennial snows, are exhibited by means of appropriate tints of colour.

These maps will, I trust, facilitate the study of geographical
distribution as a science, by showing, in some cases, an adequate
cause in the nature of the terrestrial surface for the actual distribution of certain groups of animals. As it is hoped they will
be constantly referred to, double folding has been avoided, and
they are consequently rather small; but Mr. Stanford, and his
able assistant in the map department, Mr. Bolton, have taken
great care in working out the details from the latest observations;
and this, combined with the clearness and the beauty of their
execution, will I trust render them both interesting and instructive.

In order to make the book more intelligible to those readers
who have no special knowledge of systematic zoology, and to
whom most of the names with which its pages are often crowded
must necessarily be unmeaning, I give a series of twenty plates,
each one illustrating at once the physical aspect and the special
zoological character of some well-marked division of a region.
Great care has been taken to associate in the pictures, such species
only as do actually occur together in nature; so that each plate
represents a scene which is, at all events, not an impossible one.
The species figured all belong to groups which are either peculiar to, or very characteristic of, the region whose zoology they
illustrate; and it is hoped that these pictures will of themselves
serve to convey a notion of the varied types of the higher
animals in their true geographical relations. The artist, Mr. J.
B. Zwecker, to whose talent as a zoological draughtsman and
great knowledge both of animal and vegetable forms we are
indebted for this set of drawings, died a few weeks after he

had put the final touches to the proofs. He is known to many
readers by his vigorous illustrations of the works of Sir Samuel
Baker, Livingstone, and many other travellers,—but these, his
last series of plates, were, at my special request, executed with
a care, delicacy, and artistic finish, which his other designs
seldom exhibit. It must, however, be remembered, that the
figures of animals here given are not intended to show specific
or generic characters for the information of the scientific zoologist, but merely to give as accurate an idea as possible, of some
of the more remarkable and more restricted types of beast and
bird, amid the characteristic scenery of their native country;—
and in carrying out this object there are probably few artists
who would have succeeded better than Mr. Zwecker has
done.

The general arrangement of the separate parts of which the
work is composed, has been, to some extent, determined by
the illustrations and maps, which all more immediately belong
to Part III. It was at first intended to place this part last, but
as this arrangement would have brought all the illustrations
into the second volume, its place was changed,—perhaps in
other respects for the better, as it naturally follows Part II.
Yet for persons not well acquainted with zoology, it will perhaps be advisable to read the more important articles of Part
IV. (and especially the observations at the end of each order)
after Part II., thus making Part III. the conclusion of the
work.

Part IV. is, in fact, a book of reference, in which the distribution of all the families and most of the genera of the higher
animals, is given in systematic order. Part III. is treated
somewhat more popularly; and, although it is necessarily
crowded with scientific names (without which the inferences

and conclusions would have nothing solid to rest on), these may be omitted by the non-scientific reader, or merely noted as a certain number or proportion of peculiar generic types. Many English equivalents to family and generic names are, however, given; and, assisted by these, it is believed that any reader capable of understanding Lyell's "Principles," or Darwin's "Origin," will have no difficulty in following the main arguments and appreciating the chief conclusions arrived at in the present work.

To those who are more interested in facts than in theories, the book will serve as a kind of dictionary of the geography and affinities of animals. By means of the copious Index, the native country, the systematic position, and the numerical extent of every important and well established genus of land-animal may be at once discovered;—information now scattered through hundreds of volumes.

In the difficult matters of synonymy, and the orthography of generic names, I have been guided rather by general utility than by any fixed rules. When I have taken a whole family group from a modern author of repute, I have generally followed his nomenclature throughout. In other cases, I use the names which are to be found in a majority of modern authors, rather than follow the strict rule of priority in adopting some newly discovered appellation of early date. In orthography I have adopted all such modern emendations as seem coming into general use, and which do not lead to inconvenience ; but where the alteration is such as to completely change the pronunciation and appearance of a well-known word, I have not adopted it. I have also thought it best to preserve the initial letter of well-known and old-established names, for convenience of reference to the Indices of established works. As an example I may refer to *Enicurus*,—a name which has been in use nearly half a century, and which is to be found under the letter *E*, in Jerdon's Birds of India, Blyth's Catalogue, Bonaparte's Conspectus, and the Proceedings of the Zoological Society of London down to 1865. Classicists now write *Henicurus* as the correct form; but this seems to me one of those cases in which orthographical accuracy should give way to priority, and still more to convenience.

In combining and arranging so much detail from such varied sources, many errors and omissions must doubtless have occurred. Owing to my residence at a distance from the scientific libraries of the metropolis, I was placed at a great disadvantage; and I could hardly have completed the work at all, had I not been permitted to have a large number of volumes at once, from the library of the Zoological Society of London, and to keep them for months together;—a privilege for which I return my best thanks to Mr. Sclater the Secretary, and to the Council.

Should my book meet with the approval of working naturalists, I venture to appeal to them, to assist me in rendering any future editions more complete, by sending me (to the care of my publishers) notes of any important omissions, or corrections of any misstatements of fact; as well as copies of any of their papers or essays, and especially of any lists, catalogues, and monographs, containing information on the classification or distribution of living or extinct animals.

To the many friends who have given me information or assistance I beg to tender my sincere thanks. Especially am I indebted to Professor Newton, who not only read through much of my rough MSS., but was so good as to make numerous corrections and critical notes. These were of great value to me, as they often contained or suggested important additional matter, or pointed out systematic and orthographical inaccuracies.

Professor Flower was so good as to read over my chapters on extinct animals, and to point out several errors into which I had fallen.

Dr. Günther gave me much valuable information on the classification of reptiles, marking on my lists the best established and most natural genera, and referring me to reliable sources of information.

I am also greatly indebted to the following gentlemen for detailed information on special subjects :—

To Sir Victor Brooke, for a MS. arrangement of the genera of Bovidæ, with the details of their distribution:

To Mr. Dresser, for lists of the characteristic birds of Northern and Arctic Europe:

To Dr. Hooker, for information on the colours and odours of New Zealand plants:

To Mr. Kirby, for a list of the butterflies of Chili:

To Professor Mivart, for a classification of the Batrachia, and an early proof of his article on "Apes" in the Encyclopedia Britannica:

To Mr. Salvin, for correcting my list of the birds of the Galapagos, and for other assistance:

To Mr. Sharpe, for MS. lists of the birds of Madagascar and the Cape Verd Islands:

To Canon Tristram, for a detailed arrangement of the difficult family of the warblers,—Sylviidæ:

To Viscount Walden, for notes on the systematic arrangement of the Pycnonotidæ and Timaliidæ, and for an early proof of his list of the birds of the Philippine Islands.

I also have to thank many naturalists, both in this country and abroad, who have sent me copies of their papers; and I trust they will continue to favour me in the same manner.

An author may easily be mistaken in estimating his own work. I am well aware that this first outline of a great subject is, in parts, very meagre and sketchy; and, though perhaps overburthened with some kinds of detail, yet leaves many points most inadequately treated. It is therefore with some hesitation that I venture to express the hope that I have made some approach to the standard of excellence I have aimed at;—which was, that my book should bear a similar relation to the eleventh and twelfth chapters of the " Origin of Species," as Mr. Darwin's " Animals and Plants under Domestication " does to the first chapter of that work. Should it be judged worthy of such a rank, my long, and often wearisome labours, will be well repaid.

MARCH, 1876.

CONTENTS OF THE FIRST VOLUME.

PART I.

THE PRINCIPLES AND GENERAL PHENOMENA OF DISTRIBUTION.

CHAPTER VII.

EXTINCT MAMMALIA OF THE NEW WORLD.

CHAPTER VIII.

VARIOUS EXTINCT ANIMALS ;—AND ON THE ANTIQUITY OF THE GENERA OF INSECTS AND LAND-MOLLUSCA.

PART III.

ZOOLOGICAL GEOGRAPHY :   A REVIEW OF THE CHIEF FORMS OF LIFE IN THE SEVERAL REGIONS AND SUB-REGIONS, WITH THE INDICATIONS THEY AFFORD OF GEOGRAPHICAL MUTATIONS.

CHAPTER IX.

THE ORDER OF SUCCESSION OF THE REGIONS.—COSMOPOLITAN GROUPS OF ANIMALS.—TABLES OF DISTRIBUTION.

CHAPTER XIII.

THE AUSTRALIAN REGION.

MAPS AND ILLUSTRATIONS IN VOL. I.

ERRATA IN VOL. I.

I have detected several misprints and small errors in the final impression, and Dr. Meyer, who has translated the work into German, has kindly communicated all that he has noticed.   It is not thought necessary to give here all the smaller orthographical errors, most of which will be corrected in the Index. The following seem, however, to be of sufficient importance to justify me in asking my readers to correct them in their copies.

Page 93, 12 lines from foot, *for* Hocco *read* Hoazin.
,, 97, line 2, *for* Hocco *read* Hoazin.
,, 147, 13 lines from foot, *for* three-handed *read* three-banded.
,, 177, line 6, *for* Lycænidæ *read* Zygænidæ.
,, 183, line 20, *for* third *read* fourth.
,, 238, line 18, *for* Spirigidea *read* Sphingidea.
,, 242, *insert* | 92a | Tamias | 1 | All Northern Asia | N. America.
,, 245, last line, *insert in 2nd column* (6).
,, 309, line 20, *for* Motacilla *read* Budytes.
,, 327, 12 lines from foot, *after* Hindostan *read* and.
,, 331, last line, for *Icthyopsis* read *Icthyophis*.
,, 340, line 15, for *Edolius* read *Bhringa*.
,, 348, line 17, *for* Flores *read* New Guinea.
,, 371, 11 lines from foot, *for* and Borneo *read* Borneo and Philippines.
,, 391, 10 lines from foot, *after* Celebes *add* and the Papuan Islands.
,, 391, 9 lines from foot, *omit* New Guinea or.
,, 414, 6 lines from foot, for *Epimachus* read *Seleucides*.
,, 415, line 10   *for*   ditto   *read*   ditto.
,, 427, line 20, *after* Celebes *add* and on some of the Philippine Islands.
,, 427, 5 lines from foot, *for* tusks *read* jaw.
,, 462, 15 lines from foot, *for* p. 156 *read* p. 166.
,, 474, 9 lines from foot, *after* Celebes *add* Papua.

GEOGRAPHICAL DISTRIBUTION
OF ANIMALS.

PART I.

THE PRINCIPLES AND GENERAL PHENOMENA
OF DISTRIBUTION.

CHAPTER I.

IT is a fact within the experience of most persons, that the various species of animals are not uniformly dispersed over the surface of the country. If we have a tolerable acquaintance with any district, be it a parish, a county, or a larger extent of territory, we soon become aware that each well-marked portion of it has some peculiarities in its animal productions. If we want to find certain birds or certain insects, we have not only to choose the right season but to go to the right place. If we travel beyond our district in various directions we shall almost certainly meet with something new to us; some species which we were accustomed to see almost daily will disappear, others which we have never seen before will make their appearance. If we go very far, so as to be able to measure our journey by degrees of latitude and longitude and to perceive important changes of climate and vegetation, the differences in the forms of animal life will become greater; till at length we shall come to a country where almost everything will be new, all the familiar creatures of our own district being replaced by others more or less differing from them.

If we have been observant during our several journeys, and have combined and compared the facts we have collected, it will become apparent that the change we have witnessed has been of two distinct kinds. In our own and immediately surrounding districts, particular species appeared and disappeared because

the soil, the aspect, or the vegetation, was adapted to them or
the reverse. The marshes, the heaths, the woods and forests,
the chalky downs, the rocky mountains, had each their peculiar
inhabitants, which reappeared again and again as we came to
tracts of country suitable for them. But as we got further away
we began to find that localities very similar to those we had
left behind were inhabited by a somewhat different set of species;
and this difference increased with distance, notwithstanding
that almost identical external conditions might be often met
with. The first class of changes is that of *stations*; the second
that of *habitats*. The one is a *local*, the other a *geographical*
phenomenon. The whole area over which a particular animal
is found may consist of any number of *stations*, but rarely of
more than one *habitat*. Stations, however, are often so extensive
as to include the entire range of many species. Such are the
great seas and oceans, the Siberian or the Amazonian forests,
the North African deserts, the Andean or the Himalayan
highlands.

There is yet another difference in the nature of the change
we have been considering. The new animals which we meet
with as we travel in any direction from our starting point, are
some of them very much like those we have left behind us,
and can be at once referred to familiar types; while others
are altogether unlike anything we have seen at home. When
we reach the Alps we find another kind of squirrel, in Southern Italy a distinct mole, in Southern Europe fresh warblers
and unfamiliar buntings. We meet also with totally new
forms; as the glutton and the snowy owl in Northern, the genet
and the hoopoe in Southern, and the saiga antelope and
collared pratincole in Eastern Europe. The first series are
examples of what are termed *representative species*, the second
of distinct groups or *types* of animals. The one represents a
comparatively recent modification, and an origin in or near the
locality where it occurs; the other is a result of very ancient
changes both organic and inorganic, and is connected with some
of the most curious and difficult of the problems we shall have
to discuss.

Having thus defined our subject, let us glance at the opinions
that have generally prevailed as to the nature and causes of
the phenomena presented by the geographical distribution of
animals.

It was long thought, and is still a popular notion, that the
manner in which the various kinds of animals are dispersed
over the globe is almost wholly due to diversities of climate and
of vegetation. There is indeed much to favour this belief. The
arctic regions are strongly characterised by their white bears
and foxes, their reindeer, ermine, and walruses, their white
ptarmigan, owls, and falcons; the temperate zone has its foxes
and wolves, its rabbits, sheep, beavers, and marmots, its sparrows
and its song birds; while tropical regions alone produce apes and
elephants, parrots and peacocks, and a thousand strange quadrupeds and brilliant birds which are found nowhere in the cooler
regions. So the camel, the gazelle and the ostrich live in the
desert; the bison on the prairie; the tapir, the deer, and the
jaguar in forests. Mountains and marshes, plains and rocky
precipices, have each their animal inhabitants; and it might well
be thought, in the absence of accurate inquiry, that these and
other differences would sufficiently explain why most of the
regions and countries into which the earth is popularly divided
should have certain animals peculiar to them and should want
others which are elsewhere abundant.

A more detailed and accurate knowledge of the productions of
different portions of the earth soon showed that this explanation
was quite insufficient; for it was found that countries exceedingly similar in climate and all physical features may yet have
very distinct animal populations. The equatorial parts of Africa
and South America, for example, are very similar in climate
and are both covered with luxuriant forests, yet their animal life
is widely different; elephants, apes, leopards, guinea-fowls
and touracos in the one, are replaced by tapirs, prehensile-
tailed monkeys, jaguars, curassows and toucans in the other.
Again, parts of South Africa and Australia are wonderfully
similar in their soil and climate; yet one has lions, antelopes,
zebras and giraffes; the other only kangaroos, wombats, phalan-

gers and mice. In like manner parts of North America and Europe are very similar in all essentials of soil climate and vegetation, yet the former has racoons, opossums, and humming-birds; while the latter possesses moles, hedgehogs and true fly-catchers. Equally striking are the facts presented by the distribution of many large and important groups of animals. Marsupials (opossums, phalangers &c.) are found from temperate Van Diemen's land to the tropical islands of New Guinea and Celebes, and in America from Chili to Virginia. No crows exist in South America, while they inhabit every other part of the world, not excepting Australia. Antelopes are found only in Africa and Asia; the sloths only in South America; the true lemurs are confined to Madagascar, and the birds-of-paradise to New Guinea.

If we examine more closely the distribution of animals in any extensive region, we find that different, though closely allied species, are often found on the opposite sides of any considerable barrier to their migration. Thus, on the two sides of the Andes and Rocky Mountains in America, almost all the mammalia, birds, and insects are of distinct species. To a less extent, the Alps and Pyrenees form a similar barrier, and even great rivers and river plains, as those of the Amazon and Ganges, separate more or less distinct groups of animals. Arms of the sea are still more effective, if they are permanent; a circumstance in some measure indicated by their depth. Thus islands far away from land almost always have very peculiar animals found nowhere else; as is strikingly the case in Madagascar and New Zealand, and to a less degree in the West India islands. But shallow straits, like the English Channel or the Straits of Malacca, are not found to have the same effect, the animals being nearly or quite identical on their opposite shores. A change of climate or a change of vegetation may form an equally effective barrier to migration. Many tropical and polar animals are pretty accu-rately limited by certain isothermal lines; and the limits of the great forests in most parts of the world strictly determine the ranges of many species.

Naturalists have now arrived at the conclusion, that by some slow process of development or transmutation, all animals have been produced from those which preceded them ; and the old notion that every species was specially created as they now exist, at a particular time and in a particular spot, is abandoned as opposed to many striking facts, and unsupported by any evidence. This modification of animal forms took place very slowly, so that the historical period of three or four thousand years has hardly produced any perceptible change in a single species. Even the time since the last glacial epoch, which on the very lowest estimate must be from 50,000 to 100,000 years, has only served to modify a few of the higher animals into very slightly different species. The changes of the forms of animals appear to have accompanied, and perhaps to have depended on, changes of physical geography, of climate, or of vegetation ; since it is evident that an animal which is well adapted to one condition of things will require to be slightly changed in constitution or habits, and therefore generally in form, structure, or colour, in order to be equally well adapted to a changed condition of surrounding circumstances. Animals multiply so rapidly, that we may consider them as continually trying to extend their range ; and thus any new land raised above the sea by geological causes becomes immediately peopled by a crowd of competing inhabitants, the strongest and best adapted of which alone succeed in maintaining their position.

If we keep in view these facts—that the minor features of the earth's surface are everywhere slowly changing; that the forms, and structure, and habits of all living things are also slowly changing; while the great features of the earth, the continents, and oceans, and loftiest mountain ranges, only change after very long intervals and with extreme slowness; we must see that the present distribution of animals upon the several parts of the earth's surface is the final product of all these wonderful revolutions in organic and inorganic nature. The greatest and most radical differences in the productions of any part of the globe must be dependent on isolation by the most effectual and most permanent barriers. That ocean which has remained broadest and deepest from the most remote geological epoch

will separate countries the productions of which most widely
and radically differ; while the most recently-depressed seas,
or the last-formed mountain ranges, will separate countries
the productions of which are almost or quite identical. It
will be evident, therefore, that the study of the distribution
of animals and plants may add greatly to our knowledge
of the past history of our globe. It may reveal to us, in a
manner which no other evidence can, which are the oldest
and most permanent features of the earth's surface, and which
the newest. It may indicate the existence of islands or continents now sunk beneath the ocean, and which have left no
record of their existence save the animal and vegetable productions which have migrated to adjacent lands. It thus
becomes an important adjunct to geology, which can rarely do
more than determine what lands have been raised above the
waters, under what conditions and at what period; but can
seldom ascertain anything of the position or extent of those
which have sunk beneath it. Our present study may often
enable us, not only to say where lands must have recently
disappeared, but also to form some judgment as to their extent, and the time that has elapsed since their submersion.

Having thus briefly sketched the nature and objects of the
subject we have to study, it will be necessary—before entering
on a detailed examination of the zoological features of the
different parts of the earth, and of the distribution of the orders,
families, and genera of animals—to examine certain preliminary
facts and principles essential for our guidance. We must first
inquire what are the powers of multiplication and dispersal of
the various groups of animals, and the nature of the barriers
that most effectually limit their range. We have then to
consider the effects of changes in physical geography and in
climate; to examine the nature and extent of such changes as
have been known to occur; to determine what others are possible
or probable; and to ascertain the various modes in which such
changes affect the structure, the distribution, or the very existence of animals.

Two subjects of a different nature must next engage our attention. We have to deal with two vast masses of facts, each involving countless details, and requiring subdivision and grouping to be capable of intelligible treatment. All the continents and their chief subdivisions, and all the more important islands of the globe, have to be compared as regards their various animal forms. To do this effectively we require a natural division of the earth especially adapted to our purpose; and we shall have to discuss at some length the reasons for the particular system adopted,—a discussion which must to some extent anticipate and summarize the conclusions of the whole work. We have also to deal with many hundreds of families and many thousands of genera of animals, and here too a true and natural classification is of great importance. We must therefore give a connected view of the classification adopted in the various classes of animals dealt with.

And lastly, as the existing distribution of animals is the result and outcome of all preceding changes of the earth and of its inhabitants, we require as much knowledge as we can get of the animals of each country during past geological epochs, in order to interpret the facts we shall accumulate. We shall, therefore, enter upon a somewhat detailed sketch of the various forms of extinct animals that have lived upon the earth during the Tertiary period; discuss their migrations at various epochs, the changes of physical geography that they imply, and the extent to which they enable us to determine the birthplace of certain families and genera.

The preliminary studies above enumerated will, it is believed, enable us to see the bearing of many facts in the distribution of animals that would otherwise be insoluble problems; and, what is hardly less valuable, will teach us to estimate the comparative importance of the various groups of animals, and to avoid the common error of cutting the gordian knot of each difficulty by vast hypothetical changes in existing continents and oceans —probably the most permanent features of our globe.

CHAPTER II.

THE MEANS OF DISPERSAL AND THE MIGRATIONS OF ANIMALS.

ALL animals are capable of multiplying so rapidly, that if a single pair were placed in a continent with abundance of food and no enemies, they might fully stock it in a very short time. Thus, a bird which produces ten pairs of young during its lifetime (and this is far below the fertility of many birds) will, if we take its life at five years, increase to a hundred millions in about forty years, a number sufficient to stock a large country. Many fishes and insects are capable of multiplying several thousandfold each year, so that in a few years they would reach billions and trillions. Even large and slow breeding mammals, which have only one at a birth but continue to breed from eight to ten successive years, may increase from a single pair to ten millions in less than forty years.

But as animals rarely have an unoccupied country to breed in, and as the food in any one district is strictly limited, their natural tendency is to roam in every direction in search of fresh pastures, or new hunting grounds. In doing so, however, they meet with many obstacles. Rocks and mountains have to be climbed, rivers or marshes to be crossed, deserts or forests to be traversed; while narrow straits or wider arms of the sea separate islands from the main land or continents from each other. We have now to inquire what facilities the different classes of animals have for overcoming these obstacles, and what kind of barriers are most effectual in checking their progress.

Means of Dispersal of Mammalia.—Many of the largest mammalia are able to roam over whole continents and are hardly stopped by any physical obstacles. The elephant is almost equally at home on plains and mountains, and it even climbs to the highest summit of Adam's Peak in Ceylon, which is so steep and rocky as to be very difficult of ascent for man. It traverses rivers with great ease and forces its way through the densest jungle. There seems therefore to be no limit to its powers of wandering, but the necessity of procuring food and its capacity of enduring changes of climate. The tiger is another animal with great powers of dispersal. It crosses rivers and sometimes even swims over narrow straits of the sea, and it can endure the severe cold of North China and Tartary as well as the heats of the plains of Bengal. The rhinoceros, the lion, and many of the ruminants have equal powers of dispersal; so that wherever there is land and sufficient food, there are no limits to their possible range. Other groups of animals are more limited in their migrations. The apes, lemurs, and many monkeys are so strictly adapted to an arboreal life that they can never roam far beyond the limits of the forest vegetation. The same may be said of the squirrels, the opossums, the arboreal cats, and the sloths, with many other groups of less importance. Deserts or open country are equally essential to the existence of others. The camel, the hare, the zebra, the giraffe and many of the antelopes could not exist in a forest country any more than could the jerboas or the prairie marmots.

There are other animals which are confined to mountains, and could not extend their range into lowlands or forests. The goats and the sheep are the most striking group of this kind, inhabiting many of the highest mountains of the globe; of which the European ibex and mouflon are striking examples. Rivers are equally necessary to the existence of others, as the beaver, otter, water-vole and capybara; and to such animals high mountain-ranges or deserts must form an absolutely impassable barrier.

Climate as a Limit to the Range of Mammals.—Climate appears to limit the range of many animals, though there is some reason to believe that in many cases it is not the climate itself so much as the change of vegetation consequent on climate which produces the effect. The quadrumana appear to be limited by climate, since they inhabit almost all the tropical regions but do not range more than about 10° beyond the southern and 12° beyond the northern tropic, while the great bulk of the species are found only within an equatorial belt about 30° wide. But as these animals are almost exclusively fruit-eaters, their distribution depends as much on vegetation as on temperature; and this is strikingly shown by the fact that the *Semnopithecus schistaceus* inhabits the Himalayan mountains to a height of 11,000 feet, where it has been seen leaping among fir-trees loaded with snow-wreaths! Some northern animals are bounded by the isothermal of 32°. Such are the polar bear and the walrus, which cannot live in a state of nature far beyond the limits of the frozen ocean; but as they live in confinement in temperate countries, their range is probably limited by other conditions than temperature.

We must not therefore be too hasty in concluding, that animals which we now see confined to a very hot or a very cold climate are incapable of living in any other. The tiger was once considered a purely tropical animal, but it inhabits permanently the cold plains of Manchuria and the Amoor, a country of an almost arctic winter climate. Few animals seem to us more truly inhabitants of hot countries than the elephants and rhinoceroses; yet in Post-tertiary times they roamed over the whole of the northern continents to within the arctic circle; and we know that the climate was then as cold as it is now, from their entire bodies being preserved in ice. Some change must recently have occurred either in the climate, soil, or vegetation of Northern Asia which led to the extinction of these forerunners of existing tropical species; and we must always bear in mind that similar changes may have acted upon other species which we now find restricted within narrow limits, but which may once have roamed over a wide and varied territory.

Valleys and Rivers as Barriers to Mammals.—To animals which thrive best in dry and hilly regions, a broad level and marshy valley must often prove an effectual barrier. The difference of vegetation and of insect life together with an unhealthy atmosphere, no doubt often checks migration if it is attempted. Thus

many animals are restricted to the slopes of the Himalayas
or to the mountains of Central India, the flat valley of the
Ganges forming a limit to their range. In other cases, however,
it is the river rather than the valley which is the barrier. In
the great Amazonian plains many species of monkeys, birds, and
even insects are found up to the river banks on one side but do
not cross to the other. Thus in the lower part of the Rio Negro
two monkeys, the *Jacchus bicolor* and the *Brachiurus couxiou*, are
found on the north bank of the river but never on the south,
where a red-whiskered *Pithecia* is alone found. Higher up *Ateles
paniscus* extends to the north bank of the river while *Lagothrix
humboldtii* comes down to the south bank ; the former being a
native of Guiana, the latter of Ecuador. The range of the birds
of the genus *Psophia* or trumpeters, is also limited by the rivers
Amazon, Madeira, Rio Negro and some others ; so that in these
cases we are able to define the limits of distribution with an
unusual degree of accuracy, and there is little doubt the same
barriers also limit a large number of other species.

Arms of the Sea as Barriers to Mammals.—Very few mammals
can swim over any considerable extent of sea, although many can
swim well for short distances. The jaguar traverses the widest
streams in South America, and the bear and bison cross the
Mississippi ; and there can be no doubt that they could swim over
equal widths of salt water, and if accidentally carried out to sea
might sometimes succeed in reaching islands many miles distant.
Contrary to the common notion pigs can swim remarkably well.
Sir Charles Lyell tells us in his " Principles of Geology " that
during the floods in Scotland in 1829, some pigs only six months
old that were carried out to sea, swam five miles and got on
shore again. He also states, on the authority of the late Edward
Forbes, that a pig jumped overboard to escape from a terrier in
the Grecian Archipelago, and swam safely to shore many miles
distant. These facts render it probable that wild pigs, from
their greater strength and activity, might under favourable circumstances cross arms of the sea twenty or thirty miles wide ;
and there are facts in the distribution of this tribe of animals
which seem to indicate that they have sometimes done so. Deer

take boldly to the water and can swim considerable distances, but we have no evidence to show how long they could live at sea or how many miles they could traverse. Squirrels, rats, and lemmings often migrate from northern countries in bands of thousands and hundreds of thousands, and pass over rivers, lakes and even arms of the sea, but they generally perish in the salt-water. Admitting, however, the powers of most mammals to swim considerable distances, we have no reason to believe that any of them could traverse without help straits of upwards of twenty miles in width, while in most cases a channel of half that distance would prove an effectual barrier.

Ice-floes and Driftwood as Aiding the Dispersal of Mammals.— In the arctic regions icebergs originate in glaciers which descend into the sea, and often bear masses of gravel, earth, and even some vegetation on their surfaces; and extensive level ice-fields break away and float southwards. These might often carry with them such arctic quadrupeds as frequent the ice, or even on rare occasions true land-animals, which might sometimes be stranded on distant continents or islands. But a more effectual because a more wide-spread agent, is to be found in the uprooted trees and rafts of driftwood often floated down great rivers and carried out to sea. Such rafts or islands are sometimes seen drifting a hundred miles from the mouth of the Ganges with living trees erect upon them; and the Amazon, the Orinoco, Mississippi, Congo, and most great rivers produce similar rafts. Spix and Martius declare that they saw at different times on the Amazon, monkeys, tiger-cats, and squirrels, being thus carried down the stream. On the Parana, pumas, squirrels, and many other quadrupeds have been seen on these rafts; and Admiral W. H. Smyth informed Sir C. Lyell that among the Philippine islands after a hurricane, he met with floating masses of wood with trees growing upon them, so that they were at first mistaken for islands till it was found that they were rapidly drifting along. Here therefore, we have ample means for carrying all the smaller and especially the arboreal mammals out to sea; and although in most cases they would perish there, yet in some favourable instances strong winds or unusual tidal currents might carry them safely to shores per haps several hundred miles from their native country. The fact of green trees so often having been seen erect on these rafts is most important; for they would act as a sail by which the raft might be propelled in one direction for several days in succession, and thus at last reach a shore to which a current alone would never have carried it.

There are two groups of mammals which have quite exceptional means of dispersal—the bats which fly, and the cetacea, seals, &c., which swim. The former are capable of traversing considerable spaces of sea, since two North American species either regularly or occasionally visit the Bermudas, a distance of 600 miles from the mainland. The oceanic mammals (whales and porpoises) seem to have no barrier but temperature; the polar species being unable to cross the equator, while the tropical forms are equally unfitted for the cold polar waters. The shore-feeding manatees, however, can only live where they find food; and a long expanse of rocky coast would probably be as complete a barrier to them as a few hundred miles of open ocean. The amphibious seals and walruses seem many of them to be capable of making long sea journeys, some of the species being found on islands a thousand miles apart, but none of the arctic are identical with the antartic species.

The otters with one exception are freshwater animals, and we have no reason to believe they could or would traverse any great distances of salt water. In fact, they would be less liable to dispersal across arms of the sea than purely terrestrial species, since their powers of swimming would enable them to regain the shore if accidentally carried out to sea by a sudden flood.

Means of Dispersal of Birds.—It would seem at first sight that no·barriers could limit the range of birds, and that they ought to be the most ubiquitous of living things, and little fitted therefore to throw any light on the laws or causes of the geographical distribution of animals. This, however, is far from being the case; many groups of birds are almost as strictly limited by barriers as the mammalia; and from their larger numbers and the avidity with which they have been collected, they furnish

materials of the greatest value for our present study. The
different groups of birds offer remarkable contrasts in the extent
of their range, some being the most cosmopolite of the higher
animals, while others are absolutely confined to single spots on
the earth's surface. The petrels (*Procellariidæ*) and the gulls
(*Laridæ*) are among the greatest wanderers; but most of the
species are confined to one or other of the great oceans, or to the
arctic or antarctic seas, a few only being found with scarcely
any variation over almost the whole globe. The sandpipers and
plovers wander along the shores as far as do the petrels over the
ocean. Great numbers of them breed in the arctic regions and
migrate as far as India and Australia, or down to Chili and
Brazil; the species of the old and new worlds, however, being
generally distinct. In striking contrast to these wide ranges
we find many of the smaller perching birds, with some of the
parrots and pigeons, confined to small islands of a few square
miles in extent, or to single valleys or mountains on the mainland.

Dispersal of Birds by Winds.—Those groups of birds which
possess no powers of flight, such as the ostrich, cassowary, and
apteryx, are in exactly the same position as mammalia as regards
their means of dispersal, or are perhaps even inferior to them;
since, although they are able to cross rivers by swimming, it is
doubtful if they could remain so long in the water as most land
quadrupeds. A very large number of short-winged birds, such
as toucans, pittas, and wrens, are perhaps worse off; for they can
fly very few miles at a time, and on falling into the water would
soon be drowned. It is only the strong-flying species that can
venture to cross any great width of sea; and even these rarely do
so unless compelled by necessity to migrate in search of food, or
to a more genial climate. Small and weak birds are, however,
often carried accidentally across great widths of ocean by violent
gales. This is well exemplified by the large numbers of
stragglers from North America, which annually reach the
Bermudas. No less than sixty-nine species of American birds
have occurred in Europe, most of them in Britain and Heligo-
land. They consist chiefly of migratory birds which in autumn

return along the eastern coasts of the United States, and often
fly from point to point across bays and inlets. They are then
liable to be blown out to sea by storms, which are prevalent at
this season; and it is almost always at this time of year that
their occurrence has been noted on the shores of Europe. It
may, however, be doubted whether this is not an altogether
modern phenomenon, dependent on the number of vessels constantly on the Atlantic which afford resting-places to the wanderers; as it is hardly conceivable that such birds as titlarks,
cuckoos, wrens, warblers, and rails, could remain on the wing
without food or rest, the time requisite to pass over 2,000 miles
of ocean. It is somewhat remarkable that no European birds
reach the American coast but a few which pass by way of
Iceland and Greenland; whereas a considerable number do
reach the Azores, fully half way across; so that their absence
can hardly be due to the prevailing winds being westerly. The
case of the Azores is, however, an argument for the unassisted
passage of birds for that distance ; since two of the finches are
peculiar 'species,' but closely allied to European forms, so that
their progenitors must, probably, have reached the islands before
the Atlantic was a commercial highway.

Barriers to the Dispersal of Birds.—We have seen that, as a
rule, wide oceans are an almost absolute barrier to the passage of
most birds from one continent to another; but much narrower
seas and straits are also very effectual barriers where the habits
of the birds are such as to preserve them from being carried
away by storms. All birds which frequent thickets and forests,
and which feed near or on the ground, are secure from such
accidents; and they are also restricted in their range by the
extent of the forests they inhabit. In South America a large
number of the birds have their ranges determined by the extent of the forest country, while others are equally limited to the
open plains. Such species are also bounded by mountain ranges
whenever these rise above the woody region. Great rivers, such
as the Amazon, also limit the range of many birds, even when
there would seem to be no difficulty in their crossing them. The
supply of food, and the kind of vegetation, soil, and climate

C

best suited to a bird's habits, are probably the causes which mark
out the exact limits of the range of each species; to which must
be added the prevalence of enemies of either the parent birds,
the eggs, or the young. In the Malay Archipelago pigeons abound
most where monkeys do not occur; and in South America the
same birds are comparatively scarce in the forest plains where
monkeys are very abundant, while they are plentiful on the open
plains and campos, and on the mountain plateaux, where these
nest-hunting quadrupeds are rarely found. Some birds are
confined to swamps, others to mountains; some can only live on
rocky streams, others on deserts or grassy plains.

The Phenomena of Migration.—The term "migration" is often
applied to the periodical or irregular movements of all animals;
but it may be questioned whether there are any regular migrants but birds and fishes. The annual or periodical movements
of mammalia are of a different class. Monkeys ascend the
Himalayas in summer to a height of 10,000 to 12,000 feet, and
descend again in winter. Wolves everywhere descend from the
mountains to the lowlands in severe weather. In dry seasons
great herds of antelopes move southwards towards the Cape of
Good Hope. The well-known lemmings, in severe winters, at
long intervals, move down from the mountains of Scandinavia in
immense numbers, crossing lakes and rivers, eating their way
through haystacks, and surmounting every obstacle till they
reach the sea, whence very few return. The alpine hare, the
arctic fox, and many other animals, exhibit similar phenomena
on a smaller scale; and generally it may be said, that whenever
a favourable succession of seasons has led to a great multiplication of any species, it must on the pressure of hunger seek
food in fresh localities. For such movements as these we have
no special term. The summer and winter movements best
correspond to true migration, but they are always on a small
scale, and of limited extent; the other movements are rather
temporary incursions than true migrations.

The annual movements of many fishes are more strictly
analogous to the migration of birds, since they take place
in large bodies and often to considerable distances, and are

immediately connected with the process of reproduction. Some, as the salmon, enter rivers; others, as the herring and mackerel, approach the coast in the breeding season; but the exact course of their migrations is unknown, and owing to our complete ignorance of the area each species occupies in the ocean, and the absence of such barriers and of such physical diversities as occur on the land, they are of far less interest and less connected with our present study than the movements of birds, to which we shall now confine ourselves.

Migrations of Birds.—In all the temperate parts of the globe there are a considerable number of birds which reside only a part of the year, regularly arriving and leaving at tolerably fixed epochs. In our own country many northern birds visit us in winter, such as the fieldfare, redwing, snow-bunting, turnstone, and numerous ducks and waders; with a few, like the black red-start, and (according to Rev. C. A. Johns) some of the woodcocks from the south. In the summer a host of birds appear—the cuckoo, the swifts and swallows, and numerous warblers, being the most familiar,—which stay to build their nests and rear their young, and then leave us again. These are true migrants; but a number of other birds visit us occasionally, like the waxwing, the oriole, and the bee-eater, and can only be classed as stragglers, which, perhaps from too rapid multiplication one year and want of food the next, are driven to extend their ordinary range of migration to an unusual degree. We will now endeavour to sketch the chief phenomena of migration in different countries.

Europe.—It is well ascertained that most of the birds that spend their spring and summer in the temperate parts of Europe pass the winter in North Africa and Western Asia. The winter visitants, on the other hand, pass the summer in the extreme north of Europe and Asia, many of them having been found to breed in Lapland. The arrival of migratory birds from the south is very constant as to date, seldom varying more than a week or two, without any regard to the weather at the time; but the departure is less constant, and more dependent on the weather. Thus the swallow always comes to us about the middle

of April, however cold it may be, while its departure may take
place from the end of September to late in October, and is said by
Forster to occur on the first N. or N.E. wind after the 20th of
September.

Almost all the migratory birds of Europe go southward to
the Mediterranean, move along its coasts east or west, and cross
over in three places only; either from the south of Spain, in the
neighbourhood of Gibraltar, from Sicily over Malta, or to the
east by Greece and Cyprus. They are thus always in sight of
land. The passage of most small birds (and many of the larger
ones too) takes place at night; and they only cross the Mediterranean when the wind is steady from near the east or west,
and when there is moonlight.

It is a curious fact, but one that seems to be well authenti-
cated, that the males often leave before the females, and both
before the young birds, which in considerable numbers migrate
later and alone. These latter, however, seldom go so far as the
old ones; and numbers of young birds do not cross the Mediterranean, but stay in the south of Europe. The same rule applies
to the northward migration; the young birds stopping short
of the extreme arctic regions, to which the old birds migrate.[1]
When old and young go together, however, the old birds take
the lead. In the south of Europe few of the migratory birds
stay to breed, but pass on to more temperate zones; thus, in the
south of France, out of 350 species only 60 breed there. The
same species is often sedentary in one part of Europe and migratory in another; thus, the chaffinch is a constant resident in
England, Germany, and the middle of France; but a migrant in
the south of France and in Holland: the rook visits the south
of France in winter only: the *Falco tinnunculus* is both a
resident and a migrant in the south of France, according to
M. Marcel de Serres, there being two regular passages every
year, while a certain number always remain.

[1] Marcel de Serres states this as a general fact for wading and swimming
birds. He says that the old birds arrive in the extreme north almost alone,
the young remaining on the shores of the Baltic, or on the lakes of Austria,
Hungary, and Russia. See his prize essay, *Des Causes des Migrations*, &c.
2nd. ed., Paris, 1845, p. 121.

We see, then, that migration is governed by certain intelligible laws; and that it varies in many of its details, even in the same species, according to changed conditions. It may be looked upon as an exaggeration of a habit common to all locomotive animals, of moving about in search of food. This habit is greatly restricted in quadrupeds by their inability to cross the sea or even to pass through the highly-cultivated valleys of such countries as Europe; but the power of flight in birds enables them to cross every kind of country, and even moderate widths of sea; and as they mostly travel at night and high in the air, their movements are difficult to observe, and are supposed to be more mysterious than they perhaps are. In the tropics birds move about to different districts according as certain fruits become ripe, certain insects abundant, or as flooded tracts dry up. On the borders of the tropics and the temperate zone extends a belt of country of a more or less arid character, and liable to be parched at the summer solstice. In winter and early spring its northern margin is verdant, but it soon becomes burnt up, and most of its birds necessarily migrate to the more fertile regions to the north of them. They thus follow the spring or summer as it advances from the south towards the pole, feeding on the young flower buds, the abundance of juicy larvæ, and on the ripening fruits; and as soon as these become scarce they retrace their steps homewards to pass the winter. Others whose home is nearer the pole are driven south by cold, hunger, and darkness, to more hospitable climes, returning northward in the early summer. As a typical example of a migratory bird, let us take the nightingale. During the winter this bird inhabits almost all North Africa, Asia Minor, and the Jordan Valley. Early in April it passes into Europe by the three routes already mentioned, and spreads over France, Britain, Denmark, and the south of Sweden, which it reaches by the beginning of May. It does not enter Brittany, the Channel Islands, or the western part of England, never visiting Wales, except the extreme south of Glamorganshire, and rarely extending farther north than Yorkshire. It spreads over Central Europe, through Austria and Hungary to Southern Russia and the warmer parts of Siberia, but it nevertheless breeds in the Jordan Valley, so that in some places it is only the surplus population that migrates. In August and September, all who can return to their winter quarters.

Migrations of this type probably date back from at least the period when there was continuous land along the route passed over; and it is a suggestive fact that this land connection is known to have existed in recent geological times. Britain was connected with the Continent during, and probably before, the glacial epoch; and Gibraltar, as well as Sicily and Malta, were also recently united with Africa, as is proved by the fossil elephants and other large mammalia found in their caverns, by the comparatively shallow water still existing in this part of the Mediterranean while the remainder is of oceanic profundity, and by the large amount of identity in the species of land animals still inhabiting the opposite shores of the Mediterranean. The submersion of these two tracts of land (which were perhaps of considerable extent) would be a slow process, and from year to year the change might be hardly perceptible It is easy to see how the migration that had once taken place over continuous land would be kept up, first over lagoons and marshes, then over a narrow channel, and subsequently over a considerable sea, no one generation of birds ever perceiving any difference in the route.

There is, however, no doubt that the sea-passage is now very dangerous to many birds. Quails cross in immense flocks, and great numbers are drowned at sea whenever the weather is unfavourable. Some individuals always stay through the winter in the south of Europe, and a few even in England and Ireland; and were the sea to become a little wider the migration would cease, and the quail, like some other birds, would remain divided between south Europe and north Africa. Aquatic birds are observed to follow the routes of great rivers and lakes, and the shores of the sea. One great body reaches central Europe by way of the Danube from the shores of the Black Sea; another ascends the Rhone Valley from the Gulf of Lyons.

India and China.—In the peninsula of India and in China great numbers of northern birds arrive during September and October, and leave from March to May. Among the smaller birds are wagtails, pipits, larks, stonechats, warblers, thrushes, buntings, shrikes, starlings, hoopoes, and quails. Some species of cranes and storks, many ducks, and great numbers of *Scolo-pacidæ* also visit India in winter; and to prey upon these come a band of rapacious birds—the peregrine falcon, the hobby, kestrel, common sparrowhawk, harrier, and the short-eared owl. These birds are almost all natives of Europe and Western Asia; they spread over all northern and central India, mingling with the sedentary birds of the oriental fauna, and give to the ornithology of Hindostan at this season quite a European aspect. The peculiar species of the higher Himalayas do not as a rule descend to the plains in winter, but merely come lower down the mountains; and in southern India and Ceylon comparatively few of these migratory birds appear.

In China the migratory birds follow generally the coast line, coming southwards in winter from eastern Siberia and northern Japan; while a few purely tropical forms travel northwards in summer to Japan, and on the mainland as far as the valley of the Amoor.

North America.—The migrations of birds in North America have been carefully studied by resident naturalists, and present some interesting features   The birds of the eastern parts of North America are pre-eminently migratory, a much smaller proportion being permanent residents than in corresponding latitudes in Europe. Thus, in Massachusetts there are only about 30 species of birds which are resident all the year, while the regular summer visitors are 106   Comparing with this our own country, though considerably further north, the proportions are reversed; there being 140 residents and 63 summer visitors. This difference is clearly due to the much greater length and severity of the winter, and the greater heat of summer, in America than with us. The number of permanent residents increases pretty regularly as we go southward; but the number of birds at any locality during the breeding season seems to increase as we go northward as far as Canada, where, according to Mr. Allen, more species breed than in the warm Southern States. Even in the extreme north, beyond the limit of forests, there are no less than 60 species which breed; in Canada about 160; while in Carolina there are only 135, and in Louisiana, 130. The extent of the migration varies greatly, some species only going a few degrees north and south, while others migrate annually from the tropics to the extreme north of the continent; and every gradation occurs between these extremes. Among those which migrate furthest are the species of *Dendrœca*, and other American flycatching warblers (*Mniotiltidæ*), many of which breed on the shores of Hudson's Bay, and spend the winter in Mexico or the West Indian islands.

The great migratory movement of American birds is almost wholly confined to the east coast; the birds of the high central plains and of California being for the most part sedentary, or only migrating for short distances. All the species which reach South America, and most of those which winter in Mexico and Guatemala, are exclusively eastern species; though a few Rocky Mountain birds range southward along the plateaux of Mexico and Guatemala, but probably not as regular annual migrants.

In America as in Europe birds appear in spring with great regularity, while the time of the autumnal return is less constant. More curious is the fact, also observed in both hemispheres, that they do not all return by the same route followed in going northwards, some species being constant visitors to certain localities in spring but not in autumn, others in autumn but not in spring.

Some interesting cases have been observed in America of a gradual alteration in the extent of the migration of certain birds. A Mexican swallow (*Hirundo lunifrons*) first appeared in Ohio in 1815. Year by year it increased the extent of its range till by 1845 it had reached Maine and Canada; and it is now quoted by American writers as extending its annual migrations to Hudson's Bay. An American wren (*Troglodytes ludovicianus*) is another bird which has spread considerably northwards since the time of the ornithologist Wilson ; and the rice-bird, or " Bob-o'-link," of the Americans, continually widens its range as rice and wheat are more extensively cultivated. This bird winters in Cuba and other West Indian Islands, and probably also in Mexico. In April it enters the Southern States and passes northward, till in June it reaches Canada and extends west to the Saskatchewan River in 54° north latitude.

South Temperate America.—The migratory birds of this part of the world have been observed by Mr. Hudson at Buenos Ayres. As in Europe and North America, there are winter and summer visitors, from Patagonia and the tropics respectively. Species of *Pyrocephalus, Milvulus,* swallows, and a humming-bird, are among the most regular of the summer visitors. They are all insectivorous birds. From Patagonia species of *Tæni-optera, Cinclodes,* and *Centrites,* come in winter, with two gulls, two geese, and six snipes and plovers. Five species of swallows appear at Buenos Ayres in spring, some staying to breed, others passing on to more temperate regions farther south. As a rule the birds which come late and leave early are the most regular. Some are very irregular in their movements, the *Molothrus bona-riensis,* for example, sometimes leaves early in autumn, some-times remains all the winter. Some resident birds also move in winter to districts where they are never seen in summer.

General Remarks on Migration.—The preceding summary of the main facts of migration (which might have been almost in-definitely extended, owing to the great mass of detailed infor-mation that exists on the subject) appears to accord with the view already suggested, that the " instinct " of migration has arisen from the habit of wandering in search of food common to all animals, but greatly exaggerated in the case of birds by their powers of flight and by the necessity for procuring a large amount of soft insect food for their unfledged young. Migra-tion in its simple form may be best studied in North America, where it takes place over a continuous land surface with a con-siderable change of climate from south to north. We have here (as probably in Europe and elsewhere) every grade of migration, from species which merely shift the northern and southern limits of their range a few hundred miles, so that in the central parts of the area the species is a permanent resident, to others which move completely over 1,000 miles of latitude, so that in all the intervening districts they are only known as birds of passage. Now, just as the rice-bird and the Mexican swallow have extended their migrations, owing to favourable conditions induced by human agency; so we may presume that large numbers of species would extend their range where favourable conditions arose through natural causes. If we go back only as far as the height of the glacial epoch, there is reason to believe that all North America, as far south as about 40° north latitude, was covered with an almost continuous and perennial ice-sheet. At this time the migratory birds would extend up to this barrier (which would probably terminate in the midst of luxuriant vegetation, just as the glaciers of Switzerland now often terminate amid forests and corn-fields), and as the cold decreased and the ice retired almost imperceptibly year by year, would follow it up farther and farther according as the peculiarities of vegetation and insect-food were more or less suited to their several constitutions. It is an ascertained fact that many individual birds return year after year to build their nests in the same spot. This shows a strong local attachment, and is, in fact, the faculty or feeling on which their very existence probably depends. For were they to wander at random each year, they would almost certainly not meet with places so well suited to them, and might even get into districts where they or their young would inevitably perish. It is also a curious fact that in so many cases the old birds migrate first, leaving the young ones behind, who follow some short time later, but do not go so far as their parents. This is very strongly opposed to the notion of an imperative instinct. The old birds have been before, the young have not; and it is only when the old ones have all or nearly all gone that the young go too, probably following some of the latest stragglers. They wander, however, almost at random, and the majority are destroyed before the next spring. This is proved by the fact that the birds which return in spring are as a rule not more numerous than those which came the preceding spring, whereas those which went away in autumn were two or three times as numerous. Those young birds that do get back, however, have learnt by experience, and the next year they take care to go with the old ones. The most striking fact in favour of the "instinct" of migration is the "agitation," or excitement, of confined birds at the time when their wild companions are migrating. It seems probable, however, that this is what may be called a social excitement, due to the anxious cries of the migrating birds; a view supported by the fact stated by Marcel de Serres, that the black swan of Australia, when domesticated in Europe, sometimes joins wild swans in their northward migration. We must remember too that migration at the proper time is in many cases absolutely essential to the existence of the species; and it is therefore not improbable that some strong social emotion should have been gradually developed in the race, by the circumstance that all who for want of such emotion did not join their fellows inevitably perished.

The mode by which a passage originally overland has been converted into one over the sea offers no insuperable difficulties, as has already been pointed out. The long flights of some birds without apparently stopping on the way is thought to be inexplicable, as well as their finding their nesting-place of the previous year from a distance of many hundreds or even a thousand miles. But the observant powers of animals are very great; and birds flying high in the air may be guided by the physical features of the country spread out beneath them in a way that would be impracticable to purely terrestrial animals.

It is assumed by some writers that the breeding-place of a species is to be considered as its true home rather than that to which it retires in winter; but this can hardly be accepted as a rule of universal application. A bird can only breed successfully where it can find sufficient food for its young; and the reason probably why so many of the smaller birds leave the warm southern regions to breed in temperate or even cold latitudes, is because caterpillars and other soft insect larvæ are there abundant at the proper time, while in their winter home the larvæ have all changed into winged insects. But this favourable breeding district will change its position with change of climate; and as the last great change has been one of increased warmth in all the temperate zones, it is probable that many of the migratory birds are comparatively recent visitors. Other changes may however have taken place, affecting the vegetation and consequently the insects of a district; and we have seldom the means of determining in any particular case in what direction the last extension of range occurred. For the purposes of the study of geographical distribution therefore, we must, except in special cases, consider the true range of a species to comprise all the area which it occupies regularly for any part of the year, while all those districts which it only visits at more or less distant intervals, apparently driven by storms or by hunger, and where it never regularly or permanently settles, should not be included as forming part of its area of distribution.

Means of Dispersal of Reptiles and Amphibia.—If we leave out of consideration the true marine groups—the turtles and sea-snakes—reptiles are scarcely more fitted for traversing seas and oceans than are mammalia. We accordingly find that in those oceanic islands which possess no indigenous mammals, land reptiles are also generally wanting. The several groups of these animals, however, differ considerably both in their means of dispersal and in their power of resisting adverse conditions. Snakes are most dependent on climate, becoming very scarce in temperate and cold climates and entirely ceasing at 62° north latitude, and they do not ascend very lofty mountains, ceasing at 6,000 feet elevation in the Alps. Some inhabit deserts, others swamps and marshes, while many are adapted for a life in forests. They swim rivers easily, but apparently have no means of passing the sea, since they are very rarely found on oceanic islands. Lizards are also essentially tropical, but they go somewhat farther north than snakes, and ascend higher on the mountains, reaching 10,000 feet in the Alps. They possess too some unknown means (probably in the egg-state) of passing over the ocean, since they are found to inhabit many islands where there are neither mammalia nor snakes.

The amphibia are much less sensitive to cold than are true reptiles, and they accordingly extend much farther north, frogs being found within the arctic circle. Their semi-aquatic life also gives them facilities for dispersal, and their eggs are no doubt sometimes carried by aquatic birds from one pond or stream to another. Salt water is fatal to them as well as to their eggs, and hence it arises that they are seldom found in those oceanic islands from which mammalia are absent. Deserts and oceans would probably form the most effectual barriers to their dispersal ; whereas both snakes and lizards abound in deserts, and have some means of occasionally passing the ocean which frogs and salamanders do not seem to possess.

Means of Dispersal of Fishes.—The fact that the same species of freshwater fish often inhabit distinct river systems, proves that they have some means of dispersal over land. The many authentic accounts of fish falling from the atmosphere, indicate one of the means by which they may be transferred from one river basin to another, viz., by hurricanes and whirlwinds, which often carry up considerable quantities of water and with it fishes of small size. In volcanic countries, also, the fishes of subterranean streams may sometimes be thrown up by volcanic explosions, as Humboldt relates happened in South America. Another mode by which fishes may be distributed is by their eggs being occasionally carried away by aquatic birds ; and it is stated by Gmelin that geese and ducks during their migrations feed on the eggs of fish, and that some of these pass through their bodies with their vitality unimpaired.[1] Even water-beetles flying from one pond to another might occasionally carry with them some of the smaller eggs of fishes. But it is probable that fresh-water fish are also enabled to migrate by changes of level causing streams to alter their course and carry their waters into adjacent basins. On plateaux the sources of distinct river systems often approach each other, and the same thing occurs with lateral tributaries on the lowlands near their mouths. Such changes, although small in extent, and occurring only at long intervals, would

[1] Quoted in Lyell's *Principles of Geology* (11th ed. vol. ii. p. 374), from *Amœn. Acad. Essay* 75.

act very powerfully in modifying the distribution of fresh-water fish.

Sea fish would seem at first sight to have almost unlimited means of dispersal, but this is far from being the case. Temperature forms a complete barrier to a large number of species, cold water being essential to many, while others can only dwell in the warmth of the tropics. Deep water is another barrier to large numbers of species which are adapted to shores and shallows; and thus the Atlantic is quite as impassable a gulf to most fishes as it is to birds. Many sea fishes migrate to a limited extent for the purpose of depositing their spawn in favourable situations. The herring, an inhabitant of the deep sea, comes in shoals to our coast in the breeding season; while the salmon quits the northern seas and enters our rivers, mounting upwards to the clear cold water near their sources to deposit its eggs. Keeping in mind the essential fact that changes of temperature and of depth are the main barriers to the dispersal of fish, we shall find little difficulty in tracing the causes that have determined their distribution.

Means of Dispersal of Mollusca.—The marine, fresh-water, and land mollusca are three groups whose powers of dispersal and consequent distribution are very different, and must be separately considered. The *Pteropoda*, the *Ianthina*, and other groups of floating molluscs, drift about in mid-ocean, and their dispersal is probably limited chiefly by temperature, but perhaps also by the presence of enemies or the scarcity of proper food. The univalve and bivalve mollusca, of which the whelk and the cockle may be taken as types, move so slowly in their adult state, that we should expect them to have an exceedingly limited distribution; but the young of all these are free swimming embryos, and they thus have a powerful means of dispersal, and are carried by tides and currents so as ultimately to spread over every shore and shoal that offers conditions favourable for their development. The fresh water molluscs, which one might at first suppose could not range beyond their own river-basin, are yet very widely distributed in common with almost all other fresh water productions; and Mr. Darwin has shown that this is

due to the fact, that ponds and marshes are constantly frequented
by wading and swimming birds which are pre-eminently wanderers, and which frequently carry away with them the seeds of
plants, and the eggs of molluscs and aquatic insects. Fresh
water molluscs just hatched were found to attach themselves to
a duck's foot suspended in an aquarium; and they would thus be
easily carried from one lake or river to another, and by the help
of different species of aquatic birds, might soon spread all over
the globe. Even a water-beetle has been caught with a small
living shell (*Ancylus*) attached to it; and these fly long distances
and are liable to be blown out to sea, one having been caught on
board the *Beagle* when forty-five miles from land. Although
fresh water molluscs and their eggs must frequently be carried
out to sea, yet this cannot lead to their dispersal, since salt
water is almost immediately fatal to them; and we are therefore
forced to conclude that the apparently insignificant and uncertain means of dispersal above alluded to are really what have
led to their wide distribution. The true land-shells offer a still
more difficult case, for they are exceedingly sensitive to the
influence of salt water; they are not likely to be carried by
aquatic birds, and yet they are more or less abundant all over
the globe, inhabiting the most remote oceanic islands. It has
been found, however, that land-shells have the power of lying
dormant a long time. Some have lived two years and a half
shut up in pill boxes; and one Egyptian desert snail came to life
after having been glued down to a tablet in the British Museum
for four years !

We are indebted to Mr. Darwin for experiments on the power
of land shells to resist sea water, and he found that when they
had formed a membranous diaphragm over the mouth of the
shell they survived many days' immersion (in one case fourteen
days); and another experimenter, quoted by Mr. Darwin, found that
out of one hundred land shells immersed for a fortnight in the sea,
twenty-seven recovered. It is therefore quite possible for them to
be carried in the chinks of drift wood for many hundred miles
across the sea, and this is probably one of the most effectual
modes of their dispersal. Very young shells would also some-

times attach themselves to the feet of birds walking or resting
on the ground, and as many of the waders often go far inland,
this may have been one of the methods of distributing species
of land shells; for it must always be remembered that nature can
afford to wait, and that if but once in a thousand years a single
bird should convey two or three minute snails to a distant island,
this is all that is required for us to find that island well stocked
with a great and varied population of land shells.

*Means of Dispersal of Insects and the Barriers which Limit
their Range.*—Winged insects, as a whole, have perhaps more
varied means of dispersal over the globe than any other highly
organised animals. Many of them can fly immense distances,
and the more delicate ones are liable to be carried by storms
and hurricanes over a wide expanse of ocean. They are often
met with far out at sea. Hawk-moths frequently fly on board
ships as they approach the shores of tropical countries, and they
have sometimes been captured more than 250 miles from the
nearest land. Dragon-flies came on board the *Adventure* frigate
when fifty miles off the coast of South America. A southerly
wind brought flies in myriads to Admiral Smyth's ship in the
Mediterranean when he was 100 miles distant from the coast of
Africa. A large Indian beetle (*Chrysochroa ocellata*) was quite
recently caught alive in the Bay of Bengal by Captain Payne of
the barque *William Mansoon*, 273 miles from the nearest land.
Darwin caught a locust 370 miles from land; and in 1844
swarms of locusts several miles in extent, and as thick as the
flakes in a heavy snowstorm, visited Madeira. These must have
come with perfect safety more than 300 miles; and as they continued flying over the island for a long time, they could evidently
have travelled to a much greater distance. Numbers of living
beetles belonging to seven genera, some aquatic and some terrestrial, were caught by Mr. Darwin in the open sea, seventeen
miles from the coast of South America, and they did not seem
injured by the salt water. Almost all the accidental causes that
lead to the dispersal of the higher animals would be still more
favourable for insects. Floating trees could carry hundreds of
insects for one bird or mammal; and so many of the larvæ, eggs,

and pupæ of insects have their abode in solid timber, that they might survive being floated immense distances. Great numbers of tropical insects have been captured in the London docks, where they have been brought in foreign timber; and some have emerged from furniture after remaining torpid for many years. Most insects have the power of existing weeks or months without food, and some are very tenacious of life. Many beetles will survive immersion for hours in strong spirit; and water a few degrees below the boiling point will not always kill them. We can therefore easily understand how, in the course of ages insects may become dispersed by means which would be quite inadequate in the case of the higher animals. The drift-wood and tropical fruits that reach Ireland and the Orkneys; the double cocoa-nuts that cross the Indian ocean from the Seychelle Islands to the coast of Sumatra; the winds that carry volcanic dust and ashes for thousands of miles; the hurricanes that travel in their revolving course over wide oceans; all indicate means by which a few insects may, at rare intervals be carried to remote regions, and become the progenitors of a group of allied forms.

But the dispersal of insects requires to be looked at from another point of view. They are, of all animals, perhaps the most wonderfully adapted for special conditions; and are so often fitted to fill one place in nature and one only, that the barriers against their permanent displacement are almost as numerous and as effective as their means of dispersal. Hundreds of species of lepidoptera, for example, can subsist in the larva state only on one species of plant; so that even if the perfect insects were carried to a new country, the continuance of the race would depend upon the same or a closely allied plant being abundant there. Other insects require succulent vegetable food all the year round, and are therefore confined to tropical regions; some can live only in deserts, others in forests; some are dependent on water-plants, some on mountain-vegetation. Many are so intimately connected with other insects during some part of their existence that they could not live without them; such are the parasitical hymenoptera and diptera, and those mimicking species whose welfare depends upon their being

D

mistaken for something else.   Then again, insects have enemies
in every stage of their existence—the egg, the larva, the pupa,
and the perfect form; and the abundance of any one of these
enemies may render their survival impossible in a country otherwise well suited to them.   Ever bearing in mind these two
opposing classes of facts, we shall not be surprised at the
enormous range of some groups of insects, and at the extreme
localization of others; and shall be able to give a rational account
of many phenomena of distribution that would otherwise seem
quite unintelligible.

CHAPTER III.

THE distribution of animals over the earth's surface, is evidently dependent in great measure upon those grand and important characteristics of our globe, the study of which is termed physical geography. The proportion of land and water; the outlines and distribution of continents; the depth of seas and oceans; the position of islands; the height, direction, and continuity of mountain chains; the position and extent of deserts, lakes, and forests; the direction and velocity of ocean currents, as well as of prevalent winds and hurricanes; and lastly, the distribution of heat and cold, of rain, snow, and ice, both in their means and in their extremes, have all to be considered when we endeavour to account for the often unequal and unsymmetrical manner in which animals are dispersed over the globe. But even this knowledge is insufficient unless we inquire further as to the evidence of permanence possessed by each of these features, in order that we may give due weight to the various causes that have led to the existing facts of animal distribution.

Land and Water.—The well-known fact that nearly three-fourths of the surface of the earth is occupied by water, and but a little more than one-fourth by land, is important as indicating the vast extent of ocean by which many of the continents and islands are separated from each other. But there is another fact

D 2

which greatly increases its importance, namely, that the mean height of the land is very small compared with the mean depth of the sea. It has been estimated by Humboldt that the mean height of all the land surface does not exceed a thousand feet, owing to the comparative narrowness of mountain ranges and the great extent of alluvial plains and valleys; the ocean bed, on the contrary, not only descends deeper than the tops of the highest mountains rise above its surface, but these profound depths are broad sunken plains, while the shallows correspond to the mountain ranges, so that its mean depth is, as nearly as can be estimated, twelve thousand feet[1] Hence, as the area of water is three times that of the land, the total cubical contents of the land, above the sea level, would be only $\frac{1}{36}$ that of the waters which are below that level. The important result follows, that whereas it is scarcely possible that in past times the amount of land surface should ever greatly have exceeded that which now exists, it is just possible that all the land may have been at some time submerged; and therefore in the highest degree probable that among the continual changes of land and sea that have been always going on, the amount of land surface has often been much less than it is now. For the same reason it is probable that there have been times when large masses of land have been more isolated from the rest than they are at present; just as South America would be if North America were submerged, or as Australia would become if the Malay Archipelago were to sink beneath the ocean. It is also very important to bear in mind the fact insisted on by Sir Charles Lyell, that the shallow parts of the ocean are almost always in the vicinity of land; and that an amount of elevation that would make little difference to the bed of the ocean, would raise up extensive tracts of dry land in the vicinity of existing continents. It is almost certain, therefore, that changes in the distribution of land and sea must have taken place more frequently by additions to, or

---

[1] This estimate has been made for me by Mr. Stanford from the materials used in delineating the contours of the ocean-bed on our general map. It embodies the result of all the soundings of the *Challenger*, *Tuscarora*, and other vessels, obtainable up to August, 1875.

modifications of pre-existing land, than by the upheaval of entirely new continents in mid-ocean. These two principles will throw light upon two constantly recurring groups of facts in the distribution of animals,—the restriction of peculiar forms to areas not at present isolated,—and on the other hand, the occurrence of allied forms in lands situated on opposite shores of the great oceans.

Continental Areas.—Although the dry land of the earth's surface is distributed with so much irregularity, that there is more than twice as much north of the equator as there is south of it, and about twice as much in the Asiatic as in the American hemisphere; and, what is still more extraordinary, that on a hemisphere of which a point in St. George's Channel between England and Ireland is the centre, the land is nearly equal in extent to the water, while in the opposite hemisphere it is in the proportion of only one-eighth,—yet the whole of the land is almost continuous. It consists essentially of only three masses: the American, the Asia-African, and the Australian. The two former are only separated by thirty-six miles of shallow sea at Behring's Straits, so that it is possible to go from Cape Horn to Singapore or the Cape of Good Hope without ever being out of sight of land; and owing to the intervention of the numerous islands of the Malay Archipelago the journey might be continued under the same conditions as far as Melbourne and Hobart Town. This curious fact, of the almost perfect continuity of all the great masses of land notwithstanding their extremely irregular shape and distribution, is no doubt dependent on the circumstances just alluded to; that the great depth of the oceans and the slowness of the process of upheaval, has almost always produced the new lands either close to, or actually connected with pre-existing lands; and this has necessarily led to a much greater uniformity in the distribution of organic forms, than would have prevailed had the continents been more completely isolated from each other.

The isthmuses which connect Africa with Asia, and North with South America, are, however, so small and insignificant compared with the vast extent of the countries they unite that we can hardly consider them to form more than a nominal connection. The Isthmus of Suez indeed, being itself a desert, and connecting districts which for a great distance are more or less desert also, does not effect any real union between the luxuriant forest-clad regions of intertropical Asia and Africa.   The Isthmus of Panama is a more effectual line of union, since it is hilly, well watered, and covered with luxuriant vegetation; and we accordingly find that the main features of South American zoology are continued into Central America and Mexico.   In Asia a great transverse barrier exists, dividing that continent into a northern and southern portion; and as the lowlands occur on the south and the highlands on the north of the great mountain range, which is situated not far beyond the tropic, an abrupt change of climate is produced; so that a belt of about a hundred miles wide, is all that intervenes between a luxuriant tropical region and an almost arctic waste.   Between the northern part of Asia, and Europe, there is no barrier of importance; and it is impossible to separate these regions as regards the main features of animal life.   Africa, like Asia, has a great transverse barrier, but it is a desert instead of a mountain chain; and it is found that this desert is a more effectual barrier to the diffusion of animals than the Mediterranean Sea; partly because it coincides with the natural division of a tropical from a temperate climate, but also on account of recent geological changes which we shall presently allude to.   It results then from this outline sketch of the earth's surface, that the primary divisions of the geographer correspond approximately with those of the zoologist.   Some large portion of each of the popular divisions forms the nucleus of a zoological region; but the boundaries are so changed that the geographer would hardly recognise them: it has, therefore, been found necessary to give them those distinct names which will be fully explained in our next chapter.

Recent Changes in the Continental Areas.—The important fact has been now ascertained, that a considerable portion of the Sahara south of Algeria and Morocco was under water at a very recent epoch.   Over much of this area sea-shells, identical with those now living in the Mediterranean, are abundantly scattered, not only in depressions below the level of the sea but up to a height of 900 feet above it. Borings for water made by the French government have shown, that these shells occur twenty feet deep in the sand; and the occurrence of abundance of salt, sometimes even forming considerable hills, is an additional proof of the disappearance of a large body of salt water. The common cockle is one of the most abundant of the shells found; and the Rev. H. B. Tristram discovered a new fish, in a salt lake nearly 300 miles inland, but which has since been found to inhabit the Gulf of Guinea. Connected with this proof of recent elevation in the Sahara, we have most interesting indications of subsidence in the area of the Mediterranean, which were perhaps contemporaneous. Sicily and Malta are connected with Africa by a submerged bank from 300 to 1,200 feet below the surface; while the depth of the Mediterranean, both to the east and west, is enormous, in some parts more than 13,000 feet; and another submerged bank with a depth of 1,000 feet occurs at the straits of Gibraltar. In caves in Sicily, remains of the living African elephant have been found by Baron Anca; and in other caves Dr. Falconer discovered remains of the *Elephas antiquus* and of two species of *Hippopotamus*. In Malta, three species of elephant have been discovered by Captain Spratt; a large one closely allied to *E. antiquus* and two smaller ones not exceeding five feet high when adult. These facts clearly indicate, that when North Africa was separated by a broad arm of the sea from the rest of the continent, it was probably connected with Europe; and this explains why zoologists find themselves obliged to place it along with Europe in the same zoological region.

Besides this change in the level of the Sahara and the Mediterranean basin, Europe has undergone many fluctuations in its physical geography in very recent times. In Wales, abundance of sea-shells of living species have been found at an elevation of 1,300 feet; and in Sardinia there is proof of an elevation of 300 feet since the human epoch; and these are only samples of many such changes of level. But these changes, though very important locally and as connected with geological problems, need not be further noticed here; as they were not of a

nature to affect the larger features of the earth's surface or to
determine the boundaries of great zoological regions.

The only other other recent change of great importance which
can be adduced to illustrate our present subject, is that which
has taken place between North and South America. The living
marine shells of the opposite coasts of the isthmus of Panama,
as well as the corals and fishes, are generally of distinct species,
but some are identical and many are closely allied; the West
Indian fossil shells and corals of the Miocene period, however,
are found to be largely identical with those of the Pacific coast.
The fishes of the Atlantic and Pacific shores of America are
as a rule very distinct; but Dr. Günther has recently shown
that a considerable number of species inhabiting the seas on
opposite sides of the isthmus are absolutely identical. These
facts certainly indicate, that during the Miocene epoch a broad
channel separated North and South America; and it seems probable that a series of elevations and subsidences have taken
place uniting and separating them at different epochs; the most
recent submersion having lasted but a short time, and thus,
while allowing the passage of abundance of locomotive fishes,
not admitting of much change in the comparatively stationary
mollusca.

The Glacial Epoch as affecting the Distribution of Animals.—
The remarkable refrigeration of climate in the northern hemisphere within the epoch, of existing species, to which the term
Glacial epoch is applied, together with the changes of level that
accompanied and perhaps assisted to produce it, has been one of
the chief agents in determining many of the details of the existing distribution of animals in temperate zones. A comparison
of the effects produced by existing glaciers with certain superficial phenomena in the temperate parts of Europe and North
America, renders it certain that between the Newer Pliocene and
the Recent epochs, a large portion of the northern hemisphere
must have been covered with a sheet of ice several thousand
feet thick, like that which now envelopes the interior of Greenland. Much further south the mountains were covered with
perpetual snow, and sent glaciers down every valley; and all the

great valleys on the southern side of the Alps poured down streams of ice which stretched far out into the plains of Northern Italy, and have left their débris in the form of huge mountainous moraines, in some cases more than a thousand feet high. In Canada and New Hampshire the marks of moving ice are found on the tops of mountains from 3,000 to 5,000 feet high; and the whole surface of the country around and to the north of the great lakes is scored by glaciers. Wherever the land was submerged during a part of this cold period, a deposit called boulder-clay, or glacial-drift has been formed. This is a mass of sand, clay, or gravel, full of angular or rounded stones of all sizes, up to huge blocks as large as a cottage; and especially characterized by these stones being distributed confusedly through it, the largest being as often near the top as near the bottom, and never sorted into layers of different sizes as in materials carried by water. Such deposits are known to be formed by glaciers and icebergs; when deposited on the land by glaciers they form moraines, when carried into water and thus spread with more regularity over a wider area they form drift. This drift is rarely found except where there is other evidence of ice-action, and never south of the 40th parallel of latitude, to which in the northern hemisphere signs of ice-action extend. In the southern hemisphere, in Patagonia and in New Zealand, exactly similar phenomena occur.

A very interesting confirmation of the reality of this cold epoch is derived from the study of fossil remains. Both the plants and animals of the Miocene period indicate that the climate of Central Europe was decidedly warmer or more equable than it is now; since the flora closely resembled that of the Southern United States, with a likeness also to that of Eastern Asia and Australia. Many of the shells were of tropical genera; and there were numbers of large mammalia allied to the elephant, rhinoceros, and tapir. At the same time, or perhaps somewhat earlier, a temperate climate extended into the arctic regions, and allowed a magnificent vegetation of shrubs and forest trees, some of them evergreen, to flourish within twelve degrees of the Pole. In the Pliocene period we find ourselves among forms implying a climate very little different from the present; and our own Crag formation furnishes evidence of a gradual refrigeration of climate; since its three divisions, the Coralline, Red, and Norwich Crags, show a decreasing number of southern, and an increasing number of northern species, as we approach the Glacial epoch. Still later than these we have the shells of the drift, almost all of which are northern and many of them arctic species. Among the mammalia indicative of cold, are the mammoth and the reindeer. In gravels and cave-deposits of Post-Pliocene date we find the same two animals, which soon disappear as the climate approached its present condition; and Professor Forbes has given a list of fifty shells which inhabited the British seas before the Glacial epoch and inhabit it still, but are all wanting in the glacial deposits. The whole of these are found in the Newer Pliocene strata of Sicily and the south of Europe, where they escaped destruction during the glacial winter.

There are also certain facts in the distribution of plants, which are so well explained by the Glacial epoch that they may be said to give an additional confirmation to it. All over the northern hemisphere within the glaciated districts, the summits of lofty mountains produce plants identical with those of the polar regions. In the celebrated case of the White Mountains in New Hampshire, United States (latitude 45°), all the plants on the summit are arctic species, none of which exist in the lowlands for near a thousand miles further north. It has also been remarked that the plants of each mountain are more especially related to those of the countries directly north of it. Thus, those of the Pyrenees and of Scotland are Scandinavian, and those of the White Mountains are all species found in Labrador. Now, remembering that we have evidence of an exceedingly mild and uniform climate in the arctic regions during the Miocene period and a gradual refrigeration from that time, it is evident that with each degree of change more and more hardy plants would be successively driven southwards; till at last the plains of the temperate zone would be inhabited by plants, which were once confined to alpine heights or to the arctic regions.

As the icy mantle gradually melted off the face of the earth these plants would occupy the newly exposed soil, and would thus necessarily travel in two directions, back towards the arctic circle and up towards the alpine peaks. The facts are thus exactly explained by a cause which independent evidence has proved to be a real one, and every such explanation is an additional proof of the reality of the cause. But this explanation implies, that in cases where the Glacial epoch cannot have so acted alpine plants should not be northern plants; and a striking proof of this is to be found on the Peak of Teneriffe, a mountain 12,000 feet high. In the uppermost 4,500 feet of this mountain above the limit of trees, Von Buch found only eleven species of plants, eight of which were peculiar; but the whole were allied to those found at lower elevations. On the Alps or Pyrenees at this elevation, there would be a rich flora comprising hundreds of arctic plants; and the absence of anything corresponding to them in this case, in which their ingress was cut off by the sea, is exactly what the theory leads us to expect.

Changes of Vegetation as affecting the Distribution of Animals. —As so many animals are dependent on vegetation, its changes immediately affect their distribution. A remarkable example of this is afforded by the pre-historic condition of Denmark, as interpreted by means of the peat-bogs and kitchen-middens. This country is now celebrated for its beech-trees; oaks and pines being scarce; and it is known to have had the same vegetation in the time of the Romans. In the peat-bogs, however, are found deposits of oak trees; and deeper still pines alone occur. Now the kitchen-middens tell us much of the natural history of Denmark in the early Stone period; and a curious confirmation of the fact that Denmark like Norway was then chiefly covered with pine forests is obtained by the discovery, that the Capercailzie was then abundant, a bird which feeds almost exclusively on the young shoots and seeds of pines and allied plants. The cause of this change in the vegetation is unknown; but from the known fact that when forests are destroyed trees of a different kind usually occupy the ground, we may suppose that some such change as a temporary submergence might cause an entirely different vegetation and a considerably modified fauna to occupy the country.

Organic Changes as affecting Distribution.—We have now briefly touched on some of the direct effects of changes in physical geography, climate, and vegetation, on the distribution of animals; but the indirect effects of such changes are probably of quite equal, if not of greater importance. Every change becomes the centre of an ever-widening circle of effects. The different members of the organic world are so bound together by complex relations, that any one change generally involves numerous other changes, often of the most unexpected kind. We know comparatively little of the way in which one animal or plant is bound up with others, but we know enough to assure us that groups the most apparently disconnected are often dependent on each other. We know, for example, that the introduction of goats into St. Helena utterly destroyed a whole flora of forest trees; and with them all the insects, mollusca, and perhaps birds directly or indirectly dependent on them. Swine, which ran wild in Mauritius, exterminated the Dodo. The same animals are known to be the greatest enemies of venomous serpents. Cattle will, in many districts, wholly prevent the growth of trees; and with the trees the numerous insects dependent on those trees, and the birds which fed upon the insects, must disappear, as well as the small mammalia which feed on the fruits, seeds, leaves, or roots. Insects again have the most wonderful influence on the range of mammalia. In Paraguay a certain species of fly abounds which destroys new-born cattle and horses; and thus neither of these animals have run wild in that country, although they abound both north and south of it. This inevitably leads to a great difference in the vegetation of Paraguay, and through that to a difference in its insects, birds, reptiles, and wild mammalia. On what causes the existence of the fly depends we do not know, but it is not improbable that some comparatively slight changes in the temperature or humidity of the air at a particular season, or the introduction of some enemy might lead to its extinction or banishment. The whole face of the country would then soon be changed: new species would come in, while many others would be unable to live there; and the immediate cause of this great alteration would probably be quite imperceptible to us, even if we could watch it in progress year by year. So, in South Africa, the celebrated Tsetse fly inhabits certain districts having well defined limits; and where it abounds no horses, dogs, or cattle can live. Yet asses, zebras, and antelopes are unaffected by it. So long as this fly continues to exist, there is a living barrier to the entrance of certain animals, quite as effectual as a lofty mountain range or a wide arm of the sea. The complex relations of one form of life with others is nowhere better illustrated than in Mr. Darwin's celebrated case of the cats and clover, as given in his *Origin of Species*, 6th ed., p. 57. He has observed that both wild heartsease and red-clover are fertilized in this country by humble-bees only, so that the production of seed depends on the visits of these insects. A gentleman who has specially studied humble-bees finds that they are largely kept down by field-mice, which destroy their combs and nests. Field-mice in their turn are kept down by cats; and probably also by owls; so that these carnivorous animals are really the agents in rendering possible the continued existence of red-clover and wild heartsease. For if they were absent, the field-mice having no enemies, would multiply to such an extent as to destroy all the humble-bees; and these two plants would then produce no seed and soon become extinct.

Mr. Darwin has also shown that one species often exterminates another closely allied to it, when the two are brought into contact. One species of swallow and thrush are known to have increased at the expense of allied species. Rats, carried all over the world by commerce, are continually extirpating other species of rats. The imported hive-bee is, in Australia, rapidly exterminating a native stingless bee. Any slight change, therefore, of physical geography or of climate, which allows allied species hitherto inhabiting distinct areas to come into contact, will often lead to the extermination of one of them; and this extermination will be effected by no external force, by no actual enemy, but merely because the one is slightly better adapted to live, to increase, and to maintain itself under adverse circumstances, than the other.

Now if we consider carefully the few suggestive facts here referred to (and many others of like import are to be found in Mr. Darwin's various works), we shall be led to conclude that the several species, genera, families, and orders, both of animals and vegetables which inhabit any extensive region, are bound together by a series of complex relations; so that the increase, diminution, or extermination of any one, may set in motion a series of actions and reactions more or less affecting a large portion of the whole, and requiring perhaps centuries of fluctuation before the balance is restored The range of any species or group in such a region, will in many cases (perhaps in most) be determined, not by physical barriers, but by the competition of other organisms. Where barriers have existed from a remote epoch, they will at first have kept back certain animals from coming in contact with each other; but when the assemblage of organisms on the two sides of the barrier have, after many ages, come to form a balanced organic whole, the destruction of the barrier may lead to a very partial intermingling of the peculiar forms of the two regions. Each will have become modified in special ways adapted to the organic and physical conditions of the country, and will form a living barrier to the entrance of animals less perfectly adapted to those conditions. Thus while the abolition of ancient barriers will always lead to much intermixture of forms, much extermination and widespread alteration in some families of animals; other important groups will be unable materially to alter their range; or they may make temporary incursions into the new territory, and be ultimately driven back to very near their ancient limits.

In order to make this somewhat difficult subject more intelligible, it may be well to consider the probable effects of certain hypothetical conditions of the earth's surface :—

1. If the dry land of the globe had been from the first continuous, and nowhere divided up by such boundaries as lofty mountain ranges, wide deserts, or arms of the sea, it seems probable that none of the larger groups (as *orders*, *tribes*, or

families,) would have a limited range; but, as is to some extent the case in tropical America east of the Andes, every such group would be represented over the whole area, by countless minute modifications of form adapted to local conditions.

2. One great physical barrier would, however, even then exist; the hot equatorial zone would divide the faunas and floras of the colder regions of the northern and southern hemispheres from any chance of intermixture. This one barrier would be more effectual than it is now, since there would be no lofty mountain ranges to serve as a bridge for the partial interchange of northern and southern forms.

3. If such a condition of the earth as here supposed continued for very long periods, we may conceive that the action and reaction of the various organisms on each other, combined with the influence of very slowly changing physical conditions, would result in an almost perfect organic balance, which would be manifested by a great stability in the average numbers, the local range, and the peculiar characteristics of every species.

4. Under such a condition of things it is not improbable that the total number of clearly differentiated specific forms might be much greater than it is now, though the number of generic and family types might perhaps be less; for dominant species would have had ample time to spread into every locality where they could exist, and would then become everywhere modified into forms best suited to the permanent local conditions.

5. Now let us consider what would be the probable effect of the introduction of a barrier, cutting off a portion of this homogeneous and well-balanced world. Suppose, for instance, that a subsidence took place, cutting off by a wide arm of the sea a large and tolerably varied island. The first and most obvious result would be that the individuals of a number of species would be divided into two portions, while others, the limits of whose range agreed approximately with the line of subsidence, would exist in unimpaired numbers on the new island or on the main land. But the species whose numbers were diminished and whose original area was also absolutely diminished by the portion now under the sea, would not be able to hold their

ground against the rival forms whose numbers were intact. Some would probably diminish and rapidly die out; others which produced favourable varieties, might be so modified by natural selection as to maintain their existence under a different form; and such changes would take place in varying modes on the two sides of the new strait.

6. But the progress of these changes would necessarily affect the other species in contact with them. New places would be opened in the economy of nature which many would struggle to obtain; and modification would go on in ever-widening circle and very long periods of time might be required to bring the whole again into a state of equilibrium.

7. A new set of factors would in the meantime have come into play. The sinking of land and the influx of a large body of water could hardly take place without producing important climatal changes. The temperature, the winds, the rains, might all be affected, and more or less changed in duration and amount. This would lead to a quite distinct movement in the organic world. Vegetation would certainly be considerably affected, and through this the insect tribes. We have seen how closely the life of the higher animals is often bound up with that of insects; and thus a set of changes might arise that would modify the numerical proportions, and even the forms and habits of a great number of species, would completely exterminate some, and raise others from a subordinate to a dominant position. And all these changes would occur differently on opposite sides of the strait, since the insular climate could not fail to differ considerably from that of the continent.

8. But the two sets of changes, as above indicated, produced by different modes of action of the same primary cause, would act and react on each other; and thus lead to such a far-spreading disturbance of the organic equilibrium as ultimately perhaps to affect in one way or another, every form of life upon the earth.

This hypothetical case is useful as enabling us better to realize how wide-spreading might be the effects of one of the simplest changes of physical geography, upon a compact mass of mutually adapted organisms. In the actual state of things, the physical changes that occur and have occurred through all geological epochs are larger and more varied. Almost every mile of land surface has been again and again depressed beneath the ocean ; most of the great mountain chains have either originated or greatly increased in height during the Tertiary period ; marvellous alterations of climate and vegetation have taken place over half the land-surface of the earth ; and all these vast changes have influenced a globe so cut up by seas and oceans, by deserts and snow-clad mountains, that in many of its more isolated land-masses ancient forms of life have been preserved, which, in the more extensive and more varied continents have long given way to higher types. How complex then must have been the actions and reactions such a state of things would bring about ; and how impossible must it be for us to guess, in most cases, at the exact nature of the forces that limit the range of some species and cause others to be rare or to become extinct ! All that we can in general hope to do is, to trace out, more or less hypothetically, some of the larger changes in physical geography that have occurred during the ages immediately pre-ceeding our own, and to estimate the effect they will probably have produced on animal distribution. We may then, by the aid of such knowledge as to past organic mutations as the geo-logical record supplies us with, be able to determine the probable birthplace and subsequent migrations of the more important genera and families ; and thus obtain some conception of that grand series of co-ordinated changes in the earth and its in-habitants, whose final result is seen in the forms and the geo-graphical distribution of existing animals.

CHAPTER IV.

ON ZOOLOGICAL REGIONS.

To the older school of Naturalists the native country of an animal was of little importance, except in as far as climates differed. Animals were supposed to be specially adapted to live in certain zones or under certain physical conditions, and it was hardly recognised that apart from these conditions there was any influence in locality which could materially affect them. It was believed that, while the animals of tropical, temperate, and arctic climates, essentially differed; those of the tropics were essentially alike all over the world. A group of animals was said to inhabit the "Indies;" and important differences of structure were often overlooked from the idea, that creatures equally adapted to live in hot countries and with certain general resemblances, would naturally be related to each other. Thus the Toucans and Hornbills, the Humming-Birds and Sun-Birds, and even the Tapirs and the Elephants, came to be popularly associated as slightly modified varieties of tropical forms of life; while to naturalists, who were acquainted with the essential differences of structure, it was a never-failing source of surprise, that under climates and conditions so apparently identical, such strangely divergent forms should be produced.

To the modern naturalist, on the other hand, the native country (or "habitat" as it is technically termed) of an animal or a group of animals, is a matter of the first importance; and, as regards the general history of life upon the globe, may be considered to be one of its essential characters. The structure, affinities, and habits of a species, now form only a part of its natural history. We require also to know its exact range at the present day and in prehistoric times, and to have some knowledge of its geological age, the place of its earliest appearance on the globe, and of the various extinct forms most nearly allied to it. To those who accept the theory of development as worked out by Mr. Darwin, and the views as to the general permanence and immense antiquity of the great continents and oceans so ably developed by Sir Charles Lyell, it ceases to be a matter of surprise that the tropics of Africa, Asia, and America should differ in their productions, but rather that they should have anything in common. Their similarity, not their diversity, is the fact that most frequently puzzles us.

The more accurate knowledge we have of late years obtained of the productions of many remote regions, combined with the greater approaches that have been made to a natural classification of the higher animals, has shown, that every continent or well-marked division of a continent, every archipelago and even every island, presents problems of more or less complexity to the student of the geographical distribution of animals. If we take up the subject from the zoological side, and study any family, order, or even extensive genus, we are almost sure to meet with some anomalies either in the present or past distribution of the various forms. Let us adduce a few examples of these problems.

Deer have a wonderfully wide range, over the whole of Europe, Asia, and North and South America; yet in Africa south of the great desert there are none. Bears range over the whole of Europe, Asia, and North America, and true pigs of the genus *Sus*, over all Europe and Asia and as far as New Guinea; yet both bears and pigs, like deer, are absent from Tropical and South Africa.

Again, the West Indian islands possess very few Mammalia, all of small size and allied to those of America, except one

E 2

genus; and that belongs to an Order, "Insectivora," entirely
absent from South America, and to a family, "Centetidæ," all
the other species of which inhabit Madagascar only. And as
if to add force to this singular correspondence we have one
Madagascar species of a beautiful day-flying Moth, *Urania*, all
the other species of which inhabit tropical America. These
insects are gorgeously arrayed in green and gold, and are quite
unlike any other Lepidoptera upon the globe.

The island of Ceylon generally agrees in its productions with
the Southern part of India; yet it has several birds which are
allied to Malayan and not to Indian groups, and a fine butterfly
of the genus *Hestia*, as well as several genera of beetles, which
are purely Malayan.

Various important groups of animals are distributed in a
way not easy to explain. The anthropoid apes in West Africa
and Borneo; the tapirs in Malaya and South America; the
camel tribe in the deserts of Asia and the Andes; the trogons
in South America and Tropical Asia, with one species in Africa;
the marsupials in Australia and America, are examples.

The cases here adduced (and they might be greatly multiplied)
are merely to show the kind of problems with which the
naturalist now has to deal; and in order to do so he requires
some system of geographical arrangement, which shall serve
the double purpose of affording a convenient subdivision of his
subject, and at the same time of giving expression to the main
results at which he has arrived. Hence the recent discussions
on "Zoological Regions," or, what are the most natural
primary divisions of the earth as regards its forms of animal
life.

The divisions in use till quite recently were of two kinds;
either those ready made by geographers, more especially the
quarters or continents of the globe; or those determined by
climate and marked out by certain parallels of latitude or by
isothermal lines. Either of these methods was better than
none at all; but from the various considerations explained in
the preceding chapters, it will be evident, that such divisions
must have often been very unnatural, and have disguised many

of the most important and interesting phenomena which a study of the distribution of animals presents to us.

The merit of initiating a more natural system, that of determining zoological regions, not by any arbitrary or *a priori* consideration but by studying the actual ranges of the more important groups of animals, is due to Mr. Sclater, who, in 1857, established six primary zoological regions from a detailed examination of the distribution of the chief genera and families of Birds. Before stating what these regions are, what objections have been made to them, what other divisions have been since proposed, and what are those which we shall adopt in this work, it will be well to consider the general principles which should guide us in the choice between rival systems.

Principles on which Zoological Regions should be formed.— It will be evident in the first place that nothing like a perfect zoological division of the earth is possible. The causes that have led to the present distribution of animal life are so varied, their action and reaction have been so complex, that anomalies and irregularities are sure to exist which will mar the symmetry of any rigid system. On two main points every system yet proposed, or that probably can be proposed, is open to objection; they are,—1stly, that the several regions are not of equal rank;—2ndly, that they are not equally applicable to all classes of animals. As to the first objection, it will be found impossible to form any three or more regions, each of which differs from the rest in an equal degree or in the same manner. One will surpass all others in the possession of peculiar families; another will have many characteristic genera; while a third will be mainly distinguished by negative characters. There will also be found many intermediate districts, which possess some of the characteristics of two well-marked regions, with a few special features of their own, or perhaps with none; and it will be a difficult question to decide in all cases which region should possess this doubtful territory, or whether it should be formed into a primary region itself. Again, two regions which have now well-marked points of difference, may be shown to have been much more alike at a comparatively recent geological epoch;

and this, it may be said, proves their fundamental unity and that they ought to form but one primary region. To obviate some of these difficulties a binary or dichotomous division is sometimes proposed; that portion of the earth which differs most from the rest being cut off as a region equal in rank to all that remains, which is subjected again and again to the same process.

To decide these various points it seems advisable that convenience, intelligibility, and custom, should largely guide us. The first essential is, a broadly marked and easily remembered set of regions; which correspond, as nearly as truth to nature will allow, with the distribution of the most important groups of animals. What these groups are we shall presently explain. In determining the number, extent, and boundaries of these regions, we must be guided by a variety of indications, since the application of fixed rules is impossible. They should evidently be of a moderate number, corresponding as far as practicable with the great natural divisions of the globe marked out by nature, and which have always been recognized by geographers. There should be some approximation to equality of size, since there is reason to believe that a tolerably extensive area has been an essential condition for the development of most animal forms; and it is found that, other things being equal, the numbers, variety and importance of the forms of animal and vegetable life, do bear some approximate relation to extent of area. Although the possession of peculiar families or genera is the main character of a primary zoological region, yet the negative character of the absence of certain families or genera is of equal importance, *when this absence does not manifestly depend on unsuitability to the support of the group,* and especially *when there is now no physical barrier preventing their entrance.* This will become evident when we consider that the importance of the possession of a group by one region depends on its absence from the adjoining regions; and if there is now no barrier to its entrance, we may be sure that there has once been one; and that the possession of the area by a distinct and well balanced set of organisms, which must have been slowly developed and adjusted, is the living barrier that now keeps out intruders.

When it is ascertained that the chief differences which now obtain between two areas did not exist in Miocene or Pliocene times, the fact is one of great interest, and enables us to speculate with some degree of probability as to the causes that have brought about the present state of things; but it is not a reason for uniting these two areas into one region. Our object is to represent as nearly as possible the main features of the distribution of existing animals, not those of any or all past geological epochs. Should we ever obtain sufficient information as to the geography and biology of the earth at past epochs, we might indeed determine approximately what were the Pliocene or Miocene or Eocene zoological regions; but any attempt to exhibit all these in combination with those of our own period, must lead to confusion.

The binary or dichotomous system, although it brings out the fundamental differences of the respective regions, is an inconvenient one in its application, and rather increases than obviates the difficulty as to equality or inequality of regions; for although a, b, c, and d, may be areas of unequal zoological rank, a being the most important, and d the least, yet this inequality will probably be still greater if we first divide them into a, on one side, and b, c, and d, on the other, and then, by another division, make b, an area of the second, and c, and d, of the third rank only.

Coming to the second objection, the often incompatible distribution of different groups of animals, affords ground for opposition to any proposed scheme of zoological regions. There is first the radical difference between land and sea animals; the most complete barriers to the dispersal of the one, sometimes offering the greatest facilities for the emigration of the other, and *vice versa*. A large number of marine animals, however, frequent shallow water only; and these, keeping near the coasts, will agree generally in their distribution with those inhabiting the land. But among land animals themselves there are very great differences of distribution, due to certain specialities in their organization or mode of life. These act mainly in two ways,—1stly, by affecting the facilities with which they can be dispersed, either voluntarily or involuntarily;—2ndly, by the conditions which enable them to multiply and establish themselves in certain areas and not in others. When both these means of diffusion are at a maximum, the dispersal of a group becomes universal, and ceases to have much interest for us. This is the case with certain groups of fungi and lichens, as well as with some of the lower animals; and in a less degree, as has been shown by Mr. Darwin, with many fresh-water plants and animals. At the other extreme we may place certain arboreal vertebrata such as sloths and lemurs, which have no means of passing such barriers as narrow straits or moderately high mountains, and whose survival in any new country they might reach, would be dependent on the presence of suitable forests and the absence of dangerous enemies. Almost equally, or perhaps even more restricted, are the means of permanent diffusion of terrestrial molluscs; since these are without any but very rare and accidental means of being safely transported across the sea; their individual powers of locomotion are highly restricted; they are especially subject to the attacks of enemies; and they often depend not only on a peculiar vegetation, but on the geological character of the country, their abundance being almost in direct proportion to the presence of some form of calcareous rocks. Between these extremes we find animals possessed of an infinite gradation of powers to disperse and to maintain themselves; and it will evidently be impossible that the limits which best define the distribution of one group, should be equally true for all others.

Which class of Animals is of most importance in determining Zoological Regions.—To decide this question we have to consider which groups of animals are best adapted to exhibit, by their existing distribution, the past changes and present physical condition of the earth's surface; and at the same time, by the abundance of their remains in the various tertiary formations will best enable us to trace out the more recent of the series of changes, both of the earth's surface and of its inhabitants, by which the present state of things has been brought about. For this purpose we require a group which shall be dependent for its means of dispersal on the distribution of land and water, and on the presence or absence of lofty mountains, desert plains or plateaux, and great forests; since these are the chief physical features of the earth's surface whose modifications at successive periods we wish to discover. It is also essential that they should not be subject to dispersal by many accidental causes; as this would inevitably in time tend to obliterate the effect of natural barriers, and produce a scattered distribution, the causes of which we could only guess at. Again, it is necessary that they should be so highly organized as not to be absolutely dependent on other groups of animals, and with so much power of adaptation as to be able to exist in one form or another over the whole globe. And lastly, it is highly important that the whole group should be pretty well known, and that a fairly natural classification, especially of its minor divisions such as families and genera, should have been arrived at; the reason for which last proviso is explained in our next chapter, on classification.

Now in every one of these points the mammalia are preemi nent; and they possess the additional advantage of being the most highly developed class of organized beings, and that to which we ourselves belong. We should therefore construct our typical or standard Zoological Regions in the first place, from a consideration of the distribution of mammalia, only bringing to our aid the distribution of other groups to determine doubtful points. Regions so established will be most closely in accordance with those long-enduring features of physical geography, on which the distribution of all forms of life fundamentally depend; and all discrepancies in the distribution of other classes of animals must be capable of being explained, either by their exceptional means of dispersion or by special conditions affecting their perpetuation and increase in each locality.

If these considerations are well founded, the objections of those who study insects or molluscs, for example,— that our regions are not true for their departments of nature—cannot be maintained. For they will find, that a careful consideration of the exceptional means of dispersal and conditions of existence of each group, will explain most of the divergences from the normal distribution of higher animals.

We shall thus be led to an intelligent comprehension of the phenomena of distribution in all groups, which would not be the case if every specialist formed regions for his own particular study. In many cases we should find that no satisfactory division of the earth could be made to correspond with the distribution even of an entire class; but we should have the coleopterist and the lepidopterist each with his own Geography. And even this would probably not suffice, for it is very doubtful if the detailed distribution of the Longicornes, so closely dependent on woody vegetation, could be made to agree with that of the Staphylinidæ or the Carabidæ which abound in many of the most barren regions, or with that of the Scarabeidæ, largely dependent on the presence of herbivorous mammalia. And when each of these enquirers had settled a division of the earth into "regions" which exhibited with tolerable accuracy the phenomena of distribution of his own group, we should have gained nothing whatever but a very complex mode of exhibiting the bare facts of distribution. We should then have to begin to work out the causes of the divergence of one group from another in this respect; but as each worker would refer to his own set of regions as the type, the whole subject would become involved in inextricable confusion. These considerations seem to make it imperative that one set of "regions" should be established as typical for Zoology; and it is hoped the reasons here advanced will satisfy most naturalists that these regions can be best determined, in the first place, by a study of the distribution of the mammalia, supplemented in doubtful cases by that of the other vertebrates. We will now proceed to a discussion of what these regions are.

Various Zoological Regions proposed since 1857.—It has already been pointed out that a very large number of birds are limited by the same kind of barriers as mammalia; it will therefore not be surprising that a system of regions formed to suit the one, should very nearly represent the distribution of the other. Mr. Sclater's regions are as follows :—

1. The Palæarctic Region; including Europe, Temperate Asia, and N. Africa to the Atlas mountains.

2. The Ethiopian Region; Africa south of the Atlas, Madagascar, and the Mascarene Islands, with Southern Arabia.

3. The Indian Region; including India south of the Himalayas, to South China, and to Borneo and Java.

4. The Australian Region; including Celebes and Lombock, eastward to Australia and the Pacific Islands.

5. The Nearctic Region; including Greenland, and N. America, to Northern Mexico.

6. The Neotropical Region; including South America, the Antilles, and Southern Mexico.

This division of the earth received great support from Dr. Günther, who, in the *Proceedings of the Zoological Society* for 1858, showed that the geographical distribution of Reptiles agreed with it very closely, the principal difference being that the reptiles of Japan have a more Indian character than the birds, this being especially the case with the snakes. In the volume for 1868 of the same work, Professor Huxley discusses at considerable length the primary and secondary zoological divisions of the earth. He gives reasons for thinking that the most radical primary division, both as regards birds and mammals, is into a Northern and Southern hemisphere (Arctogæa and Notogæa), the former, however, embracing all Africa, while the latter includes only Australasia and the Neotropical or Austro-Columbian region. Mr. Sclater had grouped his regions primarily into Palæogæa and Neogæa, the Old and New Worlds of geographers; a division which strikingly accords with the distribution of the passerine birds, but not so well with that of mammalia or reptiles. Professor Huxley points out that the Nearctic, Palæarctic, Indian, and Ethiopian regions of Mr. Sclater have a much greater resemblance to each other than any one of them has to Australia or to South America; and he further suggests that New Zealand alone has peculiarities which might entitle it to rank as a primary region

along with Australasia and South America; and that a Circumpolar Province might be conveniently recognised as of equal
rank with the Palæarctic and Nearctic provinces.

In 1866, Mr. Andrew Murray published a large and copiously
illustrated volume on the *Geographical Distribution of Mammals*, in which he maintains that the great and primary
mammalian regions are only four: 1st. The Palæarctic region
of Mr. Sclater, extended to include the Sahara and Nubia;
2nd. the Indo-African region, including the Indian and Ethiopian
regions of Mr. Sclater; 3rd. the Australian region (unaltered);
4th. the American region, including both North and South
America. These are the regions as *described* by Mr. Murray,
but his coloured map of "Great Mammalian Regions" shows
all Arctic America to a little south of the Isothermal of 32°
Fahr. as forming with Europe and North Asia one great region.

At the meeting of the British Association at Exeter in 1869,
Mr. W. T. Blanford read a paper on the Fauna of British India,
in which he maintained that a large portion of the peninsula
of India had derived its Fauna mainly from Africa; and that the
term "Indian region" of Mr. Sclater was misleading, because
India proper, if it belongs to it at all, is the least typical portion
of it. He therefore proposes to call it the "Malayan region,"
because in the Malay countries it is most highly developed.
Ceylon and the mountain ranges of Southern India have marked
Malay affinities.

In 1871 Mr. E. Blyth published in *Nature* "A suggested new
Division of the Earth into Zoological Regions," in which he
indicates seven primary divisions or regions, subdivided into
twenty-six sub-regions. The seven regions are defined as
follows: 1. The Boreal region; including the whole of the
Palæarctic and Nearctic regions of Mr. Sclater along with the
West Indies, Central America, the whole chain of the Andes,
with Chili and Patagonia. 2. The Columbian region; consisting
of the remaining part of South America. 3. The Ethiopian
region; comprising besides that region of Mr. Sclater, the valley
of the Jordan, Arabia, and the desert country towards India,
with all the plains and table lands of India and the northern

half of Ceylon. 4. The Lemurian region; consisting of Madagascar and its adjacent islands. 5. The Austral-Asian region; which is the Indian region of Mr. Sclater without the portion taken to be added to the Ethiopian region. 6. The Melanesian region; which is the Australian region of Mr. Sclater without New Zealand and the Pacific Islands, which form 7. the Polynesian region. Mr. Blyth thinks this is "a true classification of zoological regions as regards mammalia and birds."

In an elaborate paper on the birds of Eastern North America, their distribution and migrations (*Bulletin of Museum of Comparative Zoology, Cambridge, Massachusetts*, Vol. 2), Mr. J. A. Allen proposes a division of the earth in accordance with what he terms, "the law of circumpolar distribution of life in zones," as follows: 1. Arctic realm. 2. North temperate realm. 3. American tropical realm. 4. Indo-African tropical realm. 5. South American tropical realm. 6. African temperate realm. 7. Antarctic realm. 8. Australian realm. Some of these are subdivided into regions; (2) consisting of the American and the Europæo-Asiatic regions; (4) into the African and Indian regions; (8) into the tropical Australian region, and one comprising the southern part of Australia and New Zealand. The other realms each form a single region.

Discussion of proposed Regions.—Before proceeding to define the regions adopted in this work, it may be as well to make a few remarks on some of the preceding classifications, and to give the reasons which seem to render it advisable to adopt very few of the suggested improvements on Mr. Sclater's original proposal. Mr. Blyth's scheme is one of the least natural, and also the most inconvenient. There can be little use in the knowledge that a group of animals is found in the Boreal Region, if their habitat might still be either Patagonia, the West Indies, or Japan; and it is difficult to see on what principle the Madagascar group of islands is made of equal rank with this enormous region, seeing that its forms of life have marked African affinities. Neither does it seem advisable to adopt the Polynesian Region, or that comprising New Zealand alone (as hinted at by Professor Huxley and since adopted by

Mr. Sclater in his Lectures on Geographical Distribution at the
Zoological Gardens in May 1874), because it is absolutely without indigenous mammalia and very poor in all forms of life,
and therefore by no means prominent or important enough to
form a primary region of the earth.

It may be as well here to notice what appears to be a serious
objection to making New Zealand, or any similar isolated
district, one of the great zoological regions, comparable to South
America, Australia, or Ethiopia ; which is, that its claim to that
distinction rests on grounds which are liable to fail. It is
because New Zealand, in addition to its negative merits, possesses
three families of birds (Apterygidæ living, Dinornithidæ and
Palapterygidæ extinct), and a peculiar lizard-like reptile,
Hatteria, which has to be classed in a distinct order, Rhyncho-
cephalina, that the rank of a Region is claimed for it. But
supposing, what is not at all improbable, that other Rhyncho-
cephalina should be discovered in the interior of Australia or
in New Guinea, and that Apterygidæ or Palapterygidæ should
be found to have inhabited Australia in Post-Pliocene times,
(as Dinornithidæ have already been proved to have done) the
claims of New Zealand would entirely fail, and it would be
universally acknowledged to be a part of the great Australian
region. No such reversal can take place in the case of the
other regions ; because they rest, not upon one or two, but upon a
large number of peculiarities, of such a nature that there is no
room upon the globe for discoveries that can seriously modify
them. Even if one or two peculiar types, like Apterygidæ or
Hatteria, should permanently remain characteristic of New Zealand alone, we can account for these by the extreme isolation of
the country, and the absence of enemies, which have enabled
these defenceless birds and reptiles to continue their existence ;
just as the isolation and protection of the caverns of Carniola
have enabled the *Proteus* to survive in Europe. But supposing
that the *Proteus* was the sole representative of an order of
Batrachia, and that two or three other equally curious and
isolated forms occurred with it, no one would propose that these
caverns or the district containing them, should form one of the

primary divisions of the earth. Neither can much stress be laid on the negative characteristics of New Zealand, since they are found to an almost equal extent in every oceanic island.

Again, it is both inconvenient and misleading to pick out certain tracts from the midst of one region or sub-region and to place them in another, on account of certain isolated affinities which may often be accounted for by local peculiarities. Even if the resemblance of the fauna of Chili and Patagonia to that of the Palæarctic and Nearctic regions was much greater than it is, this mode of dealing with it would be objectionable; but it is still more so, when we find that these countries have a strongly marked South American character, and that the northern affinities are altogether exceptional. The Rodentia, which comprise a large portion of the mammalia of these countries, are wholly South American in type, and the birds are almost all allied to forms characteristic of tropical America.

For analogous reasons the Ethiopian must not be made to include any part of India or Ceylon; for although the Fauna of Central India has some African affinities, these do not preponderate; and it will not be difficult to show that to follow Mr. Andrew Murray in uniting bodily the Ethiopian and Indian regions of Mr. Sclater, is both unnatural and inconvenient. The resemblances between them are of the same character as those which would unite them both with the Palæarctic and Nearctic regions; and although it may be admitted, that, as Professor Huxley maintains, this group forms one of the great primary divisions of the globe, it is far too extensive and too heterogeneous to subserve the practical uses for which we require a division of the world into zoological regions.

Reasons for adopting the six Regions first proposed by Mr. Sclater. —So that we do not violate any clear affinities or produce any glaring.irregularities, it is a positive, and by no means an unimportant, advantage to have our named regions approximately equal in size, and with easily defined, and therefore easily remembered, boundaries. . All elaborate definitions of interpenetrating frontiers, as well as regions extending over three-fourths of the land surface of the globe, and including places which are

the antipodes of each other, would be most inconvenient, even
if there were not such difference of opinion about them.   There
can be little doubt, for example, that the most radical zoological
division of the earth is made by separating the Australian region from the rest; but although it is something useful and
definite to know that a group of animals is peculiar to Australia,
it is exceedingly vague and unsatisfactory to say of any other
group merely that it is extra-Australian.   Neither can it be said
that, from any point of view, these two divisions are of equal
importance.   The next great natural division that can be made
is the separation of the Neotropical Region of Mr. Sclater from
the rest of the world.   We thus have three primary divisions,
which Professor Huxley seems inclined to consider as of
tolerably equal zoological importance.   But a consideration of
all the facts, zoological and palæontological, indicates, that the
great northern division (Arctogæa) is fully as much more important than either Australia or South America, as its four component parts are less important; and if so, convenience requires
us to adopt the smaller rather than the larger divisions.

This question, of comparative importance or equivalence of
value, is very difficult to determine.   It may be considered from
the point of view of speciality or isolation, or from that of
richness and variety of animal forms.   In isolation and speciality,
determined by what they want as well as what they possess, the
Australian and Neotropical regions are undoubtedly each comparable with the rest of the earth (Arctogæa).   But in richness
and variety of forms, they are both very much inferior, and are
much more nearly comparable with the separate regions which
compose it.   Taking the families of mammalia as established by
the best authors, and leaving out the Cetacea and the Bats,
which are almost universally distributed, and about whose
classification there is much uncertainty, the number of families
represented in each of Mr. Sclater's regions is as follows:

I.	Palæarctic region has 31 families of terrestrial mammalia.
II.	Ethiopian	„	„ 40	„	„	„
III.	Indian	„	„ 31	„	„	„
IV.	Australian	„	„ 14	„	„	„
V.	Neotropical	„	„ 26	„	„	„
VI.	Nearctic	„	„ 23	„	„	„

We see, then, that even the exceedingly rich and isolated Neo-tropical region is less rich and diversified in its forms of mammalian life than the very much smaller area of the Indian region, or the temperate Palæarctic, and very much less so than the Ethiopian region; while even the comparatively poor Nearctic region, is nearly equal to it in the number of its family types.   If these were united they would possess fifty-five families, a number very disproportionate to those of the remaining two.   Another consideration is, that although the absence of certain forms of life makes a region more isolated, it does not make it zoologically more important; for we have only to suppose some five or six families, now common to both, to become extinct either in the Ethiopian or the Indian regions, and they would become as strongly differentiated from all other regions as South America, while still remaining as rich in family types.   In birds exactly the same phenomenon recurs, the family types being less numerous in South America than in either of the other tropical regions of the earth, but a larger proportion of them are restricted to it.   It will be shown further on, that the Ethiopian and Indian, (or, as I propose to call it in this work, Oriental) regions, are sufficiently differentiated by very important groups of animals peculiar to each; and that, on strict zoological principles they are entitled to rank as regions of equal value with the Neotropical and Australian.   It is perhaps less clear whether the Palæarctic should be separated from the Oriental region, with which it has undoubtedly much in common; but there are many and powerful reasons for keeping it distinct.   There is an unmistakably different facies in the animal forms of the two regions; and although no families of mammalia or birds, and not many genera, are wholly confined to the Palæarctic region, a very considerable number of both have their metropolis in it, and are very richly represented. The distinction between the characteristic forms of life in tropical and cold countries is, on the whole, very strongly marked in the northern hemisphere; and to refuse to recognise this in a subdivision of the earth which is established for the very purpose of expressing such contrasts more clearly and concisely than by ordinary geographical terminology, would be both illogical and

F

inconvenient. The one question then remains, whether the
Nearctic region should be kept separate, or whether it should
form part of the Palæarctic or of the Neotropical regions. Professor Huxley and Mr. Blyth advocate the former course ; Mr.
Andrew Murray (for mammalia) and Professor Newton (for birds)
think the latter would be more natural. No doubt much is to
be said for both views, but both cannot be right; and it will be
shown in the latter part of this chapter that the Nearctic region
is, on the whole, fully as well defined as the Palæarctic, by positive characters which differentiate it from both the adjacent
regions. More evidence in the same direction will be found in
the Second Part of this work, in which the extinct faunas of the
several regions are discussed.

A confirmation of the general views here set forth, as to the
distinctness and approximate equivalence of the six regions, is
to be found in the fact, that if any two or more of them are combined they themselves become divisions of the next lower rank,
or " sub-regions ; "—and these will be very much more important,
both zoologically and geographically, than the subdivisions of
the remaining regions. It is admitted then that these six regions
are by no means of precisely equal rank, and that some of them
are far more isolated and better characterized than others; but
it is maintained that, looked at from every point of view, they
are more equal in rank than any others that can be formed;
while in geographical equality, compactness of area, and facility
of definition, they are beyond all comparison better than any
others that have yet been proposed for the purpose of facilitat-
ing the study of geographical distribution. They may be arranged and grouped as follows, so as to exhibit their various
relations and affinities.

	Regions.		
Neogæa	NEOTROPICAL ...	Austral zone.........	Notogæa.
	NEARCTIC.........	Boreal zone	
Palæogæa	PALÆARCTIC ...		Arctogæa.
	ETHIOPIAN	Palæotropical zone	
	ORIENTAL		
	AUSTRALIAN ...	Austral zone	Notogæa.

The above table shows the regions placed in the order followed
in the Fourth Part of this work, and the reasons for which are

explained in Chapter IX. As a matter of convenience, and for other reasons adduced in the same chapter, the detailed exposition of the geographical distribution of the animals of the several regions in Part III. commences with the Palæarctic and terminates with the Nearctic region.

Objections to the system of Circumpolar Zones.—Mr. Allen's system of "realms" founded on climatic zones (given at p. 61), having recently appeared in an ornithological work of considerable detail and research, calls for a few remarks. The author continually refers to the "*law of the distribution of life in circumpolar zones,*" as if it were one generally accepted and that admits of no dispute. But this supposed "law" only applies to the smallest details of distribution—to the range and increasing or decreasing numbers of *species* as we pass from north to south, or the reverse ; while it has little bearing on the great features of zoological geography—the limitation of groups of *genera* and *families* to certain areas. It is analogous to the "*law of adaptation*" in the organisation of animals, by which members of various groups are suited for an aerial, an aquatic, a desert, or an arboreal life ; are herbivorous, carnivorous, or insectivorous ; are fitted to live underground, or in fresh waters, or on polar ice. It was once thought that these adaptive peculiarities were suitable foundations for a classification,—that whales were fishes, and bats birds ; and even to this day there are naturalists who cannot recognise the essential diversity of structure in such groups as swifts and swallows, sun-birds and humming-birds, under the superficial disguise caused by adaptation to a similar mode of life. The application of Mr. Allen's principle leads to equally erroneous results, as may be well seen by considering his separation of "the southern third of Australia" to unite it with New Zealand as one of his secondary zoological divisions. If there is one country in the world whose fauna is strictly homogeneous, that country is Australia ; while New Guinea on the one hand, and New Zealand on the other, are as sharply differentiated from Australia as any adjacent parts of the same primary zoological division can possibly be. Yet the "*law of circumpolar distribution*" leads to the division of

Australia by an arbitrary east and west line, and a union of the northern two-thirds with New Guinea, the southern third with New Zealand. Hardly less unnatural is the supposed equivalence of South Africa (the African temperate realm) to all tropical Africa and Asia, including Madagascar (the Indo-African tropical realm). South Africa has, it is true, some striking peculiarities; but they are absolutely unimportant as compared with the great and radical differences between tropical Africa and tropical Asia. On these examples we may fairly rest our rejection of Mr. Allen's scheme.

We must however say a few words on the zoo-geographical nomenclature proposed in the same paper, which seems also very objectionable. The following terms are proposed: *realm, region, province, district, fauna and flora;* the first being the highest, the last the lowest and smallest sub-division. Considering that most of these terms have been used in very different senses already, and that no means of settling their equivalence in different parts of the globe has been even suggested, such a complex system must lead to endless confusion. Until the whole subject is far better known and its first principles agreed upon, the simpler and the fewer the terms employed the better; and as "region" was employed for the primary divisions by Mr. Sclater, eighteen years ago, and again by Mr. Andrew Murray, in his Geographical Distribution of Mammals; nothing but obscurity can result from each writer using some new, and doubtfully better, term. For the sub-divisions of the regions no advantage is gained by the use of a distinct term—"province"—which has been used (by Swainson) for the primary divisions, and which does not itself tell you what rank it holds; whereas the term "sub-region" speaks for itself as being unmistakably next in subordination to region, and this clearness of meaning gives it the preference over any independent term. As to minor named sub-divisions, they seem at present uncalled for; and till the greater divisions are themselves generally agreed on, it seems better to adopt no technical names for what must, for a long time to come, be indeterminate.

Does the Arctic Fauna characterize an independent Region.—

The proposal to consider the Arctic regions as constituting one of the primary zoological divisions of the globe, has been advocated by many naturalists. Professor Huxley seems to consider it advisable, and Mr. Allen unhesitatingly adopts it, as well as an "antarctic" region to balance it in the southern hemisphere. The reason why an " Arctic Region " finds no place in this work may therefore be here stated.

No species or group of animals can properly be classed as " arctic," which does not exclusively inhabit or greatly preponderate in arctic lands. For the purpose of establishing the need of an " arctic " zoological region, we should consider chiefly such groups as are circumpolar as well as arctic; because, if they are confined to, or greatly preponderate in, either the eastern or western hemispheres, they can be at once allocated to the Nearctic or Palæarctic regions, and can therefore afford no justification for establishing a new primary division of the globe.

Thus restricted, only three genera of land mammalia are truly arctic: *Gulo, Myodes,* and *Rangifer.* Two species of widely dispersed genera are also exclusively arctic, *Ursus maritimus* and *Vulpes lagopus.*

Exclusively arctic birds are not much more numerous. Of land birds there are only three genera (each consisting of but a single species), *Pinicola, Nyctea,* and *Surnia. Lagopus* is circumpolar, but the genus has too wide an extension in the temperate zone to be considered arctic. Among aquatic birds we have the genus of ducks, *Somateria ;* three genera of Uriidæ, *Uria, Catarractes,* and *Mergulus;* and the small family Alcidæ, consisting of the genera *Alca* and *Fratercula.* Our total then is, three genera of mammalia, three of land, and six of aquatic birds, including one peculiar family.

In the southern hemisphere there is only the single genus *Aptenodytes* that can be classed as antarctic ; and even that is more properly south temperate.

In dealing with this arctic fauna we have two courses open to us; we must either group them with the other species and genera which are common to the two northern regions, or we must form a separate primary region for them. As a matter of convenience the former plan seems the best; and it is that which is in accordance with our treatment of other intermediate tracts which contain special forms of life. The great desert zone, extending from the Atlantic shores of the Sahara across Arabia to Central Asia, is a connecting link between the Palæarctic, Ethiopian, and Oriental regions, and contains a number of "desert" forms wholly or almost wholly restricted to it; but the attempt to define it as a separate region would introduce difficulty and confusion. Neither to the "desert" nor to the "arctic" regions could any defined limits, either geographical or zoological, be placed; and the attempt to determine what species or genera should be allotted to them would prove an insoluble problem. The reason perhaps is, that both are essentially unstable, to a much greater extent than those great masses of land with more or less defined barriers, which constitute our six regions. The Arctic Zone has been, within a recent geological period, both vastly more extensive and vastly less extensive than it is at present. At a not distant epoch it extended over half of Europe and of North America. At an earlier date it appears to have vanished altogether; since a luxuriant vegetation of tall deciduous trees and broad-leaved evergreens flourished within ten degrees of the Pole! The great deserts have not improbably been equally fluctuating; hence neither the one nor the other can present that marked individuality in their forms of life, which seems to have arisen only when extensive tracts of land have retained some considerable stability both of surface and climatal conditions, during periods sufficient for the development and co-adaptation of their several assemblages of plants and animals.

We must also consider that there is no geographical difficulty in dividing the Arctic Zone between the two northern regions. The only debateable lands, Greenland and Iceland, are generally admitted to belong respectively to America and Europe. Neither is there any zoological difficulty; for the land mammalia and birds are on the whole wonderfully restricted to their respective regions even in high latitudes; and the aquatic forms are, for our present purpose, of much less importance. As a primary division the " Arctic region " would be out of all proportion to the other six, whether as regards its few peculiar types or the limited number of forms and species actually inhabiting it; but it comes in well as a connecting link between two regions, where the peculiar forms of both are specially modified; and is in this respect quite analogous to the great desert zone above referred to.

I now proceed to characterize briefly the six regions adopted in the present work, together with the sub-regions into which they may be most conveniently and naturally divided, as shown in our general map.

Palæarctic Region.—This very extensive region comprises all temperate Europe and Asia, from Iceland to Behring's Straits and from the Azores to Japan. Its southern boundary is somewhat indefinite, but it seems advisable to comprise in it all the extra-tropical part of the Sahara and Arabia, and all Persia, Cabul, and Beloochistan to the Indus. It comes down to a little below the upper limit of forests in the Himalayas, and includes the larger northern half of China, not quite so far down the coast as Amoy. It has been said that this region differs from the Oriental by negative characters only; a host of tropical families and genera being absent, while there is little or nothing but peculiar species to characterize it absolutely. This however is not true. The Palæarctic region is well characterized by possessing 3 families of vertebrata peculiar to it, as well as 35 peculiar genera of mammalia, and 57 of birds, constituting about one-third of the total number it possesses. These are amply sufficient to characterize a region positively; but we must also consider the absence of many important groups of the Oriental, Ethiopian, and Nearctic regions; and we shall then find, that taking positive and negative characters together, and making some allowance for the necessary poverty of a temperate as compared with tropical regions, the Palæarctic is almost as strongly marked and well defined as any other.

Sub-divisions of the Palæarctic Region.—These are by no means so clearly indicated as in some of the other regions, and they are adopted more for convenience than because they are very natural or strongly marked.

The first, or European sub-region, comprises Central and Northern Europe as far South as the Pyrenees, the Maritime and Dinaric Alps, the Balkan mountains, the Black Sea, and the Caucasus. On the east the Caspian sea and the Ural mountains seem the most obvious limit; but it is doubtful if they form the actual boundary, which is perhaps better marked by the valley of the Irtish, where a pre-glacial sea almost certainly connected the Aral and Caspian seas with the Arctic ocean, and formed an effective barrier which must still, to some extent, influence the distribution of animals.

The next, or Mediterranean sub-region, comprises South Europe, North Africa with the extra-tropical portion of the Sahara, and Egypt to about the first or second cataracts; and eastward through Asia Minor, Persia, and Cabul, to the deserts of the Indus.

The third, or Siberian sub-region, consists of all north and central Asia north of Herat, as far as the eastern limits of the great desert plateau of Mongolia, and southward to about the upper limit of trees on the Himalayas.

The fourth, or Manchurian sub-region, consists of Japan and North China with the lower valley of the Amoor; and it should probably be extended westward in a narrow strip along the Himalayas, embracing about 1,000 or 2,000 feet of vertical distance below the upper limit of trees, till it meets an eastern extension of the Mediterranean sub-region a little beyond Simla. These extensions are necessary to avoid passing from the Oriental region, which is essentially tropical, directly to the Siberian sub-region, which has an extreme northern character; whereas the Mediterranean and Manchurian sub-regions are more temperate in climate. It will be found that between the upper limit of most of the typical Oriental groups and the Thibetan or Siberian fauna, there is a zone in which many forms occur common to temperate China. This is especially the case among the pheasants and finches.

Ethiopian Region.—The limits of this region have been indicated by the definition of the Palæarctic region. Besides Africa south of the tropic of Cancer, and its islands, it comprises the southern half of Arabia.

This region has been said to be identical in the main characters of its mammalian fauna with the Oriental region, and has therefore been united with it by Mr. A. Murray. Most important differences have however been overlooked, as the following summary of the peculiarities of the Ethiopian region will, I think, show.

It possesses 22 peculiar families of vertebrates; 90 peculiar genera of mammalia, being two-thirds of its whole number; and 179 peculiar genera of birds, being three-fifths of all it possesses. It is further characterized by the absence of several families and genera which range over the whole northern hemisphere, details of which will be found in the chapter treating of the region. There are, it is true, many points of resemblance, not to be wondered at between two tropical regions in the same hemisphere, and which have evidently been at one time more nearly connected, both by intervening lands and by a different condition of the lands that even now connect them. But these resemblances only render the differences more remarkable; since they show that there has been an ancient and long-continued separation of the two regions, developing a distinct fauna in each, and establishing marked specialities which the temporary intercommunication and immigration has not sufficed to remove. The entire absence of such wide-spread groups as bears and deer, from a country many parts of which are well adapted to them, and in close proximity to regions where they abound, would alone mark out the Ethiopian region as one of the primary divisions of the earth, even if it possessed a less number than it actually does of peculiar family and generic groups.

Sub-divisions of the Ethiopian Region.—The African continent south of the tropic of Cancer is more homogeneous in its prominent and superficial zoological features than most of the other regions, but there are nevertheless important and deep-seated local peculiarities. Two portions can be marked off as possessing many peculiar forms; the luxuriant forest district of equatorial West Africa, and the southern extremity or Cape district. The remaining portion has no well-marked divisions, and a large proportion of its animal forms range over it from Nubia and Abyssinia, to Senegal on the one side and to the Zambesi on the other; this forms our first or East-African sub-region.

The second, or West African sub-region extends along the coast from Senegal to Angola, and inland to the sources of the Shary and the Congo.

The third, or South African sub-region, comprises the Cape Colony and Natal, and is roughly limited by a line from Delagoa Bay to Walvish Bay.

The fourth, or Malagasy sub-region, consists of Madagascar and the adjacent islands, from Rodriguez to the Seychelles; and this differs so remarkably from the continent that it has been proposed to form a distinct primary region for its reception. Its productions are indeed highly interesting; since it possesses 3 families, and 2 sub-families of mammals peculiar to itself, while almost all its genera are peculiar. Of these a few show Oriental or Ethiopian affinities, but the remainder are quite isolated. Turning to other classes of animals, we find that the birds are almost as remarkable; but, as might be expected, a larger number of genera are common to surrounding countries. More than 30 genera are altogether peculiar, and some of these are so isolated as to require to be classed in separate families or sub-families. The African affinity is however here more strongly shown by the considerable number (13) of peculiar Ethiopian genera which in Madagascar have representative species. There can be no doubt therefore about Madagascar being more nearly related to the Ethiopian than to any other region; but its peculiarities are so great, that, were it not for its small size and the limited extent of its fauna, its claim to rank as a separate region might not seem unreasonable. It is true that it is not poorer in mammals than Australia; but that country is far more isolated, and cannot be so decidedly and naturally associated with any other region as Madagascar can be with the Ethiopian. It is therefore the better and more natural course to keep it as a sub-region; the peculiarities it exhibits being of exactly the same kind as those presented by the Antilles, by New Zealand, and even by Celebes and Ceylon, but in a much greater degree.

Oriental Region.—On account of the numerous objections that have been made to naming a region from the least characteristic portion of it, and not thinking " Malayan," proposed by Mr. Blanford, a good term, (as it has a very circumscribed and definite meaning, and especially because the " Malay " archipelago is half of it in the Australian region,) I propose to use the word " Oriental " instead of " Indian," as being geographically applicable to the whole of the countries included in the region and to very few beyond it ; as being euphonious, and as being free from all confusion with terms already used in zoological geography. I trust therefore that it may meet with general acceptance.

This small, compact, but rich and varied region, consists of all India and China from the limits of the Palæarctic region ; all the Malay peninsula and islands as far east as Java and Baly, Borneo and the Philippine Islands ; and Formosa. It is positively characterized by possessing 12 peculiar families of vertebrata ; by 55 genera of land mammalia, and 165 genera of land birds, altogether confined to it; these peculiar genera forming in each case about one half of the total number it possesses.

Sub-divisions of the Oriental region.—First we have the Indian sub-region, consisting of Central India from the foot of the Himalayas in the west, and south of the Ganges to the east, as far as a line drawn from Goa curving south and up to the Kistna river; this is the portion which has most affinity with Africa.

The second, or Ceylonese sub-region, consists of the southern extremity of India with Ceylon; this is a mountainous forest region, and possesses several peculiar forms as well as some Malayan types not found in the first sub-region.

Next we have the Indo-Chinese sub-region, comprising South China and Burmah, extending westward along the Himalayan range to an altitude of about 9,000 or 10,000 feet, and southward to Tavoy or Tenasserim.

The last is the Indo-Malayan sub-region, comprising the Peninsula of Malacca and the Malay Islands to Baly, Borneo, and the Philippines.

On account of the absence from the first sub-region of many of the forms most characteristic of the other three, and the number of families and genera of mammalia and birds which occur in it and also in Africa, it has been thought by some naturalists that this part of India has at least an equal claim to be classed as a part of the Ethiopian region. This question will be found fully discussed in Chapter XII. devoted to the Oriental region, where it is shown that the African affinity is far less than has been represented, and that in all its essential features Central India is wholly Oriental in its fauna.

Before leaving this region a few words may be said about Lemuria, a name proposed by Mr. Sclater for the site of a supposed submerged continent extending from Madagascar to Ceylon and Sumatra, in which the Lemuroid type of animals was developed. This is undoubtedly a legitimate and highly probable supposition, and it is an example of the way in which a study of the geographical distribution of animals may enable us to reconstruct the geography of a bygone age. But we must not, as Mr. Blyth proposed, make this hypothetical land one of our actual Zoological regions. It represents what was probably a primary Zoological region in some past geological epoch; but what that epoch was and what were the limits of the region in question, we are quite unable to say. If we are to suppose that it comprised the whole area now inhabited by Lemuroid animals, we must make it extend from West Africa to Burmah, South China, and Celebes; an area which it possibly did once occupy, but which cannot be formed into a modern Zoological region without violating much more important affinities. If, on the other hand, we leave out all those areas which undoubtedly belong to other regions, we reduce Lemuria to Madagascar and its adjacent islands, which, for reasons already stated, it is not advisable to treat as a primary Zoological region. The theory of this ancient continent and the light it may throw on existing anomalies of distribution, will be more fully considered in the geographical part of this work.

Australian Region.—Mr. Sclater's original name seems preferable to Professor Huxley's, "Austral-Asian;" the inconvenience of which alteration is sufficiently shown by the fact that Mr. Blyth proposed to use the very same term as an appropriate substitute for the "Indian region" of Mr. Sclater. Australia is the great central mass of the region; it is by far the richest in varied and highly remarkable forms of life; and it therefore seems in every way fitted to give a name to the region of which it is the essential element. The limits of this region in the Pacific are somewhat obscure, but as so many of the Pacific Islands are extremely poor zoologically, this is not of great importance.

Sub-divisions of the Australian Region.—The first sub-region is the Austro-Malayan, including the islands from Celebes and Lombock on the west to the Solomon Islands on the east. The Australian sub-region comes next, consisting of Australia and Tasmania. The third, or Polynesian sub-region, will consist of all the tropical Pacific Islands, and is characterized by several peculiar genera of birds which are all allied to Australian types. The fourth, consists of New Zealand with Auckland, Chatham, and Norfolk Islands, and must be called the New Zealand sub-region.

The extreme peculiarities of New Zealand, due no doubt to its great isolation and to its being the remains of a more extensive land, have induced several naturalists to suggest that it ought justly to form a Zoological region by itself. But the inconveniences of such a procedure have been already pointed out; and when we look at its birds as a whole (they being the only class sufficiently well represented to found any conclusion upon) we find that the majority of them belong to Australian genera, and where the genera are peculiar they are most nearly related to Australian types. The preservation in these islands of a single representative of a unique order of reptiles, is, as before remarked, of the same character as the preservation of the *Proteus* in the caverns of Carniola ; and can give the locality where it happens to have survived no claim to form a primary Zoological region, unless supported by a tolerably varied and distinctly characterized fauna, such as never exists in a very restricted and insular area.

Neotropical Region.—Mr. Sclater's original name for this. region is preserved, because change of nomenclature is always an evil; and neither Professor Huxley's suggested alteration " Austro-Columbia," nor Mr. Sclater's new term "Dendrogæa," appear to be improvements. The region is essentially a tropical one, and the extra-tropical portion of it is not important enough to make the name inappropriate. That proposed by Professor Huxley is not free from the same kind of criticism, since it would imply that the region was exclusively South American, whereas a considerable tract of North America belongs to it. This region includes South America, the Antilles and tropical North America ; and it possesses more peculiar families of vertebrates and genera of birds and mammalia than any other region.

Subdivisions of the Neotropical Region.—The great central mass of South America, from the shores of Venezuela to Paraguay and Eastern Peru, constitutes the chief division, and may be termed the Brazilian sub-region. It is on the whole a forest country; its most remarkable forms are highly developed arboreal types; and it exhibits all the characteristics of this rich and varied continent in their highest development.

The second, or Chilian sub-region, consists of the open plains, pampas, and mountains of the southern extremity of the continent; and we must include in it the west side of the Andes as far as the limits of the forest near Payta, and the whole of the high Andean plateaus as far as 4° of south latitude; which makes it coincide with the range of the Camelidæ and Chinchillidæ.

The third, or Mexican sub-region, consists of Central America and Southern Mexico, but it has no distinguishing characteristics except the absence of some of the more highly specialized Neotropical groups. It is, however, a convenient division as comprising the portion of the North American continent which belongs zoologically to South America.

The fourth, or Antillean sub-region, consists of the West India islands (except Trinidad and Tobago, which are detached portions of the continent and must be grouped in the first sub-region); and these reproduce, in a much less marked degree, the phenomena presented by Madagascar. Terrestrial mammals are almost entirely wanting, but the larger islands possess three genera which are altogether peculiar to them. The birds are of South American forms, but comprise many peculiar genera. Terrestrial molluscs are more abundant and varied than in any part of the globe of equal extent; and if these alone were considered, the Antilles would constitute an important Zoological region.

Nearctic Region.—This region comprises all temperate North America and Greenland. The arctic lands and islands beyond the limit of trees form a transitional territory to the Palæarctic region, but even here there are some characteristic species. The southern limit between this region and the Neotropical is a little uncertain; but it may be drawn at about the Rio Grande del Norte on the east coast, and a little north of Mazatlan on the west; while on the central plateau it descends much farther south, and should perhaps include all the open highlands of Mexico and Guatemala. This would coincide with the range of several characteristic Nearctic genera.

Distinction of the Nearctic from the Palæarctic Region.—The Nearctic region possesses twelve peculiar families of vertebrates or one-tenth of its whole number. It has also twenty-four peculiar genera of mammalia and fifty-two of birds, in each case nearly one-third of all it possesses. This proportion is very nearly the same as in the Palæarctic region, while the number of peculiar families of vertebrata is very much greater. It has been already seen that both Mr. Blyth and Professor Huxley are disposed to unite this region with the Palæarctic, while Professor Newton, in his article on birds in the new edition of the

Encyclopædia Britannica, thinks that as regards that class it can hardly claim to be more than a sub-region of the Neotropical. These views are mutually destructive, but it will be shown in the proper place, that on independent grounds the Nearctic region can very properly be maintained.

Subdivisions of the Nearctic Region.—The sub-regions here depend on the great physical features of the country, and have been in some cases accurately defined by American naturalists. First we have the Californian sub-region, consisting of California and Oregon—a narrow tract between the Sierra Nevada and the Pacific, but characterized by a number of peculiar species and by several genera found nowhere else in the region.

The second, or Rocky Mountain sub-region, consists of this great mountain range with its plateaus, and the central plains and prairies to about 100° west longitude, but including New Mexico and Texas in the South.

The third and most important sub-region, which may be termed the Alleghanian, extends eastward to the Atlantic, including the Mississippi Valley, the Alleghany Mountains, and the Eastern United States. This is an old forest district, and contains most of the characteristic animal types of the region.

The fourth, or Canadian sub-region, comprises all the northern part of the continent from the great lakes to the Arctic ocean; a land of pine-forests and barren wastes, characterized by Arctic types and the absence of many of the genera which distinguish the more southern portions of the region.

Observations on the series of Sub-regions.—The twenty-four sub-regions here adopted were arrived at by a careful consideration of the distribution of the more important genera, and of the materials, both zoological and geographical, available for their determination; and it was not till they were almost finally decided on, that they were found to be equal in number throughout all the regions—four in each. As this uniformity is of great advantage in tabular and diagrammatic presentations of the distribution of the several families, I decided not to disturb it unless very strong reasons should appear for adopting a greater or less number in any particular case. Such however have not arisen; and it is hoped that these divisions will prove as satis-factory and useful to naturalists in general as they have been to the author. Of course, in a detailed study of any region much more minute sub-division may be required; but even in that case it is believed that the sub-regions here adopted, will be found, with slight modifications, permanently available for ex-hibiting general results.

I give here a table showing the proportionate richness and speciality of each region as determined by its *families* of verte-brates and *genera* of mammalia and birds ; and also a general table of the regions and sub-regions, arranged in the order that seems best to show their mutual relations.

COMPARATIVE RICHNESS OF THE SIX REGIONS.

REGIONS.	VERTEBRATA.		MAMMALIA.			BIRDS.		
	Fami-lies.	Peculiar families	Genera.	Peculiar genera.	Per centage.	Genera.	Peculiar genera.	Per centage.
Palæarctic...	136	3	100	35	35	174	57	33
Ethiopian ...	174	22	140	90	64	294	179	60
Oriental......	164	12	118	55	46	340	165	48
Australian...	141	30	72	44	61	298	189	64
Neotropical..	168	44	130	103	79	683	576	86
Nearctic	122	12	74	24	32	169	52	31

TABLE OF REGIONS AND SUB-REGIONS.

Regions.	Sub-regions.	Remarks.
I. Palæarctic...	1. North-Europe.	
	2. Mediterranean (or S. Eu.)	Transition to Ethiopian.
	3. Siberia.	Transition to Nearctic.
	4. Manchuria (or Japan)	Transition to Oriental.
II. Ethiopian ...	1. East Africa.	Transition to Palæarctic.
	2. West Africa.	
	3. South Africa.	
	4. Madagascar.	

TABLE OF REGIONS AND SUB-REGIONS—*continued.*

Regions.	Sub-regions.	Remarks.
III. Oriental	1. Hindostan (or Central Ind.)	Transition to Ethiopian.
	2. Ceylon.	
	3. Indo-China (or Himalayas)	Transition to Palæarctic.
	4. Indo-Malaya.	Transition to Australian.
IV. Australian ...	1. Austro-Malaya.	Transition to Oriental.
	2. Australia.	
	3. Polynesia.	
	4. New Zealand.	Transition to Neotropical.
V. Neotropical.	1. Chili (or S. Temp. Am.)	Transition to Australian.
	2. Brazil.	
	3. Mexico (or Trop. N. Am.)	Transition to Nearctic.
	4. Antilles.	
VI. Nearctic	1. California.	
	2. Rocky Mountains.	
	3. Alleghanies (or East U. S.)	Transition to Neotropical.
	4. Canada.	Transition to Palæarctic.

CHAPTER V.

A LITTLE consideration will convince us, that no inquiry into
the causes and laws which determine the geographical distribution of animals or plants can lead to satisfactory results, unless
we have a tolerably accurate knowledge of the affinities of the
several species, genera, and families to each other; in other
words, we require a natural classification to work upon. Let us,
for example, take three animals—a, b, and c—which have a
general external resemblance to each other, and are usually
considered to be really allied; and let us suppose that a and b
inhabit the same or adjacent districts, while c is found far away
on the other side of the globe, with no animals at all resembling
it in any of the intervening countries. We should here have a
difficult problem to solve; for we should have to show that the
general laws by which we account for the main features of
distribution, will explain this exceptional case. But now, suppose some comparative anatomist takes these animals in hand,
and finds that the resemblance of c to a and b is only superficial,
while their internal structure exhibits marked and important
differences; and that c really belongs to another group of
animals, d, which inhabits the very region in which c was
found—and we should no longer have anything to explain.
This is no imaginary case. Up to a very few years ago a
curious Mexican animal, *Bassaris astuta*, was almost always
classed in the civet family (Viverridæ), a group entirely con-

fined to Africa and Asia; but it has now been conclusively shown by Professor Flower that its real affinities are with the racoons (Procyonidæ), a group confined to North and South America. In another case, however, an equally careful examination shows, that an animal peculiar to the Himalayas (*Ælurus fulgens*) has its nearest ally in the *Cercoleptes* of South America. Here, therefore, the geographical difficulty really exists, and any satisfactory theory of the causes that have led to the existing distribution of living things, must be able to account, more or less definitely, for this and other anomalies. From these cases it will be evident, that if any class or order of animals is very imperfectly known and its classification altogether artificial, it is useless to attempt to account for the anomalies its distribution may present; since those anomalies may be, to a great extent, due to false notions as to the affinities of its component species.

According to the laws and causes of distribution discussed in the preceding chapters, we should find limited and defined distribution to be the rule, universal or indefinite distribution to be the exception, in every natural group corresponding to what are usually regarded as families and genera; and so much is this the case in nature, that when we find a group of this nominal rank scattered as it were at random over the earth, we have a strong presumption that it is not natural; but is, to a considerable extent, a haphazard collection of species. Of course this reasoning will only apply, in cases where there are no unusual means of dispersal, nor any exceptional causes which might determine a scattered distribution.

From the considerations now adduced it becomes evident, that it is of the first importance for the success of our inquiry to secure a natural classification of animals, especially as regards the families and genera. The higher groups, such as classes and orders, are of less importance for our purpose; because they are almost always widely and often universally distributed, except those which are so small as to be evidently the nearly extinct representatives of a once more extensive series of forms. We now proceed to explain the classification to be adopted, as low down as the series of families. To these, equivalent English names are given wherever they exist, in order that readers possessing no technical knowledge, may form some conception of the meaning of the term " family " in zoology.

The primary divisions of the animal kingdom according to two eminent modern authorities are as follows :

HUXLEY. Classification of Animals (1869).	CARUS AND GERSTAEKER. Handbuch der Zoologie (1868).
1. Protozoa } 2. Infusoria }	1. Protozoa.
3. Cœlenterata	2. Cœlenterata.
4. Annuloida	3. Echinodermata.
5. Annulosa	{ 4. Vermes. { 5. Arthropoda.
6. Molluscoida	6. Molluscoida.
7. Mollusca	7. Mollusca.
8. Vertebrata	8. Vertebrata.

For reasons already stated it is only with the fifth, seventh, and eighth of these groups that the present work proposes to deal; and even with the fifth and seventh only partially and in the most general way.

The classes of the vertebrata, according to both the authors above quoted, are: 1. Mammalia.   2. Aves.   3. Reptilia.   4. Amphibia.   5. Pisces, in which order they will be taken here.

The sub-classes and orders of mammalia are as follows :

MAMMALIA.

	HUXLEY (1869), FLOWER (1870).		CARUS (1868).
	1. Primates		{ 1. Primates. { 5. Prosimii.
	2. Chiroptera		2. Chiroptera.
	3. Insectivora		3. Insectivora.
	4. Carnivora		{ 6. Carnivora. { 7. Pinnipedia.
Monodelphia... {	5. Cetacea } 6. Sirenia }		12. Natantia.
	7. Ungulata		{ 10. Artiodactyla. { 11. Perissodactyla.
	8. Proboscidea... ...		9. Proboscidea.
	9. Hyracoidea		8. Lamnungia.
	10. Rodentia		4. Rodentia.
	11. Edentata		13. Bruta.
Didelphia	12. Marsupialia		14. Marsupialia.
Ornithodelphia	13. Monotremata ...		15. Monotremata.

The above series of orders is arranged according to Professor Flower's *Osteology of Mammalia,* and they will follow in this succession throughout my work. Professor Huxley arranges the same orders in a different series.

In determining the manner in which the several orders shall be subdivided into families, I have been guided in my choice of classifications mainly by the degree of attention the author appears to have paid to the group, and his known ability as a systematic zoologist; and in a less degree by considerations of convenience as regards the special purposes of geographical distribution. In many cases it is a matter of great doubt whether a certain group should form several distinct families or be united into one or two; but one method may bring out the peculiarities of distribution much better than the other, and this is, in our case, a sufficient reason for adopting it.

For the Primates I follow, with some modifications, the classification of Mr. St. George Mivart given in his article "Apes" in the new edition of the *Encyclopædia Britannica,* and in his paper in the *Proceedings of the Zoological Society of London,* 1865, p. 547. It is as follows:

Order—PRIMATES, divided into two Sub-orders:
I. Anthropoidea.
II. Lemuroidea.

Sub-order—ANTHROPOIDEA.
Fam.

	Hominidæ	Man.
Simii	1. Simiidæ	Anthropoid Apes.
	2. Semnopithecidæ ...	Old-world Monkeys.
	3. Cynopithecidæ ...	Baboons and Macaques.
Cebii	4. Cebidæ	American Monkeys.
	5. Hapalidæ	Marmosets.

Sub-order—LEMUROIDEA.
Fam.
6. Lemuridæ ... ... Lemurs.
7. Tarsiidæ ... ... Tarsiers.
8. Chiromyidæ ... ... Aye-ayes.

Omitting man (for reasons stated in the preface) the three first families are considered by Professor Mivart to be sub-families of Simiidæ; but as the geographical distribution of the Old World apes is especially interesting, it is thought better to treat them as families, a rank which is claimed for the anthropoid apes by many naturalists.

As no good systematic work on the genera and species of bats has been yet published, I adopt the five families as generally used in this country, with the genera as given in the papers of Dr. J. E. Gray and Mr. Tomes. A monograph by Dr. Peters has long been promised, and his outline arrangement was published in 1865, but this will perhaps be materially altered when the work appears.

Order—CHIROPTERA.

Fam.

Frugivora		... ...	... 9. Pteropidæ	...	Fruit-eating Bats.
Insectivora	{ Istiophora	{ 10. Phyllostomidæ	...	Leaf-nosed Bats.
		{ 11. Rhinolophidæ	...	Horse-shoe Bats.
	{ Gymnorhini	{ 12. Vespertilionidæ	...	True Bats.
		{ 13. Noctilionidæ	...	Dog-headed Bats.

The genera of Chiroptera are in a state of great confusion, the names used by different authors being often not at all comparable, so that the few details given of the distribution of the bats are not trustworthy. We have therefore made little use of this order in the theoretical part of the work.

The osteology of the Insectivora has been very carefully worked out by Professor Mivart in the *Journal of Anatomy and Physiology* (Vol. ii., p. 380), and I follow his classification as given there, and in the *Proceedings of the Zoological Society* (1871).

Order—INSECTIVORA.

Fam.

14. Galeopithecidæ	...	...	...	Flying Lemurs.
15. Macroscelididæ	...	...	...	Elephant Shrews.
16. Tupaiidæ ...	...	...	...	Squirrel Shrews.
17. Erinaceidæ ...	...	...	...	Hedgehogs.
18. Centetidæ ...	...	...	...	Tenrecs.
19. Potamogalidæ	...	...	...	Otter Shrew.
20. Chrysochloridæ	...	...	...	Golden Moles.
21. Talpidæ ...	...	...	...	Moles.
22. Soricidæ ...	...	...	...	Shrews.

The next order, Carnivora, has been studied in detail by Professor Flower; and I adopt the classification given by him in the *Proceedings of the Zoological Society*, 1869, p. 4.

Order—CARNIVORA.

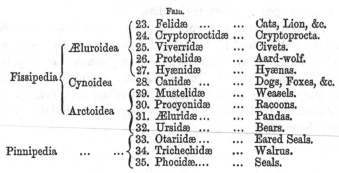

		Fam.		
Fissipedia	Æluroidea	23. Felidæ ...	...	Cats, Lion, &c.
		24. Cryptoproctidæ	...	Cryptoprocta.
		25. Viverridæ	...	Civets.
		26. Protelidæ	...	Aard-wolf.
		27. Hyænidæ	...	Hyænas.
	Cynoidea	28. Canidæ ...	...	Dogs, Foxes, &c.
	Arctoidea	29. Mustelidæ	...	Weasels.
		30. Procyonidæ	...	Racoons.
		31. Æluridæ...	...	Pandas.
		32. Ursidæ ...	...	Bears.
Pinnipedia	... ...	33. Otariidæ...	...	Eared Seals.
		34. Trichechidæ	...	Walrus.
		35. Phocidæ....	...	Seals.

The Cetacea is one of those orders the classification of which is very unsettled. The animals comprising it are so huge, and there is so much difficulty in preserving them, that only a very few species are known with anything like completeness. A considerable number of genera and species have been described or indicated; but as many of these are founded on imperfect specimens of perhaps a single individual, it is not to be wondered at that those few naturalists who occupy themselves with the study of these large animals, cannot agree as to the proper mode of grouping them into natural families. They are, however, of but little importance to us, as almost all the species inhabit the ocean, and of only a few of them can it be said that anything is accurately known of their distribution. I therefore consider it best to follow Professor Carus, who makes a smaller number of families; but I give also the arrangement of Dr. Gray in his British Museum catalogue of whales and seals, as modified subsequently in the *Proceedings of Zoological Society*, 1870, p. 772. The Zeuglodontidæ, a family of extinct tertiary whales, are classed by Professors Owen and Carus between Cetacea and Sirenia, while Professor Huxley considers them to have been carnivorous and allied to the seals.

Order—CETACEA.

	Fam. (CARUS).					Fam. (GRAY).
Sub-order I.—	Balænidæ	...	...	...	36.	Balænidæ.
Mystaceti.	Balænopteridæ	...	...	37.	Balænopteridæ.
	Catodontidæ ...	...	...	38.	Catodontidæ.
						Hyperoodontidæ.
	Hyperoodontidæ	...	...	39.	Epiodontidæ.
Sub-order II.—						Xiphiadæ.
Odontoceti.	Monodontidæ	...	...	40.	(Part of Delphinidæ.)
						Platanistidæ.
						Iniadæ.
						Delphinidæ.
	Delphinidæ ...	...	...	41.	Globiocephalidæ.
						Orcadæ.
						Belugidæ.
						Pontoporiadæ.

Extinct family   Zeuglodontidæ.

Order—SIRENIA.

The order Sirenia, comprising the sea-cows, consists of a single family:

Family 42. Manatidæ.

The extensive order Ungulata comprises the three orders Pachydermata, Solidungula, and Ruminantia of the older naturalists. The following classification is that now generally adopted, the only difference of opinion being as to whether some of the groups should be classed as families or sub-families, a matter of little importance for our purpose:

Order—UNGULATA.

		Fam.		
Perissodactyla or Odd-toed Ungulates	... ...	43. Equidæ	Horses.
		44. Tapiridæ	Tapirs.
		45. Rhinocerotidæ...	Rhinoceros.
Artiodactyla or Even-toed Ungulates	Suina	46. Hippopotamidæ	Hippopotamus.
		47. Suidæ	Swine.
	Tylopoda	48. Camelidæ ...	Camels.
	Tragulina	49. Tragulidæ ...	Chevrotains.
	Pecora	50. Cervidæ	Deer.
		51. Cameloptardidæ	Giraffes.
		52. Bovidæ	Cattle, Sheep, Antelopes, &c.

The two next orders consist of but a single family each, viz.:

Order		Fam.			
PROBOSCIDEA	...	53. Elephantidæ	...	...	Elephants.
HYRACOIDEA	...	54. Hyracidæ	...	...	Rock-rabbits.

We now come to the Rodentia, a very extensive and difficult order, in which there is still much difference of opinion as to the details of classification, although the main outlines are pretty well settled. The foundations of a true classification of this order were laid by Mr. G. R. Waterhouse more than thirty years ago, and succeeding authors have done little more than follow his arrangement with unimportant modifications. Professor Lilljeborg, of Upsala, has however made a special study of this group of animals, and has given an original and detailed classification of all the genera. (*Systematisk Öfversigt af de Gnagande Däggdjuren, Glires. Upsala,* 1866.) I follow this arrangement with a few slight modifications suggested by other naturalists, and which make it better adapted for the purposes of this work.

Order—RODENTIA.

		Fam.			
Simplicidentati	Murina (Waterhouse)	55. Muridæ	...	...	Rats.
		56. Spalacidæ	...	...	Mole-rats.
		57. Dipodidæ	...	...	Jerboas.
		58. Myoxidæ	...	...	Dormice.
		59. Saccomyidæ	...	...	Pouched Rats.
		60. Castoridæ	...	...	Beavers.
		61. Sciuridæ	...	...	Squirrels.
		62. Haploodontidæ	...		Sewellels.
	Hystricina (Waterhouse)	63. Chinchillidæ	...	...	Chinchillas.
		64. Octodontidæ	...	...	Octodons.
		65. Echimyidæ	...	...	Spiny Rats.
		66. Cercolabidæ	...	...	Tree Porcupines.
		67. Hystricidæ	...	...	Porcupines.
		68. Caviidæ	...	...	Cavies.
Duplicidentati	Leporina (Waterhouse)	69. Lagomyidæ	...	...	Pikas.
		70. Leporidæ	...	...	Hares.

The Edentata have been classified by Mr. Turner, in the *Proceedings of the Zoological Society* (1851, p. 205), by Dr. Gray in the British Museum Catalogue, and by Professor Carus in his *Handbuch.* The former takes a middle course between the numerous families of Dr. Gray, seven in number, and the two families to which Professor Carus restricts the existing species. I therefore follow Mr. Turner.

Order—EDENTATA.

	Fam.		
Bradypoda ...	71. Bradypodidæ	...	Sloths.
	72. Manididæ ...	...	Scaly Ant-eaters.
	73. Dasypodidæ	...	Armadillos.
Entomophaga {	74. Orycteropodidæ...		Ant-bears.
	75. Myrmecophagidæ		Ant-eaters.

The Marsupials have been well classified and described by Mr. Waterhouse in the first volume of his *Natural History of Mammalia,* and his arrangement is here followed. The sub-orders adopted by Professor Carus are also given.

Order—MARSUPIALIA.

	Fam.		
	76. Didelphidæ ...	...	Opossums.
Rapacia (Wagner) ... {	77. Dasyuridæ ...	...	Native Cats.
	78. Myrmecobiidæ	...	Native Ant-eater.
	79. Peramelidæ ...	...	Bandicoots.
Poephaga (Owen) ...	80. Macropodidæ	...	Kangaroos.
Carpophaga (Owen) ...	81. Phalangistidæ	...	Phalangers.
Rhizophaga (Owen) ...	82. Phascolomyidæ	...	Wombats.

Order—MONOTREMATA.

The last order, the Monotremata, consist of two families, which Professor Carus combines into one, but which it seems more natural to keep separate.

	Fam.	
83. Ornithorhynchidæ	... ...	Duckbill.
84. Echidnidæ	... ...	Echidna.

BIRDS.

Birds are perhaps the most difficult to classify of all the divisions of the vertebrata. The species and genera are exceedingly numerous, and there is such a great uniformity in general structure and even in the details of external form, that it is exceedingly difficult to find characters by which orders and families can be characterised. For a long time the system of Vigors and Swainson was followed; but this wholly ignored anatomical characters and in many cases plainly violated well-marked affinities. Characters derived from the form of the sternum, the scutellation of the tarsi, and the arrangement of the feathers, have all assisted in determining natural groups. More recently Professor Huxley has applied the variations of the bony palate to the general arrangement of birds; and still more recently Professor Garrod has studied certain leg-muscles for the same purpose. The condition of the young as regards plumage, and even the form, texture, and coloration of the egg, have also been applied to solve doubtful cases of affinity; yet the problem is not settled, and it will probably remain for another generation of ornithologists to determine with any approach to accuracy what are the most natural divisions of the class into orders and families. In a work like the present it is evidently not advisable to adopt all the recent classifications; since experience has shown that no arrangement in which one set of characters is mainly relied on, long holds its ground. Such modifications of the old system as seem to be well established will be adopted; but the older groups will be adhered to in cases where the most recent classifications are open to doubt, or seem inconvenient as separating families, which, owing to their similarity in general structure, form and habits are best kept together for the purposes of geographical distribution.

The old plan of putting the birds of prey at the head of the class, is now almost wholly given up; both because they are not the most highly organised, but only one of the most specialised forms of birds, and because their affinities are not with the Passeres, but rather with the cormorants and some other of the aquatic groups. The Passeres therefore are placed first; and the series of families is begun by the thrushes, which are certainly the most typical and generally well-organised form of birds. Instead of the Scansores and Fissirostres of the older authors, the order Picariæ, which includes them both, is adopted, but with some reluctance; as the former are, generally speaking, well marked and strongly contrasted groups, although certain families have been shown to be intermediate. In the Picariæ are included the goat-suckers, swifts, and humming-birds, sometimes separated as a distinct order, Macrochires. The parrots and the pigeons form each a separate order. The old groups of Grallæ and Anseres are preserved, as more convenient than breaking them up into widely separated parts; for though the latter plan may in some cases more strictly represent their affinities, its details are not yet established, nor is it much used by ornithologists. In accordance with these views the following is the series of orders and families of birds adopted in this work:

Class—AVES.

Orders.

1. Passeres ... { Including the great mass of the smaller birds—Crows, Finches, Flycatchers, Creepers, Honeysuckers, &c., &c.
2. Picariæ ... { Including Woodpeckers, Cuckoos, Toucans, Kingfishers, Swifts, &c., &c.
3. Psittaci ... Parrots only.
4. Columbæ ... Pigeons and the Dodo.
5. Gallinæ ... Grouse, Pheasants, Curassows, Mound-builders, &c.
6. Opisthocomi   The Hocco only.
7. Accipitres ... Eagles, Owls, and Vultures.
8. Grallæ ... Herons, Plovers, Rails, &c.
9. Anseres ... Gulls, Ducks, Divers, &c
10. Struthiones ... Ostrich, Cassowary, Apteryx, &c.

The Passeres consist of fifty families, which may be arranged and grouped in series as follows. It must however be remembered that the first family in each series is not always that which is most allied to the last family of the preceding series. All extensive natural groups consist of divergent or branching alliances, which renders it impossible to arrange the whole in one continuous series.

A.—TYPICAL OR TURDOID PASSERES.

1. Turdidæ ... ... Thrushes.
2. Sylviidæ ... ... Warblers.
3. Timaliidæ ... ... Babblers.
4. Panuridæ ... ... Reedlings.
5. Cinclidæ ... ... Dippers.
6. Troglodytidæ ... ... Wrens.
7. Chamæidæ ... ...
8. Certhiidæ ... ... Creepers.
9. Sittidæ ... ... Nuthatches.
10. Paridæ ... ... Tits.
11. Liotrichidæ ... ... Hill-tits.
12. Phyllornithidæ ... ... Green Bulbuls.
13. Pycnonotidæ ... ... Bulbuls.
14. Oriolidæ ... ... Orioles.
15. Campephagidæ ... ... Caterpillar-shrikes.
16. Dicruridæ ... ... Drongos.
17. Muscicapidæ ... ... Flycatchers.
18. Pachycephalidæ ... ... Thick-heads.
19. Laniidæ ... ... Shrikes.
20. Corvidæ ... ... Crows.
21. Paradiseidæ ... ... Paradise-birds.
22. Meliphagidæ ... ... Honey-suckers.
23. Nectarineidæ ... ... Sun-birds.

B.—TANAGROID PASSERES.

24. Dicæidæ ... ... Flower-peckers.
25. Drepanididæ ... ...
26. Cærebidæ ... ... Sugar-birds.
27. Mniotiltidæ ... ... Wood-warblers.
28. Vireonidæ ... ... Greenlets.
29. Ampelidæ ... ... Waxwings.
30. Hirundinidæ ... ... Swallows.
31. Icteridæ ... ... Hangnests.
32. Tanagridæ ... ... Tanagers.
33. Fringillidæ ... ... Finches.

C.—STURNOID PASSERES.

34. Ploceidæ ... ... Weaver-birds.
35. Sturnidæ ... ... Starlings.
36. Artamidæ ... ... Swallow-shrikes.
37. Alaudidæ ... ... Larks.
38. Motacillidæ ... ... Wagtails.

D.—FORMICAROID PASSERES.

39. Tyrannidæ ... ... Tyrants.
40. Pipridæ ... ... Manakins.
41. Cotingidæ ... ... Chatterers.
42. Phytotomidæ ... ... Plant-cutters.
43. Eurylæmidæ ... ... Broad-bills.
44. Dendrocolaptidæ ... ... American Creepers.
45. Formicariidæ ... ... Ant-thrushes.

D.—FORMICAROID PASSERES—*continued.*

46. Pteroptochidæ ...　　...　　...
47. Pittidæ　　　...　　...　　...　　Pittas.
48. Paictidæ　　　...　　...　　...

E.—ANOMALOUS PASSERES.

49. Menuridæ　　...　　...　　...　　Lyre-birds.
50. Atrichidæ　　...　　...　　...　　Scrub-birds.

The preceding arrangement is a modification of that proposed by myself in the *Ibis* (1874, p. 406). The principal alterations are adding the families Panuridæ and Sittidæ in series A, commencing series B with Dicæidæ; bringing Vireonidæ next to the allied American family Mniotiltidæ; and placing Motacillidæ in series C next to Alaudidæ. At the suggestion of Professor Newton I place Menuridæ and Atrichidæ apart from the other Passeres, as they both possess striking peculiarities of anatomical structure.

The heterogeneous families constituting the order Picariæ may be conveniently arranged as follows:

	51. Picidæ ...	...	Woodpeckers.
	52. Yungidæ ...	...	Wrynecks.
	53. Indicatoridæ	...	Honey-guides.
Sub-order—Scansores.	54. Megalæmidæ	...	Barbets.
	55. Rhamphastidæ	...	Toucans.
	56. Musophagidæ	...	Plantain-eaters.
	57. Coliidæ ...	...	Colies.
	58. Cuculidæ ...	...	Cuckoos.
Intermediate ...	59. Leptosomidæ	...	The Leptosoma.
	60. Bucconidæ ...	...	Puff-birds.
	61. Galbulidæ ...	...	Jacamars.
	62. Coraciidæ ...	...	Rollers.
	63. Meropidæ ...	...	Bee-eaters.
	64. Todidæ ...	...	Todies.
	65. Momotidæ ...	...	Motmots.
	66. Trogonidæ ...	...	Trogons.
Sub-order—Fissirostres.	67. Alcedinidæ ...	...	Kingfishers.
	68. Bucerotidæ ...	...	Hornbills.
	69. Upupidæ ...	...	Hoopoes.
	70. Irrisoridæ ...	...	Promerops.
	71. Podargidæ ...	...	Frog-mouths.
	72. Steatornithidæ	...	The Guacharo.
	73. Caprimulgidæ	...	Goatsuckers.
	74. Cypselidæ ...	...	Swifts.
	75. Trochilidæ ...	...	Humming-birds.

The Psittaci or parrot tribe are still in a very unsettled state of classification; that recently proposed by Professor Garrod differing widely from the arrangement adopted in Dr. Finsch's monograph of the order. Taking advantage of the researches of these and other authors, the following families are adopted as the most convenient in the present state of our knowledge:

76. Cacatuidæ ...	...	The Cockatoos.
77. Platycercidæ	...	The Broad-tailed Paroquets of Australia.
78. Palæornithidæ	...	The Oriental Parrots and Paroquets.
79. Trichoglossidæ	...	The Brush-tongued Paroquets and Lories.
80. Conuridæ ...	...	The Macaws and their allies.
81. Psittacidæ ...	...	The African and South American Parrots.
82. Nestoridæ ...	...	The Nestors of New Zealand.
83. Stringopidæ	...	The Owl-parrots of New Zealand.

The Columbæ, or pigeons, are also in a very unsatisfactory state as regards a natural classification. The families, sub-families, and genera proposed by various authors are very numerous, and often quite irreconcilable. I therefore adopt only two families; and generally follow Mr. G. R. Gray's hand-list for the genera, except where trustworthy authorities exist for a different arrangement. The families are:

84. Columbidæ	...	Pigeons and Doves.
85. Dididæ ...	...	The extinct Dodo and allies.

The Gallinæ, or game-birds, may be divided into seven families:

Fam.	Sub-fam.			
86. Pteroclidæ	...	...	...	... Sand-grouse.
87. Tetraonidæ	...	...	...	... Partridges and Grouse.
88. Phasianidæ ...	Pavoninæ ...		...	Peafowl.
	Lophophorinæ		...	Tragopans, &c.
	Phasianinæ ...		...	Pheasants.
	Euplocaminæ		...	Fire-backed Pheasants, &c.
	Gallinæ ...		...	Jungle-fowl.
	Meleagrinæ...		...	Turkeys.
	Numidinæ ...		...	Guinea-fowl.
89. Turnicidæ	...	...	...	... Hemipodes.
90. Megapodiidæ ...	...	...	...	Mound-makers.
91. Cracidæ ...	Cracinæ ...		...	Curassows.
	Penelopinæ ...		...	Guans.
	Oreophasinæ		...	Mountain-pheasant.
92. Tinamidæ	...	...	...	... Tinamous.

The Opisthocomi consist of one family containing a single species, the "Hocco" of Guiana.

Family 93. Opisthocomidæ.

The Accipitres, or birds of prey, which were long considered to be the highest and most perfect order of birds, are now properly placed lower down in the series, their affinities being more with the aquatic than with the perching birds. The following is the arrangement adopted by Mr. Sharpe in his recently published British Museum catalogue of diurnal birds of prey :—

Sub-orders.	Fam.		Sub-families.		
Falcones	94. Vulturidæ	...	Vulturinæ	Vultures.	
			Sarcorhamphinæ	Turkey-buzzards.	
	95. Serpentariidæ ...		Polyborniæ ...	Caracaras.	
			Accipitrinæ ...	Hawks.	
	96. Falconidæ	...	Buteoninæ ...	Buzzards.	
			Aquilinæ	Eagles.	
			Falconinæ ...	Falcons.	
Pandiones...97. Pandionidæ	...	... ...	...	Fishing-eagles.	
Striges ...98. Strigidæ	...	... ...	...	Owls.	

The Grallæ or Grallatores are in a very unsettled state. The following series of families is in accordance with the views of some of the best modern ornithologists :

99. Rallidæ ...	...	...	...	Rails, &c.
100. Scolopacidæ	...	...	...	Sandpipers and Snipes
101. Chionididæ	...	...	...	Sheath-bills.
102. Thinocoridæ	...	...	...	Quail-snipes.
103. Parridæ ...	...	...	...	Jacanas.
104. Glareolidæ	...	...	...	Pratincoles.
105. Charadriidæ	...	...	...	Plovers.
106. Otididæ ...	...	...	...	Bustards.
107. Gruidæ ...	...	...	...	Cranes.
108. Cariamidæ	...	...	...	Cariamas.
109. Aramidæ...	...	...	...	Guaraunas.
110. Psophiidæ	...	...	...	Trumpeters.
111. Eurypygidæ	...	...	...	Sun-bitterns.
112. Rhinochœtidæ	...	...	...	Kagus.
113. Ardeidæ ...	...	...	...	Herons.
114. Plataleidæ	...	...	..	Spoonbills and Ibis.
115. Ciconiidæ	...	...	...	Storks.
116. Palamedeidæ	...	...	...	Screamers.
117. Phænicopteridæ ...	...	...	Flamingoes.

H

The Anseres or Natatores are almost equally unsettled. The flamingoes are usually placed in this order, but their habits best assort with those of the waders.

	Fam.				
118.	Anatidæ	...	...	...	Duck and Geese.
119.	Laridæ	...	...	...	Gulls.
120.	Procellariidæ	...	...	...	Petrels.
121.	Pelecanidæ ...	...	...	...	Pelicans.
122.	Spheniscidæ	...	...	...	Penguins.
123.	Colymbidæ	...	...	...	Divers.
124.	Podicipidæ ...	...	...	...	Grebes.
125.	Alcidæ	...	...	...	Auks.

The last order of birds is the Struthiones or Ratitæ, considered by many naturalists to form a distinct sub-class. It consists of comparatively few species, either living or recently extinct.

		Fam.		
Living	126.	Struthionidæ	...	Ostriches.
	127.	Casuariidæ ...	...	Cassowaries.
	128.	Apterygidæ ...	...	Apteryx.
Extinct	129.	Dinornithidæ	...	Dinornis.
	130.	Palapterygidæ	...	Palapteryx.
	131.	Æpyornithidæ	...	Æpyornis.

REPTILES.

In reptiles I follow the classification of Dr. Günther as given in the *Philosophical Transactions*, vol. clvii., p. 625. He divides the class into five orders as follows :—

Sub-classes.		Orders.		
I. Squamata ...	1.	Ophidia	... ...	Serpents.
	2.	Lacertilia	... ...	Lizards.
	3.	Rhyncocephalina	...	The Hatteria.
II. Loricata ...	4.	Crocodilia	... ...	Crocodiles.
III. Cataphracta	5.	Chelonia	... ...	Tortoises.

In the arrangement of the families comprised in each of these orders I also follow the arrangement of Dr. Günther and Dr. J. E. Gray, as given in the British Museum Catalogue, or as modified by the former gentleman who has kindly given me much personal information.

The Ophidia, or Snakes, form the first order and are classified as follows :—

<table>
<tr><td colspan="2" align="center">Fam.</td><td></td></tr>
<tr><td rowspan="19">Innocuous Snakes</td><td>1. Typhlopidæ ...</td><td rowspan="4">Burrowing Snakes.</td></tr>
<tr><td>2. Tortricidæ ... ...</td></tr>
<tr><td>3. Xenopeltidæ ...</td></tr>
<tr><td>4. Uropeltidæ ...</td></tr>
<tr><td>5. Calamaridæ ...</td><td>Dwarf ground-snakes.</td></tr>
<tr><td>6. Oligodontidæ.</td><td></td></tr>
<tr><td>7. Colubridæ ... ...</td><td>Colubrine Snakes.</td></tr>
<tr><td>8. Homalopsidæ ...</td><td>Fresh-water Snakes.</td></tr>
<tr><td>9. Psammophidæ ...</td><td>Desert-snakes.</td></tr>
<tr><td>10. Rachiodontidæ.</td><td></td></tr>
<tr><td>11. Dendrophidæ ...</td><td>Tree-snakes.</td></tr>
<tr><td>12. Dryiophidæ ...</td><td>Whip-snakes.</td></tr>
<tr><td>13. Dipsasidæ ... ...</td><td>Nocturnal tree-snakes.</td></tr>
<tr><td>14. Scytalidæ.</td><td></td></tr>
<tr><td>15. Lycodontidæ ...</td><td>Fanged ground-snakes.</td></tr>
<tr><td>16. Amblycephalidæ</td><td>Blunt-heads.</td></tr>
<tr><td>17. Pythonidæ ...</td><td>Pythons.</td></tr>
<tr><td>18. Erycidæ ... ...</td><td>Sand-snakes.</td></tr>
<tr><td>19. Acrochordidæ ...</td><td>Wart-snakes.</td></tr>
<tr><td rowspan="4">Venomous Colubrine Snakes ... ...</td><td>20. Elapidæ ... ...</td><td>Cobras, &c.</td></tr>
<tr><td>21. Dendraspididæ.</td><td></td></tr>
<tr><td>22. Atractaspididæ.</td><td></td></tr>
<tr><td>23. Hydrophidæ ...</td><td>Sea-snakes.</td></tr>
<tr><td rowspan="2">Viperine Snakes ...</td><td>24. Crotalidæ ... ...</td><td>Pit-vipers.</td></tr>
<tr><td>25 Viperidæ ... ...</td><td>True vipers</td></tr>
</table>

The second order, Lacertilia, are arranged as follows :—

<table>
<tr><td>Fam.</td><td></td><td></td><td></td></tr>
<tr><td>26. Trogonophidæ ...</td><td>...</td><td>...</td><td rowspan="4">Amphisbænians.</td></tr>
<tr><td>27. Chirotidæ ...</td><td>...</td><td>...</td></tr>
<tr><td>28. Amphisbænidæ</td><td>...</td><td>...</td></tr>
<tr><td>29. Lepidosternidæ</td><td>...</td><td>...</td></tr>
<tr><td>30. Varanidæ ...</td><td>...</td><td>...</td><td>Water Lizards.</td></tr>
<tr><td>31. Helodermidæ.</td><td></td><td></td><td></td></tr>
<tr><td>32. Teidæ ...</td><td>...</td><td>...</td><td>Teguexins.</td></tr>
<tr><td>33. Lacertidæ ...</td><td>...</td><td>...</td><td rowspan="2">Land Lizards</td></tr>
<tr><td>34. Zonuridæ ...</td><td>...</td><td>...</td></tr>
<tr><td>35. Chalcidæ.</td><td></td><td></td><td></td></tr>
<tr><td>36. Anadiadæ.</td><td></td><td></td><td></td></tr>
<tr><td>37. Chirocolidæ.</td><td></td><td></td><td></td></tr>
<tr><td>38. Iphisadæ.</td><td></td><td></td><td></td></tr>
<tr><td>39. Cercosauridæ.</td><td></td><td></td><td></td></tr>
<tr><td>40. Chamæsauridæ.</td><td></td><td></td><td></td></tr>
<tr><td>41. Gymnopthalmidæ</td><td>...</td><td>...</td><td>Gape-eyed Scinks.</td></tr>
<tr><td>42. Pygopodidæ ...</td><td>...</td><td>...</td><td>Two-legged Lizards.</td></tr>
<tr><td>43. Aprasiadæ.</td><td></td><td></td><td></td></tr>
</table>

H 2

Fam.
44. Lialidæ.
45. Scincidæ ... ... Scinks.
46. Ophiomoridæ ... ... Snake-lizards.
47. Sepidæ ... ... Sand-lizards.
48. Acontiadæ.
49. Geckotidæ ... ... Geckoes.
50. Iguanidæ ... ... Iguanas.
51. Agamidæ ... ... Fringed Lizards.
52. Chameleonidæ ... ... Chameleons.

The third order, Rhyncocephalina consists of a single family:—

53. Rhyncocephalidæ ... ... The Hatteria of New Zealand.

The fourth order, Crocodilia or Loricata, consists of three families :—

54. Gavialidæ ... ... Gavials.
55. Crocodilidæ ... ... Crocodiles.
56. Alligatoridæ ... ... Alligators.

The fifth order, Chelonia, consists of four families :—

57. Testudinidæ ... ... Land and fresh-water Tortoises.
58. Chelydidæ ... ... Fresh-water Turtles.
59. Trionychidæ ... ... Soft Turtles.
60. Cheloniidæ ... ... Sea Turtles.

AMPHIBIA.

In the Amphibia I follow the classification of Professor Mivart, as given for a large part of the order in the *Proceedings of the Zoological Society* for 1869. For the remainder I follow Dr. Strauch, Dr. Günther, and a MSS. arrangement kindly furnished me by Professor Mivart.

The class is first divided into three groups or orders, and then into families as follows:—

Order I.—PSEUDOPHIDIA.

Fam.
1. Cæciliadæ ... ... Cæcilia.

Order II.—BATRACHIA URODELA.

2. Sirenidæ ... ... Siren.
3. Proteidæ ... ... Proteus.
4. Amphiumidæ ... Amphiuma.
5. Menopomidæ ... Giant Salamanders.
6. Salamandridæ ... Salamanders and Newts.

Order III. BATRACHIA ANOURA.

Fam.			Fam.		
7. Rhinophrynidæ			16. Pelodryadæ ...		
8. Phryniscidæ ...			17. Hylidæ		Tree Frogs.
9. Hylaplesidæ ...		Toads.	18. Polypedatidæ ...		
10. Bufonidæ... ...			19. Ranidæ		Frogs.
11. Xenorhinidæ ...			20. Discoglossidæ ...		
12. Engystomidæ ...			21. Pipidæ		Tongueless
13. Bombinatoridæ			22. Dactylethridæ ...		Toads.
14. Plectromantidæ		Frogs.			
15. Alytidæ					

FISHES.

These are arranged according to the classification of Dr. Günther, whose great work "The British Museum Catalogue of Fishes," has furnished almost all the material for our account of the distribution of the class.

In that work all existing fishes are arranged in six sub-classes and thirteen orders. A study of the extraordinary *Ceratodus* from Australia has induced Dr. Günther to unite three of his sub-classes; but as his catalogue will long remain a handbook for every student of fishes, it seems better to follow the arrangement there given, indicating his later views by bracketing together the groups he now thinks should be united.

Sub-class.	Order.	Families	Remarks.
Teleostei	1. Acanthopterygii ...	47	Gasterosteidæ to Notacanthi.
	2. Do. Pharyncognathi	5	Pomacentridæ to Chromidæ.
	3. Acanthini	6	Gadopsidæ to Pleurouectidæ.
	4. Physostomi	29	Siluridæ to Pegasidæ.
	5. Lophobranchii ...	2	Solenostomidæ and Syngnathidæ.
	6. Plectognathi ...	2	Sclerodermi and Gymnodontes.
Dipnoi ...	7. Sirenoidei	1	Sirenoidei.
Ganoidei	8. Holostei	3	Amiidæ to Lepidosteidæ.
	9. Chondrostei.	2	Accipenseridæ and Polydontidæ.
Chondropterygii	10. Holocephala	1	Chimæridæ.
	11. Plagiostomata ...	15	Carchariidæ to Myliobatidæ.
Cyclostomata	12. Marsipobranchii ...	2	Petromyzontidæ and Myxinidæ.
Leptocardii	13. Cirrhostomi	1	Cirrhostomi.
	Total ...	116	families.

INSECTS.

The families and genera of insects are so immensely numerous, probably exceeding fifty-fold those of all other land animals, that for this cause alone it would be impossible to enter fully into their distribution. It is also quite unnecessary, because many of the groups are so liable to be transported by accidental causes, that they afford no useful information for our subject; while others are so obscure and uninteresting that they have been very partially collected and studied, and are for this reason equally ineligible. I have therefore selected a few of the largest and most conspicuous families, which have been so assiduously collected in every part of the globe, and so carefully studied at home, as to afford valuable materials for comparison with the vertebrate groups, when we have made due allowance for the dependence of many insects on peculiar forms of vegetation, and the facility with which many of them are transported either in the egg, larva, or perfect state, by winds, currents, and other less known means.

I confine myself then, almost exclusively, to the sixteen families of Diurnal Lepidoptera or butterflies, and to six of the most extensive, conspicuous, and popular families of Coleoptera.

The number of species of Butterflies is about the same as that of Birds, while the six families of Coleoptera selected, comprise more than twenty thousand species, far exceeding the number of all other vertebrates. These families have all been recently catalogued, so that we have very complete information as to their arrangement and distribution.

LEPIDOPTERA DIURNA, OR BUTTERFLIES.

Fam.	Fam.
1. Danaidæ.	9. Libythæidæ.
2. Satyridæ.	10. Nemeobiidæ.
3. Elymniidæ.	11. Eurygonidæ.
4. Morphidæ.	12. Erycinidæ.
5. Brassolidæ.	13. Lycænidæ.
6. Acræidæ.	14. Pieridæ.
7. Heliconidæ.	15. Papilionidæ.
8. Nymphalidæ.	16. Hesperidæ.

COLEOPTERA, OR BEETLES.

Fam.		Fam.	
1. Cicindelidæ...	Tiger-beetles.	4. Cetoniidæ ...	Rose-chafers.
2. Carabidæ ...	Ground-beetles.	5. Buprestidæ ...	Metallic Beetles.
3. Lucanidæ ...	Stag-beetles.	6. Longicornia ...	Long-horned Beetles.

The above families comprise the extensive series of ground beetles (Carabidæ) containing about 9,000 species, and the Longicorns, which are nearly as numerous and surpass them in variety of form and colour as well as in beauty. The Cetoniidæ and Buprestidæ are among the largest and most brilliant of beetles ; the Lucanidæ are pre-eminent for remarkable form, and the Cicindelidæ for elegance; and all the families are especial favourites with entomologists, so that the whole earth has been ransacked to procure fresh species.

Results deduced from a study of these will, therefore, fairly represent the phenomena of distribution of Coleoptera, and, as they are very varied in their habits, perhaps of insects in general.

MOLLUSCA.

The Mollusca are usually divided into five classes as follows:—

Classes.		
I. Cephalopoda	Cuttle-fish.
II. Gasteropoda	Snails and aquatic Univalves.
III. Pteropoda	Oceanic Snails.
IV. Brachiopoda	Symmetrical Bivalves.
V. Conchifera	Unsymmetrical Bivalves.

The Gasteropoda and Conchifera alone contain land and freshwater forms, and to these we shall chiefly confine our illustrations of the geographical distribution of the Mollusca. The classification followed is that of Dr. Pfeiffer for the Operculata and Dr. Von Martens for the Helicidæ. The families chiefly referred to are:—

Class II.—GASTEROPODA.

Order 2.—Pulmonifera.

Fam.

In-operculata ... ...
1. Helicidæ.
2. Limacidæ.
3. Oncidiadæ.
4. Limnæidæ.
5. Auriculidæ.
6. Aciculidæ.
7. Diplommatinidæ.

Operculata ... ...
8. Cyclostomidæ.
9. Helicinidæ.

PART II.

ON THE DISTRIBUTION OF EXTINCT ANIMALS.

CHAPTER VI.

ALTHOUGH it may seem somewhat out of place to begin the systematic treatment of our subject with extinct rather than with living animals, it is necessary to do so in order that we may see the meaning and trace the causes of the existing distribution of animal forms. It is true, that the animals found fossil in a country are very generally allied to those which still inhabit it; but this is by no means universally the case. If it were, the attempt to elucidate our subject by Palæontology would be hopeless, since the past would show us the same puzzling diversities of faunas and floras that now exist. We find however very numerous exceptions to this rule, and it is these exceptions which tell us of the past migrations of whole groups of animals. We are thus enabled to determine what portion of the existing races of animals in a country are descendants of its ancient fauna, and which are comparatively modern immigrants; and combining these movements of the forms of life with known or probable changes in the distribution of land and sea, we shall sometimes be able to trace approximately the long series of changes which have resulted in the actual state of things. To gain this knowledge is our object in studying the " Geographical Distribution of Animals," and our plan of study must be determined, mainly, by the facilities it affords us for attaining this object. In discussing the countless details of distribution we shall meet with in our survey of the zoological regions, we shall often find it useful to refer to the evidence we possess of the range of the group in question in past times ; and when we attempt to generalise the phenomena on a large scale, with the details fresh in our memory, we shall find a reference to the extinct faunas of various epochs to be absolutely necessary.

The degree of our knowledge of the Palæontology of various parts of the world is so unequal, that it will not be advisable to treat the subject under each of our six regions.   Yet some sub-division must be made, and it seems best to consider separately the extinct animals of the Old and of the New Worlds. Those of Europe and Asia are intimately connected, and throw light on the past changes which have led to the establishment of the three great continental Old World regions, with their various subdivisions.   The wonderful extinct fauna recently discovered in North America, with what was previously known from South temperate America, not only elucidates the past history of the whole continent, but also gives indications of the mutual relations of the eastern and western hemispheres.

The materials to be dealt with are enormous; and it will be necessary to confine ourselves to a general summary, with fuller details on those points which directly bear upon our special subject.   The objects of most interest to the pure zoologist and to the geologist—those strange forms which are farthest removed from any now living—are of least interest to us, since we aim at tracing the local origin or birthplace of existing genera and families; and for this purpose animals whose affinities with living forms are altogether doubtful, are of no value whatever.

The great mass of the vertebrate fossils of the tertiary period consist of mammalia, and this is precisely the class which is of most value in the determination of zoological regions.   The animals of the secondary period, though of the highest interest to the zoologist are of little importance to us ; both because of their very uncertain affinities for any existing groups, and also because we can form no adequate notion of the distribution of land and sea in those remote epochs.   Our great object is to trace back, step by step, the varying distribution of the chief forms of life ; and to deduce, wherever possible, the physical changes which must have accompanied or caused such changes.

The natural division of our subject therefore is into geological periods. We first go back to the Post-Pliocene period, which includes that of the caves and gravels of Europe containing flint implements, and extends back to the deposit of the glacial drift in the concluding phase of the glacial epoch. Next we have the Pliocene period, divided into its later portion (the Newer Pliocene) which includes the Glacial epoch of the northern hemisphere; and its earlier portion (the Older Pliocene), represented by the red and coralline crag of England, and deposits of similar age in the continent. During this earlier epoch the climate was not very dissimilar from that which now prevails; but we next get evidence of a still earlier period, the Miocene, when a warmer climate prevailed in Europe, and the whole fauna and flora were very different. This is perhaps the most interesting portion of the tertiary deposits, and furnishes us with the most valuable materials for our present study. Further back still we have the Eocene period, with apparently an almost tropical climate in Europe; and here we find a clue to some of the most puzzling facts in the distribution of living animals. Our knowledge of this epoch is however very imperfect; and we wait for discoveries that will elucidate some of the mystery that still hangs over the origin and migrations of many important families. Beyond this there is a great chasm in the geological record as regards land animals; and we have to go so far back into the past, that when we again meet with mammalia, birds, and land-reptiles, they appear under such archaic forms that they cease to have any local or geographical significance, and we can only refer them to wide-spread classes and orders. For the purpose of elucidating geographical distribution, therefore, it is, in the present state of our knowledge, unnecessary to go back beyond the tertiary period of geology.

The remains of Mammalia being so much more numerous and important than those of other classes, we shall at first confine ourselves almost exclusively to these. What is known of the birds, reptiles, and fishes of the tertiary epoch will be best indicated by a brief connected sketch of their fossils in all parts of the globe, which we shall give in a subsequent chapter.

Historic Period.—In tracing back the history of the organic world we find, even within the limits of the historical period, that some animals have become extinct, while the distribution of others has been materially changed.   The *Rytina* of the North Pacific, the dodo of Mauritius, and the great auk of the North Atlantic coasts, have been exterminated almost in our own times.   The kitchen-middens of Denmark contain remains of the capercailzie, the *Bos primigenius*, and the beaver.   The first still abounds farther north, the second is extinct, and the third is becoming so in Europe.   The great Irish elk, a huge-antlered deer, probably existed almost down to historic times.

Pleistocene or Post-Pliocene Period.—We first meet with proofs of important changes in the character of the European fauna, in studying the remains found in the caverns of England and France, which have recently been so well explored.   These cave-remains are probably all subsequent to the Glacial epoch, and they all come within the period of man's occupation of the country.   Yet we find clear proofs of two distinct kinds of change in the forms of animal life.   First we have a change clearly trace-able to a difference of climate.   We find such arctic forms as the rein-deer, the musk-sheep, the glutton, and the lemming, with the mammoth and the woolly rhinoceros of the Siberian ice-cliffs, inhabiting this country and even the south of France. This is held to be good proof that a sub-arctic climate pre-vailed over all Central Europe ; and this climate, together with the continental condition of Britain, will sufficiently explain such a southward range of what are now arctic forms.

But together with this change we have another that seems at first sight to be in an exactly opposite direction.   We meet with numerous animals which now only inhabit Africa, or South Europe, or the warmer parts of Asia.   Such are, large felines— some closely related to the lion (*Felis spelœa*), others of alto-gether extinct type (*Machairodus*) and forming the extreme de-velopment of the feline race ;—hyænas ; horses of two or more species ; and a hippopotamus.   If we go a little further back, to the remains furnished by the gravels and brick-earths, we still find the same association of forms.   The reindeer, the glutton,

the musk-sheep, and the woolly rhinoceros, are associated with
several other species of rhinoceros and elephant; with numerous civets, now abundant only in warm countries; and with
antelopes of several species. We also meet here with a great
extension of range of forms now limited to small areas. The
Saiga antelope of Eastern Europe occurs in France, where wild
sheep and goats and the chamois were then found, together with
several species of deer, of bear, and of hyæna. A few extinct
genera even come down to this late period, such as the great
sabre-toothed tiger, *Machairodus; Galeotherium*, a form of Viverridæ; *Palæospalax*, allied to the mole; and *Trogontherium*, a
gigantic form of beaver.

We find then, that even at so early a stage of our inquiries we
meet with a problem in distribution by no means easy to solve.
How are we to explain the banishment from Europe in so short
a space of time (geologically speaking) of so many forms of life
now characteristic of warmer countries, and this too during a
period when the climate of Central Europe was itself becoming
warmer? Such a change must almost certainly have been due
to changes of physical geography, which we shall be better able
to understand when we have examined the preceding Pliocene
period. We may here notice, however, that so far as we yet
know, this great recent change in the character of the fauna is
confined to the western part of the Palæarctic region. In caves
in the Altai Mountains examined by Prof. Brandt, a great collection of fossil bones was discovered. These comprised the
Siberian rhinoceros and mammoth, and the cave hyæna; but all
the others, more than thirty distinct species, are now living in
or near the same regions. We may perhaps impute this difference to the fact that the migration of Southern types into
this part of Siberia was prevented by the great mountain and
desert barrier of the Central Asiatic plateau; whereas in Europe
there was at this time a land connection with Africa. Post-
pliocene deposits and caverns in Algeria have yielded remains
resembling the more southern European types of the Post-
pliocene period, but without any admixture of Arctic forms;
showing, as we might expect, that the glacial cold did not

extend so far south. We have here remains of *Equus, Bos, Antilope, Hippopotamus, Elephas, Rhinoceros, Ursus, Canis,* and *Hyæna,* together with *Phacochærus,* an African type of swine which has not occurred in the European deposits.

It is perhaps to the earlier portion of this period that the *Merycotherium* of the Siberian drift belongs. This was an animal related to the living camel, thus supporting the view that the *Camelidæ* are essentially denizens of the extra-tropical zone.

PLIOCENE PERIOD.

Primates.—We here first meet with evidence of the existence of monkeys in Central Europe. Species of *Macacus* have left remains not only in the Newer Pliocene of the Val d'Arno in Italy, but in beds of the same age at Grays in Essex ; while *Semnopithecus* and *Cercopithecus,* genera now confined to the Oriental and Ethiopian regions respectively, have been found in the Pliocene deposits of the South of France and Italy.

Carnivora.—Most of the genera which occurred in the Post Pliocene are found here also, and many of the same species. Few new forms appear, except *Hyænarctos,* a large bear with characters approaching the hyænas, and *Pristiphoca,* a new form of seal, both from the Older Pliocene of France ; and *Galecynus,* a fox-like animal intermediate between *Canis* and *Viverra,* from the Pliocene of Œninghen in Switzerland.

Cetacea.—Species of *Balæna, Physeter,* and *Delphinus* occur in the Older Pliocene of England and France, and with these the remains of many extinct forms, *Balænodon* and *Hoplocetus* (Balænidæ) ; *Belemnoziphius* and *Choneziphius* (Hyperoodontidæ), and *Halitherium,* an extinct form of the next order—Sirenia, now confined to the tropics, although the recently extinct *Rytina* of the N. W. Pacific shows that it is also adapted for temperate climates.

Ungulata.—The Pliocene deposits are not very rich in this order. The horses (*Equidæ*) are represented by the genus *Equus ;* and here we first meet with *Hipparion,* in which small lateral toes appear. Both genera occur in British deposits of this age.

A more interesting fact for us is the occurrence of the genus *Tapirus* in the Newer Pliocene of France and in the older beds of both France and England, since this genus is now isolated in the remotest parts of the eastern and western tropics. The genera *Rhinoceros*, *Hippopotamus*, and *Sus*, occur here as in the preceding epoch.

We next come to the deer genus (*Cervus*), which appears to have been at its maximum in this period, no less than eight species occurring in the Norwich Crag, and Forest-beds. Among the Bovidæ, the antelopes, ox, and bison, are the only forms represented here, as in the Post-Pliocene period. Passing on to the Proboscidea, we find three species of elephants and two of *Mastodon* preserved in European beds of this period, all distinct from those of Post-Pliocene times.

Rodentia.—In this order we find representatives of many living European forms; as *Cricetus* (hamster), *Arvicola* (vole), *Castor* (beaver), *Arctomys* (marmot), *Hystrix* (porcupine), *Lepus* (hare), and *Lagomys* (pika); and a few that are extinct, the most important being *Chalicomys*, allied to the beaver; and *Issiodromys*, said to come nearest to the remarkable *Pedetes* of South Africa, both found in the Pliocene formations of France.

General Conclusions as to Pliocene and Post-Pliocene Faunas of Europe.—This completes the series of fossil forms of the Pliocene deposits of Europe. They show us that the presence of numerous large carnivora and ungulates (now almost wholly tropical) in the Post-Pliocene period, was due to no exceptional or temporary cause, but was the result of a natural succession from similar races which had inhabited the same countries for long preceding ages. In order to understand the vast periods of time covered by the Pliocene and Post-Pliocene formations, the works of Sir Charles Lyell must be studied. We shall then come to see, that the present condition of the fauna of Europe is wholly new and exceptional. For a long succession of ages, various forms of monkeys, hyænas, lions, horses, hipparions, tapirs, rhinoceroses, hippopotami, elephants, mastodons, deer, and antelopes, together

I

with almost all the forms now living, produced a rich and varied fauna such as we now see only in the open country of tropical Africa. During all this period we have no reason to believe that the climate or other physical conditions of Europe were more favourable to the existence of these animals than now. We must look upon them, therefore, as true indigenes of the country, and their comparatively recent extinction or banishment as a remarkable phenomenon for which there must have been some adequate cause. What this cause was we can only conjecture; but it seems most probable that it was due to the combined action of the Glacial period, and the subsidence of large areas of land once connecting Europe with Africa. The existence, in the small island of Malta, of no less than three extinct species of elephant (two of very small stature), of a gigantic dormouse, an extinct hippopotamus, and other mammalia, together with the occurrence of remains of hippopotamus in the caves of Gibraltar, indicate very clearly that during the Pliocene epoch, and perhaps during a considerable part of the Post-Pliocene, a connection existed between South Europe and North Africa in at least these two localities. At the same time we have every reason to believe that Britain was united to the Continent, what is now the German Ocean constituting a great river-valley. During the height of the Glacial epoch, these large animals would probably retire into this Mediterranean land and into North Africa, making annual migrations northwards during the summer. But as the connecting land sank and became narrower and narrower, the migrating herds would diminish, and at last cease altogether; and when the glacial cold had passed away would be altogether prevented from returning to their former haunts.

MIOCENE PERIOD.

We now come to a period which was wonderfully rich in all forms of life, and of which the geological record is exceptionally complete. Various lacustrine, estuarine, and other deposits in Europe, North India, and North America, have furnished such a vast number of remains of extinct mammalia, as to solve many zoological problems, and to throw great light on the early distribution and centres of dispersal of various groups of animals. In order to show the bearing of these remains on our special subject, we will first give an account of the extinct fauna of Greece, of the Upper Miocene period; since this, being nearest to Africa and Asia, best exhibits the relations of the old European fauna to those countries. We shall then pass to the Miocene fauna of France and Central Europe ; and conclude with the remarkable Siwalik and other Indian extinct faunas, which throw an additional light on the early history of the animal life of the great old-world continents.

Extinct Animals of Greece.

These are from the Upper Miocene deposits at Pikermi, near Athens, and were collected by M. Gaudry a few years ago. They comprise ten living and eighteen extinct genera of mammalia, with a few birds and reptiles.

Primates.—These are represented by *Mesopithecus*, a genus believed to be intermediate between the two Indian genera of monkeys, *Semnopithecus* and *Macacus.*

Carnivora.—These were abundant. Of *Felis* there were four species, ranging from the size of a cat to that of a jaguar, a large *hyæna*, and a large weasel (*Mustela*). Besides these there were the huge *Machairodus*, larger than any existing lion or tiger, and with enormously developed canine teeth ; *Hyænictis* and *Lycæna*, extinct forms of Hyænidæ ; *Thalassictis=Ictitherium*, an extinct genus of Viverridæ but with resemblances to the hyænas, represented by three species, some of which were larger than any existing Viverridæ; *Promephytis*, an extinct form of Mustelidæ, having resemblances to the European marten, to the otters, and to the S. African *Zorilla* ; and lastly, *Simocyon*, an extraordinary carnivore of the size of a small panther, but having the canines of a cat, the molars of a dog, and the jaws shaped like those of a bear.

Ungulata.—These are numerous and very interesting. The Equidæ are represented by the three-toed *Hipparion*, which con-

I 2

tinued to exist till the Older Pliocene period. There are three large species of *Rhinoceros*, as well as a species of the extinct genus *Leptodon* of smaller size. Remains of a very large wild boar (*Sus*) were found. Very interesting is the occurrence of a species of giraffe (*Camelopardalis*) as tall as the African species but more slender; and also an extinct genus *Helladotherium*, not quite so tall as the giraffe but much more robust, and showing some approach to the Antilopidæ in its dentition. Antelopes were abundant, ranging from the size of the gazelle to that of the largest living species. Three or four seem referable to living genera, but the majority are of extinct types, and are classed in the genera *Palæotragus*, *Palæoryx*, *Tragocerus*, and *Palæoreas;* while *Dremotherium* is an ancient generalized form of *Cervidæ* or deer.

Proboscidea.—These are represented by two species of *Mastodon*, and two of *Dinotherium*, an extraordinary extinct form supposed to be, to some extent, intermediate between the elephants and the aquatic manatees (*Sirenia.*)

Rodentia.—This order is represented by a species of *Hystrix*, larger than living porcupines.

Edentata.—This order, now almost confined to South America, was represented in the Miocene period by several European species. *Ancylotherium* and *Macrotherium*, belonging to an extinct family but remotely allied to the African ant-bear (*Orycteropus*), occur in Greece.

Birds.—Species of *Phasianus* and *Gallus* were found; the latter especially interesting as being now confined to India.

Reptiles.—These are few and unimportant, consisting of a tortoise (*Testudo*) and a large lizard allied to *Varanus*.

Summary of the Miocene Fauna of Greece.—Although we cannot consider that the preceding enumeration gives us by any means a complete view of the actual inhabitants of this part of Europe during the later portion of the Miocene period, we yet obtain some important information. The resemblance that appeared in the Pliocene fauna of Europe, to that of the open country of tropical Africa, is now still more remarkable. We

CHAP. VI.]   MAMMALIA OF THE OLD WORLD.   117

not only find great felines, surpassing in size and destructive power the lions and leopards of Africa, with hyænas of a size and in a variety not to be equalled now, but also huge rhinoceroses and elephants, two forms of giraffes, and a host of antelopes, which, from the sample here obtained, were probably quite as numerous and varied as they now are in Africa. Joined with this abundance of antelopes we have the absence of deer, which probably indicates that the country was open and somewhat of a desert character, since there were deer in other parts of Europe at this epoch. The occurrence of but a single species of monkey is also favourable to this view, since a well-wooded country would most likely have supplied many forms of these animals.

Miocene Fauna of Central and Western Europe.

We have now to consider the Miocene fauna of Europe generally, of which we have very full information from numerous deposits of this age in France, Switzerland, Italy, Germany, and Hungary.

Primates.—Three distinct forms of monkeys have been found in Europe—in the South of France, in Switzerland, and Wurtemberg ; one was very like *Colobus* or *Semnopithecus;* the others— *Pliopithecus* and *Dryopithecus*—were of higher type, and belonged to the anthropomorphous apes, being nearest to the genus *Hylobates* or gibbons. Both have occurred in the South of France. The *Dryopithecus* was a very large animal (equal to the gorilla), and M. Lartet considers that in the character of its dentition it approached nearer to man than any of the existing anthropoid apes.

Insectivora.—These small animals are represented by numerous remains belonging to four families and a dozen genera. Of *Erinaceus* (hedgehog) several species are found in the Upper Miocene ; and in the Lower Miocene of Auvergne two extinct genera of the same family—*Amphechinus* and *Tetracus*—have been discovered. Several species of *Talpa* (mole) occur in the Upper Miocene of France, while the extinct *Dinylus* is from Germany, and *Palæospalax* from the Lower Miocene of the Isle of

Wight. The Malayan family Tupaiidæ or squirrel-shrews, is believed to be represented by *Oxygomphus*, a fossil discovered in South Germany (Wiesenau) by H. von Meyer. The Soricidæ or shrews, are represented by several extinct genera—*Plesiosorex*, *Mysarachne* and *Galeospalax;* as well as by *Amphisorex* and *Myogale* still living. *Echinogale*, a genus of Centetidæ now confined to Madagascar, is said to occur in the Lower Miocene of Auvergne, a most interesting determination, if correct, as it would form a transition to the *Solenodon* of the Antilles belonging to the same family; but I am informed by Prof. Flower that the affinities of the animals described under this name are very doubtful.

Carnivora.—Besides *Felis* and *Machairodus*, which extend back to the Upper Miocene, there are two other genera of Felidæ, *Pseudælurus* in the Upper Miocene of France, and *Hyænodon*, which occurs in the Upper and Lower Miocene of France, named from some resemblance in its teeth to the hyænas, and considered by some Palæontologists to form a distinct family, Hyænodontidæ. The Viverridæ, or civets, were very numerous, consisting of the living genus *Viverra*, and three extinct forms—*Thalassictis=Ictitherium*, as large as a panther, and *Soricictis*, a smaller form, occurring both in France and Hungary. Of *Hyænidæ*, there was the living genus *Hyæna*, and the extinct *Hyænictis*, which has occurred in Hungary as well as in Greece. The Canidæ, or wolf and fox family, were represented by *Pseudocyon*, near to *Canis;* *Hemicyon*, intermediate between dogs and gluttons; and *Amphicyon*, of which several species occur in the Upper and Lower Miocene of France, some of them larger than a tiger. The Mustelidæ, or weasels, were represented by five genera, the existing genera *Lutra* (otter) and *Mustela* (weasel); *Potamotherium*, an extinct form of otter; *Taxodon*, allied to the badger and otter; *Palæomephitis* in Germany, and the *Promephytis* (already noticed) in Greece. The bears were represented only by *Hyænarctos*, which has been noticed as occurring in the Pliocene, and first appears in the Upper Miocene of France. Seals are represented by a form resembling the Antarctic *Otaria*, remains of which occur in the Upper Miocene of France.

Cetacea (whales).—These occur frequently in the Miocene deposits, four living, and five extinct genera having been described; but these marine forms are not of much importance for our purpose.

Sirenia (sea-cows).—These are represented by two extinct genera, *Halitherium* and *Trachytherium*. Several species of the former have been discovered, but the latter has occurred in France only, and its affinities are doubtful.

Ungulata.—Horses are represented by *Hipparion* and *Anchitherium*, the latter occurring in both Upper and Lower Miocene and Eocene; while *Hipparion*, which is more nearly allied to living horses, first appears in the Upper Miocene and continues in the Pliocene.

Hippotherium, in the Upper Miocene of the Vienna basin, forms a transition to *Paloplotherium*, an Eocene genus of Tapiridæ or Palæotheridæ. Tapirs, allied to living forms, occur in both Upper and Lower Miocene. Rhinoceroses are still found in the Upper Miocene, and here first appear the four-toed hornless rhinoceros, *Acerotherium*. The Suidæ (swine) are rather numerous. *Sus* (wild boar) continued as far back as the Upper Miocene; but now there first appear a number of extinct forms which have been named *Hyotherium*, *Palæochœrus*, *Chœromorus*, all of a small or moderate size; *Hyopotamus*, nearly as large as a tapir; and *Anthracotherium*, nearly the size of a hippopotamus and, according to Dr. Leidy, the type of a distinct family. *Listriodon*, from the Upper Miocene of the Vienna basin, is sometimes classed with the tapirs.

We now come to a well-marked new family of Artiodactyle or even-toed Ungulata, the *Anoplotheriidæ*, which consisted of more slender long-tailed animals, allied to the swine but with indications of a transition towards the camels. The only genera that appear in the Miocene formation are, *Chalicotherium*, nearly as large as a rhinoceros, of which three species have been found in Germany and France; and *Synaphodus*, known only from its teeth, which differ somewhat from those of the *Anoplotherium* which appears earlier in the Eocene formation. Another extinct family, *Amphimericidæ* or *Xiphodontidæ*, is represented by two

genera, *Cainotherium* and *Microtherium*, in the Miocene of
France. They were of very small size, and are supposed to be
intermediate between the Suidæ and Tragulidæ.

The Camelopardalidæ, or giraffes, were represented in Europe
in Miocene times by the gigantic *Helladotherium*, which has
been found in the south of France, and in Hungary, as well as
in Greece. The chevrotains (Tragulidæ) are represented by
the extinct genus *Hyomoschus*.

The Cervidæ do not seem to have appeared in Europe before
the Upper Miocene epoch, when they were represented by
Dorcatherium and *Amphimoschus*, allied to *Moschus*, and also by
true *Cervus*, as well as by small allied forms, *Dremotherium*,
Amphitragalus (in the Lower Miocene), *Micromeryx*, *Palæomeryx*,
and *Dicrocerus*.

The Bovidæ, or hollow-horned ruminants, were not well
represented in Central Europe in Miocene times. There were
no sheep, goats, or oxen, and only a few antelopes of the genus
Tragocerus, and one allied to *Hippotragus;* and these all lived
in the Upper Miocene period, as did the more numerous
forms of Greece.

Proboscidea.—The true elephants do not extend back to the
Miocene period, but they are represented by the Mastodons,
which had less complex teeth. These first appear in the Upper
Miocene of Europe, five species being known from France,
Germany, Switzerland, and Greece. *Dinotherium*, already
noticed as occurring in Greece, extended also to Germany and
France, where remains of three species have been found.

Rodentia.—A considerable number of generic forms of this
order have been obtained from the Miocene strata. The principal genera are *Cricetodon*, allied to the hamsters, numerous in
both the Upper and Lower Miocene period of France; *Myoxus*
(the dormice) in France, and an allied genus, *Brachymys*, in Germany. The beavers were represented by the still living genus
Castor, and the extinct *Steneofiber* in France. The squirrels by
the existing *Scuirus* and *Spermophilus;* and by extinct forms,
Lithomys and *Aulacodon*, in Germany, the latter resembling the
African genus *Aulacodes*. The hares, by *Lagomys* and an

extinct form *Titanomys*. Besides these, remains referred to the South American genera, *Cavia* (cavy) and *Dasyprocta* (agouti), have been found, the former in the Upper Miocene of Switzerland, the latter in the Lower Miocene of Auvergne. *Palœomys*, allied to the West Indian *Capromys*, has been found in the same deposits ; as well as *Theridomys*, said by Gervais to be allied to *Anomalurus* and *Echimys*, the former now living in W. Africa, the latter in S. America.

Edentata.—These are only represented by the *Macrotherium* and *Ancylotherium* of the Grecian deposits, the former occurring also in France and Germany in Upper Miocene strata.

Marsupials.—These consist of numerous species related to the opossums (*Didelphys*), but separated by Gervais under the name *Peratherium*. They occur in both Upper and Lower Miocene beds.

Upper Miocene Deposits of the Siwalik Hills and other Localities in N.W. India.

These remarkable fresh-water deposits form a range of hills at the foot of the Himalayas, a little south of Simla. They were investigated for many years by Sir P. Cautley and Dr. Falconer, and add greatly to our knowledge of the early fauna of the Old World continent.

Primates.—Remains of the genera *Semnopithecus* and *Macacus* were found, with other forms of intermediate character; and some teeth indicated animals allied to the orang-utan of Borneo, and of similar size.

Carnivora.—These consisted of species of *Felis* and *Machairodus* of large size ; *Hyœna, Canis, Mellivora*, and an allied genus *Ursitaxus; Ursus*, in the deposits of the Nerbudda valley (of Pliocene age) ; *Hyœnarctus* as large as the cave bear ; *Amphicyon* of the size of a polar bear (in the deposits of the Indus valley, west of Cashmere) ; *Lutra*, and an extinct allied genus *Enhydrion*.

Ungulata.— These are very numerous, and constitute the most important feature of this ancient fauna. Horses are represented by a species of *Equus* from the Siwalik Hills and the Irawaddy deposits in Burmah, and by two others from the Pliocene of the Nerbudda Valley; while *Hippotherium*—a slender, antelope-like animal, found in the Siwalik Hills and in Europe—is supposed to form a transition from the Equidæ to the Tapiridæ. These latter are found in the Upper Indus deposits, where there is a species of *Tapirus*, and one of an extinct genus *Antelotherium*. Of *Rhinoceros*, five extinct species have been found—in the Siwalik Hills, in Perim Island, and one at an elevation of 16,000 feet in the deserts of Thibet. *Hippopotamus* occurs in the Pliocene of the Nerbudda, and is represented in the older Miocene deposits by *Hexaprotodon*, of which three species have been found in various parts of India. Another remarkable genus, *Merycopotamus*, connects *Hippopotamus* with *Anthracotherium*, one of the extinct European forms allied to the swine. These last are represented by several large species of *Sus*, and by the extinct European genus *Chærotherium*.

The extinct Anoplotheridæ are represented by a species of the European genus *Chalicotherium*, larger than a horse.

An extinct camel, larger than the living species, was found in the Siwalik Hills.

Three species of deer (*Cervus*) have been found in the Siwaliks, and one in the Nerbudda deposits.

A large and a small species of giraffe (*Camelopardalis*) were found in the Siwalik Hills and at Perim Island.

The Bovidæ are represented by numerous species of *Bos*, and by the extinct genera *Hemibos* and *Amphibos*. There are also three species of antelopes, one of which is allied to the African *Alcephalus*.

We now come to an extraordinary group of extinct animals, probably forming a new family intermediate between the antelope and the giraffe. The *Sivatherium* was an enormous four-horned ruminant, larger than a rhinoceros. It had a short trunk like a tapir, the lower horns on the forehead were simple, the upper pair palmated. The *Bramatherium*, an allied form from Perim Island, showed somewhat more affinity for the giraffe.

Proboscidea.—No less than seven species of elephants and four

of mastodons ranged over India, their remains being found in all
the deposits from the Siwalik Hills to Burmah. A large *Dinotherium* has also been found at Perim Island.

Reptiles.—Many remains of birds were found, but these have
not been determined. Reptiles were numerous and interesting,
the most remarkable being the huge tortoise, *Colossochelys*, whose
shell was twelve feet long and head and neck eight feet more.
Other small tortoises of the genera *Testudo*, *Emys*, *Trionyx*
and *Emydida* were found, the *Emys* being a living species.
There were three extinct and one living species of crocodile,
and one of them was larger than any now living. The only
other reptile of importance was a large lizard of the genus
Varanus.

*General Observations on the Miocene faunas of Europe and
Asia.*—Comparing the three faunas of approximately the same
period, and allowing for the necessarily imperfect record of
each, we find a wonderful similarity of general type over the
enormous area between France on the west and the Irawaddy
river in Burmah on the east. We may even extend our comparison to Northern China, where remains of *Hyæna*, *Tapir*,
Rhinoceros, *Chalicotherium*, and *Elephas*, have been recently
found, closely resembling those from the Miocene or Pliocene
deposits of Europe or India, and showing that the Palæarctic
region had then the same great extent from west to east
that it has now. Of about forty genera comprised in the
Indian Miocene fauna, no less than twenty-seven inhabited
Central and Western Europe during the same epoch. The Indian
Miocene fossils are much what we should expect as the forerunners of the existing fauna, the giraffes and hippopotami
being the only additions from the present Ethiopian fauna.
The numerous forms of the restricted bovine type, show that
these probably originated in India; while the monkeys appear
to be altogether of Oriental types.

In Europe, however, we meet with a totally different assemblage of animals from those that form the existing fauna. We
find apes and monkeys, many large Felidæ, numerous civets

and hyænas, tapirs, rhinoceros, hippopotamus, elephants, giraffes, and antelopes, such as now characterise the tropics of Africa and Asia.   Along with these we meet with less familiar types, showing relations with the Centetidæ of Madagascar, the Tupaiidæ of the Malay Islands, the *Capromys*, of the West Indies, and the *Echimys* of South America.   And besides all these living types we have a host of extinct forms,—ten or twelve genera allied to swine; nine genera of tapir-like animals; four of horses; nine of wolves; with many distinct forms of the long-extinct families of Anoplotheridæ, Xiphodontidæ, and the edentate Macrotheridæ.   It is almost certain that during the Miocene period Europe was not only far richer than it is now in the higher forms of life, but not improbably richer than any part of the globe now is, not excepting tropical Africa and tropical Asia.

Eocene Period.

The deposits of Eocene age are less numerous, and spread over a far more limited area, than those of the Miocene period, and only restricted portions of them furnish any remains of land animals.   Our knowledge of the Eocene mammalian fauna is therefore very imperfect and will not occupy us long, as most of the new types it furnishes are of more interest to the zoologist than to the student of distribution.   Some of the Eocene mammalia of Europe are, however, of interest in comparison with those of North America of the same age ; while others show that ancestral types of groups now confined to Australia or to South America, then inhabited Europe.

Primates.—The only undoubted Eocene examples of this order, are the *Cœnopithecus lemuroides* from the Jura, which has points of resemblance to the South American marmosets and howlers, and also to the Lemuridæ ; and a cranium recently discovered in the Department of Lot (S.W. France), undoubtedly belonging to the Lemuridæ, and which most resembles that of the West African " Potto" (*Perodicticus*).   This discovery has led to another, for it is now believed that remains formerly referred to the Anoplotheridæ (*Adapis* and *Aphelotherium* from the Upper Eocene of Paris) were also Lemurs. Some remains from the Lower Eocene of Suffolk were at first supposed to be allied to *Macacus*, but were subsequently referred to the Ungulate, *Hyracotherium*. There is still, however, some doubt as to its true affinities.

Chiroptera.—In the Upper Eocene of Paris remains of bats have been found, so closely resembling living forms as to be referred to the genus *Vespertilio*.

Carnivora.—The only feline remains, are those of *Hyænodon* in the Upper Eocene of Hampshire, and *Pterodon*, an allied form from beds of the same age in France ; with *Ælurogale*, found in the South of France in deposits of phosphate of lime of uncertain age, but probably belonging to this period. Viverridæ (civets) are represented by two genera, *Tylodon*, the size of a glutton from the Upper Eocene, and *Palæonyctis*, allied to *Viverra*, from the Middle Eocene of France. The Canidæ (wolves and foxes) appear to have been the most ancient of the existing types of Carnivora, five genera being represented by Eocene remains. Of these, *Galethylax* and *Cyotherium* were small, and with the existing genus *Canis* are found in the Upper Eocene of France. *Arctocyon*, about the size of a wolf, is a very ancient and generalised form of carnivore which can not be placed in any existing family. It is found in the Lower Eocene of France, and is thus the oldest known member of the Carnivora.

Ungulata.—These are more numerous. Equidæ (horses) are represented by the Miocene *Anchitherium* in the Lower, and by a more ancient form, *Anchilophus*, in the Middle Eocene of France. Tapiridæ and Palæotheridæ were very numerous. *Palæotherium* and the allied genus *Paloplotherium*, were abundant in France and England in Upper Eocene times. They somewhat resembled the tapir, with affinities for the horse and rhinoceros. A new genus, *Cadurcotherium*, allied to the rhinoceros and equally large, has been found in the same deposits of phosphate of lime as the lemur and *Ælurogale*. In the Middle Eocene of both England and France are found *Lophiodon* allied to the tapir, but in some of the species reaching a larger size; *Propalæotherium* and *Pachynolophus* of smaller size and having affinities for the other genera named; and *Plagiolophus*, a small, slender animal which Professor Huxley thinks may have been a direct ancestor of the horse. In the Lower Eocene we meet with *Coryphodon*, much larger than the tapir, and armed with large canine teeth; *Pliolophus*, a generalised type, allied to the tapir and horse; and *Hyracotherium*, a small animal from the Lower Eocene of England, remotely allied to the tapir.

Among the Artiodactyla, or even-toed ungulates, the swine are represented by several extinct genera, of moderate or small size—*Acotherium*, *Chæropotamus*, *Cebochœrus* and *Dichobune*, all from the Upper and the last also from the Middle Eocene of France; but *Eutelodon*, from the phosphate of lime deposits is large. The *Dichobune* was the most generalised type, presenting the characters of many of the other genera combined, and was believed by Dr. Falconer to approach the musk-deer. The *Cainotherium* of the Miocene also occurs here, and an allied genus *Plesiomeryx* from the same deposits as *Euteledon*.

The Eocene Anoplotheridæ were numerous. The *Anoplotherium* was a two-toed, long-tailed Pachyderm, ranging from the size of a hog to that of an ass; the allied *Eurytherium* was four-toed; and there are one or two others of doubtful affinity. All are from the Upper Eocene of France and England.

Rodentia.—Remains referred to the genera *Myoxus* (dormouse) and *Sciurus* (squirrel) have been found in the Upper Eocene of France; as well as *Plesiarctomys*, an extinct genus between the marmots and squirrels. The Miocene *Theridomys* is also found here.

Marsupials.—The *Didelphys* (opossum) of Cuvier, now referred to an extinct genus *Peratherium*, is found in the Upper Eocene of France and England.

General Considerations on the Extinct Mammalian Fauna of Europe.—It is a curious fact that no family, and hardly a genus, of European mammalia occurs in the Pliocene deposits, without extending back also into those of Miocene age. There are, however, a few groups which seem to be late developments or recent importations into the Palæarctic region, as they occur only in Post-Pliocene deposits. The most important of these are the badger, glutton, elk, reindeer, chamois, goat, and sheep, which only occur in caves and other deposits of Post-Pliocene age. Camels only occur in the Post-Pliocene of Siberia (*Merycotherium*), although a true *Camelus* of large size appears to have inhabited some part of Central Asia in the Upper Miocene period, being found in the Siwalik beds. The only exclusively Pliocene genera in Europe are *Ursus, Equus, Hippopotamus, Bos, Elephas, Arvicola, Trogontherium, Arctomys, Hystrix* and *Lepus;* but of these *Equus, Hippopotamus, Bos,* and *Elephas* are found in the Miocene deposits of India. Owing, no doubt, in part to the superior productiveness of the various Miocene beds, large numbers of groups appear to have their origin or earliest appearance here. Such are Insectivora, Felidæ, Hyænidæ, Mustelidæ, *Ursus,* Equidæ, *Tapirus,* Rhinocerotidæ, Hippopotamidæ, Anthracotheridæ (extinct), *Sus,* Camelopardidæ, Tragulidæ, Cervidæ, Bovidæ, Elephantidæ, and Edentata.

Groups which go back to the Eocene period, are, Primates allied to South American monkeys, as well as some of the Lemuridæ; bats of the living genus *Vespertilio;* Hyænodontidæ, an ancestral form of Carnivore; Viverridæ; Canidæ (to the Upper Eocene), and the ancestral Arctocyonidæ to the Lower Eocene; *Hyænarctos,* an ancestral type of bears and hyænas; Anchitheridæ, ancestral horses, to the Middle Eocene; Palæotheridæ, comprising numerous generalised forms, ancestors of the rhinoceros, horse, and tapir; Suidæ, with numerous generalised forms, to the Middle Eocene; Anoplotheridæ and Xiphodontidæ, ancestral families of even-toed Ungulates, connecting the ruminants with the swine; and lastly, several groups of Rodents, and a Marsupial, in the Upper Eocene. We thus find all the great types of Mammalia well developed in the earliest portion of the tertiary period; and the occurrence of Quadrumana, of the highly specialized bats (*Vespertilio*), of various forms of Carnivora, and of Ungulates, clearly differentiated into the odd and even-toed series, associated with such lower forms as

Lemurs and Marsupials—proves, that we have here hardly made an approach towards the epoch when the mammalian type itself began to diverge into its various modifications. Some of the Carnivora and Ungulates do, indeed, exhibit a less specialised structure than later forms; yet so far back as the Upper Miocene the most specialised of all carnivora, the great sabre-toothed *Machairodus*, makes its appearance.

The Miocene is, for our special study, the most valuable and instructive of the Tertiary periods, both on account of its superior richness, and because we here meet with many types now confined to separate regions. Such facts as the occurrence in Europe during this period of hippopotami, tapirs, giraffes, Tragulidæ, Edentata, and Marsupials—will assist us in solving many of the problems we shall meet with in reviewing the actual distribution of living forms of those groups. Still more light will, however, be thrown on the subject by the fossil forms of the American continent, which we will now proceed to examine.

CHAPTER VII.

THE discoveries of very rich deposits of mammalian remains in various parts of the United States have thrown great light on the relations of the faunas of very distant regions. North America now makes a near approach to Europe in the number and variety of its extinct mammalia, and in no part of the world have such perfect specimens been discovered. In what are called the "Mauvaises terres" of Nebraska (the dried-up mud of an ancient lake), thousands of entire crania and some almost entire skeletons of ancient animals have been found, their teeth absolutely perfect, and altogether more resembling the preparations of the anatomist, than time-worn fossils such as we are accustomed to see in the museums of Europe. Other deposits have been discovered in Oregon, California, Virginia, South Carolina, Texas, and Utah, ranging over all the Tertiary epochs, from Post-Pliocene to Eocene, and furnishing a remarkable picture of the numerous strange mammalia which inhabited the ancient North American continent.

NORTH AMERICA—POST-PLIOCENE PERIOD.

Insectivora.—The only indications of this order yet discovered, consists of a single tooth of some insectivorous animal found in Illinois, but which cannot be referred to any known group.

Carnivora.—These are fairly represented. Two species of *Felis* as large as a lion; the equally large extinct *Trucifelis*, found only in Texas; four species of *Canis*, some of them larger

K

than wolves; two species of *Galera*, a genus now confined to the
Neotropical region; two bears, and an extinct genus, *Arctodus;*
an extinct species of racoon (*Procyon*), and an allied extinct
genus, *Myxophagus*—show, that at a very recent period North
America was better supplied with Carnivora than it is now.
Remains of the walrus (*Trichechus*) have also been found as far
south as Virginia.

Cetacea.—Three species of dolphins belonging to existing
genera, have been found in the Eastern States; and two species
of *Manatus*, or sea-cow, in Florida and South Carolina.

Ungulata.—Six extinct horses (*Equus*), and one *Hipparion;*
the living South American tapir, and a larger extinct species; a
Dicotyles, or peccary, and an allied genus, *Platygonus;* a species
of the South American llamas (*Auchenia*), and one of a kind of
camel, *Procamelus;* two extinct bisons; a sheep, and two musk-
sheep (*Ovibos*); with three living and one extinct deer (*Cervus*),
show an important increase in its Herbivora.

Proboscidea.—Two elephants and two mastodons, added to this
remarkable assemblage of large vegetable-feeding quadrupeds.

Rodentia.—These consist mainly of genera and species still
living in North America; the only important exceptions being a
species of the South American capybara (*Hydrochœrus*) in South
Carolina; and *Praotherium*, an extinct form of hare, found in a
bone cave in Pennsylvania.

Edentata.—Here we meet with a wonderful assemblage, of six
species belonging to four extinct genera, mostly of gigantic size.
A species of *Megatherium*, three of *Megalonyx*, and one of
Mylodon—huge terrestrial sloths as large as the rhinoceros
or even as the largest elephants—ranged over the Southern
States to Pennsylvania, the latter (*Mylodon*) going as far as the
great lakes and Oregon. Another form, *Ereptodon*, has been
found in the Mississippi Valley.

Marsupialia.—The living American genus of opossums, *Didelphys*, has been found in deposits of this age in South Carolina.

Remarks on the Post-Pliocene fauna of North America.—The
assemblage of animals proved, by these remains, to have

inhabited North America at a comparatively recent epoch, is most remarkable. In Europe, we found a striking change in the fauna at the same period; but that consisted almost wholly in the presence of animals now inhabiting countries immediately to the north or south. Here we have the appearance of two new assemblages of animals, the one now confined to the Old World—horses, camels, and elephants; the other exclusively of South American type—llamas, tapirs, capybaras, *Galera*, and gigantic Edentata. The age of the various deposits in which these remains are found is somewhat uncertain, and probably extends over a considerable period of time, inclusive of the Glacial epoch, and perhaps both anterior and subsequent to it. We have here, as in Europe, the presence and apparent co-existence in the same area, of Arctic and Southern forms—the walrus and the manatee—the musk-sheep and the gigantic sloths. Unfortunately, as we shall see, the immediately preceding Pliocene deposits of North America are rather poor in organic remains; yet it can hardly be owing to the imperfection of the record of this period, that *not one* of the South American types above numerated occurs there, while a considerable number of Old World forms are represented. Neither in the preceding wonderfully rich Miocene or Eocene periods, does any *one* of these forms occur; or, with the exception of *Morotherium*, from Pliocene deposits *west* of the Rocky Mountains, any apparent ancestor of them! We have here unmistakable evidence of an extensive immigration from South into North America, not very long before the beginning of the Glacial epoch. It was an immigration of types altogether new to the country, which spread over all the southern and central portions of it, and established themselves sufficiently to leave abundance of remains in the few detached localities where they have been discovered. How such large yet defenceless animals as tapirs and great terrestrial sloths, could have made their way into a country abounding in large felines equal in size and destructiveness to the lion and the tiger, with numerous wolves and bears of the largest size, is a great mystery. But it is nevertheless certain that they did so ; and the fact that no such

K 2

migration had occurred for countless preceding ages, proves that some great barrier to the entrance of terrestrial mammalia which had previously existed, must for a time have been removed. We must defer further discussion of this subject till we have examined the relations of the existing faunas of North and South America.

TERTIARY PERIOD.

When we get to remains of the Tertiary age, especially those of the Miocene and Eocene epochs, we meet with so many interesting and connected types, and such curious relations with living forms in Europe, that it will be clearer to trace the history of each order and family throughout the Tertiary period, instead of considering each of the subdivisions of that period separately.

It will be well however first to note the few American Post-Pliocene or living genera that are found in the Pliocene beds. These consist of several species of *Canis*, from the size of a fox to that of a large wolf; a *Felis* as large as a tiger; an Otter (*Lutra*); several species of *Hipparion;* a peccary (*Dicotyles*); a deer (*Cervus*); several species of *Procamelus;* a mastodon; an elephant; and a beaver (*Castor*). It thus appears that out of nearly forty genera found in the Post-Pliocene deposits, only ten are found in the preceding Pliocene period. About twelve additional genera, however, appear there, as we shall see in going over the various orders.

Primates.—Among the vast number of extinct mammalia discovered in the Tertiary deposits of North America, no example of this order had been recognized up to 1872, when the discovery of more perfect remains showed, that a number of small animals of obscure affinities from the Lower Eocene of Wyoming, were really allied to the lemurs and perhaps also to the marmosets, the lowest form of American monkeys, but having a larger number of teeth than either. A number of other remains of small animals from the same formation, previously supposed to be allied to the Ungulata, are now shown to belong to the Primates; so that no less than twelve genera of these animals are recognized by Mr. Marsh, who classes them in two families—Limnotheridæ, comprising the genera *Limnotherium*, (which had larger canine teeth), *Thinolestes, Telmatolestes, Mesacodon, Bathrodon,* and *Antiacodon* of Marsh, with *Notharctos, Hipposyus, Microsyops,* and *Palæacodon* previously described by Leidy;—and Lemuravidæ, consisting of the genera *Lemuravus* (Marsh) and *Hyopsodus* (Leidy). The animals of the latter family were most allied to existing lemurs, but were a more generalized form, *Lemuravus* having forty-four teeth, the greatest number known in the order. These numerous forms ranged from the size of a small squirrel to that of a racoon. It is especially interesting to find these peculiar lemuroid forms in America, just when a lemur has been discovered of about the same age in Europe; and as the American forms are said to show an affinity with the South American marmosets, while the European animal is most allied to a West African group, we have evidently not yet got back far enough to find the primeval or ancestral type from which all the Primates sprang.

About the same time, in the succeeding Miocene formation, true monkeys were discovered. Mr. Marsh describes *Laopithecus* as an animal nearly the size of the largest South American monkeys, and allied both to the Cebidæ and the Eocene Limnotheridæ. Mr. Cope has described *Menotherium* from the Miocene of Colorado, as a lemuroid animal, the size of a cat, and perhaps allied to *Limnotherium*. More Miocene remains will, no doubt, be discovered, by which we shall be enabled to trace the origin of some of the existing forms of South American monkeys; and perhaps help to decide the question (now in dispute among anatomists) whether the lemurs are really Primates, or form an altogether distinct and isolated order of mammalia.

Insectivora.—This order is represented by comparatively few forms in the tertiary beds, and these are all very different from existing types. In the Upper Miocene of Dakota are found remains indicating two extinct genera, *Lepictis* and *Ictops.* In the Miocene of Colorado, Professor Cope has recently discovered four new genera, *Isacis;* allied to the preceding, but as large as a

Mephitis or skunk; *Herpetotherium*, near the moles; *Embasis*, more allied to the shrews; and *Dommina*, of uncertain affinities. Two others have been found in the Eocene of Wyoming; *Amomys*, having some resemblance to hedgehogs and to the Eastern *Tupaia;* and *Washakius*, of doubtful affinities.

Far back in the Triassic coal of North Carolina has been found the jaw of a small mammal (*Dromotherium*), the teeth of which somewhat resemble those of the Australian *Myrmecobius*, and may belong either to the Insectivora or Marsupials; if indeed, at that early period these orders were differentiated.

Carnivora.—The most ancient forms of this order are some remains found in the Middle Eocene of Wyoming, and others recently described by Professor Cope (1875) from the Eocene of New Mexico, of perhaps earlier date. The former consist of three genera, *Patriofelis, Uintacyon,* and *Sinopa*,—animals of large size but which cannot be classed in any existing family; and two others, *Mesonyx* and *Synoplotherium*, believed by Mr. Cope to be allied to *Hyænodon.* The latter consist of four genera,— *Oxyæna*, consisting of several species, some as large as a jaguar, was allied to *Hyænodon* and *Pterodon ;* *Pachyæna*, allied to the last; *Prototomus*, allied to *Amphicyon* and the Viverridæ; and *Limnocyon*, a civet-like carnivore with resemblances to the Canidæ.

In the Miocene formations we find the Feline type well developed. The wonderful *Machairodus,* which in Europe lived down to Post-Pliocene times, is found in the Upper Miocene of Dakota; and perfect crania have been discovered, showing that the chin was lengthened downwards to receive and protect the enormous canines. *Dinyctis* was allied both to *Machairodus* and to the weasels. Three new genera have been lately described by Professor Cope from the Miocene of Colorado,—*Bunœlurus*, with characters of both cats and weasels; *Daptophilus*, allied to *Dinyctis;* and *Hoplophoneus*, more allied to *Machairodus.* The Canidæ are represented by *Amphicyon*, which occurs in deposits of the same age in Europe; and by *Canis*, four species of which genus are recorded by Professor Cope from the Miocene of Colorado, and it also occurs in the Pliocene. The *Hyænodon* is represented by three species in the Miocene of Dakota and Colorado. It occurs also in the European Miocene and Upper Eocene formations, and constitutes a distinct family Hyænodontidæ, allied, according to Dr. Leidy, to wolves, cats, hyænas and weasels. The Ursidæ are represented by only one species of an extinct genus, *Leptarchus*, from the Pliocene of Nebraska. From the Pliocene of Colorado, Prof. Cope has recently described *Tomarctos*, as a "short-faced type of dog;" as well as species of *Canis* and *Martes*.

Ungulata.—The animals belonging to this order being usually of large size and accustomed to feed and travel in herds, are liable to wholesale destruction by floods, bogs, precipices, drought or hunger. It is for these reasons, probably, that their remains are almost always more numerous than those of other orders of mammalia. In America they are especially abundant; and the number of new and intermediate types about whose position there is much difference of opinion among Palæontologists, renders it very difficult to give a connected summary of them with any approach to systematic accuracy.

Beginning with the Perissodactyla, or odd-toed ungulates, we find the Equine animals remarkably numerous and interesting. The true horses of the genus *Equus*, so abundant in the Post-Pliocene formations, are represented in the Pliocene by several ancestral forms. The most nearly allied to *Equus* is *Pliohippus*, consisting of animals about the size of an ass, with the lateral toes not externally developed, but with some differences of dentition. Next come *Protohippus* and *Hipparion*, in which the lateral toes are developed but are small and functionless. Then we have the allied genera, *Anchippus*, *Merychippus*, and *Hyohippus*, related to the European *Hippotherium*, which were all still smaller animals, *Protohippus* being only 2½ feet high. In the older deposits we come to a series of forms, still unmistakably equine, but with three or more toes used for locomotion and with numerous differentiations in form, proportions, and dentition. These constitute the family Anchitheridæ. In the Miocene we have the genera *Anchitherium* (found also in the European Miocene), *Miohippus* and *Mesohippus*, all with three toes on each foot, and about the size of a sheep or large goat. In the Eocene of

Utah and Wyoming, we get a step further back, several species having been discovered about the size of a fox with four toes in front and three behind. These form the genus *Orohippus*, and are the oldest ancestral horse known. Prof. Marsh points out the remarkably perfect series of forms in America, which, beginning with this minute ancient type, is gradually modified by gaining increased size, increased speed by concentration of the limb-bones, elongation of the head and neck, the canine teeth decreased in size, the molars becoming longer and being coated with cement— till we at last come to animals hardly distinguishable, specifically, from the living horse.

Allied to these, are a series of forms showing a transition to the tapirs, and to the *Palæotherium* of the European Eocene. In the Pliocene we have *Parahippus;* in the Miocene *Lophiodon*, found in the same formation and in the Eocene of Europe, and allied to the tapir; and in the Eocene, *Palæosyops*, as large as a rhinoceros, which had large canines and was allied to the tapir and *Palæotherium ;* *Limnohyus*, forming the type of a family Limnohyidæ, which included the last genus and some others mentioned further on; and *Hyrachyus*, allied to *Lophiodon*, and to *Hyracodon* an extinct form of rhinoceros. Besides these we have *Lophiotherium* (also from the Eocene of Europe); *Diplacodon* allied to *Limnohyus*, but with affinities to modern Perissodactyla and nearly as large as a rhinoceros ; and *Colonoceras*, also belonging to the Limnohyidæ, an animal which was the size of a sheep, and had divergent protuberances or horns on its nose. A remarkable genus, *Bathmodon*, lately described by Professor Cope, and of which five species have been found in the Eocene of New Mexico and Wyoming, is believed to form the type of a new family, having some affinity to *Palæosyops* and to the extinct Brontotheridæ. It had large canine tusks but no horns.

The Rhinocerotidæ are represented in America by the genus *Rhinoceros* in the Pliocene and Miocene, and by *Aceratherium* and *Hyracodon* in the Miocene. Both the latter were hornless, and *Hyracodon* was allied to the Eocene *Hyrachyus*, one of the Lophiodontidæ. In the Eocene and Miocene deposits of Utah, and Oregon, several remarkable extinct rhinoceroses have been recently discovered, forming the genus *Diceratherium.* These had a pair of nasal horns placed side by side on the snout, not behind each other as in existing two-horned rhinoceroses, the rest of their skeleton resembling the hornless *Aceratherium.* They were of rather small size.

Next to these extinct rhinoceroses come the Brontotheridæ, an extraordinary family of large mammalia, some of which exceeded in bulk the largest living rhinoceros. They had four toes to the front and three to the hind feet, with a pair of large divergent horns on the front of the head, in both sexes. Professor Marsh and Dr. Leidy have described four genera, *Brontotherium,* *Titanotherium, Megacerops,* and *Anisacodon,* distinguished by peculiarities of dentition. Though most nearly allied to the rhinoceroses, they show some affinity for the gigantic Dinocerata of the Eocene to be noticed further on. Professor Cope has since described another genus, *Symborodon,* from the Miocene of Colorado, with no less than seven species, one nearly the size of an elephant. He thinks they had a short tapir-like proboscis. The species differ greatly in the form of the cranium and development of the horn-bearing processes.

We commence the Artiodactyla, or even-toed Ungulates, with the hog tribe. These are represented by species of peccaries, (*Dicotyles*) from the Pliocene of Nebraska and Oregon ; and by an allied form *Thinohyus,* very like *Dicotyles,* but having an additional premolar tooth and a much smaller brain-cavity. From the Miocene are three allied genera, *Nanohyus, Leptochœrus,* and *Perchœrus.* Professor Cope, however, thinks *Leptochœrus* may be Lemuroid, and allied to *Menotherium.* The Anthracotheridæ, a family which connects the Hippopotamidæ and Ruminants, and which occurs in the Miocene of Europe and India, are represented in America by the genus *Hyopotamus* from the Miocene of Dakota, and *Elotherium* from the Miocene of Oregon and the Eocene of Wyoming; the latter genus being sometimes classed with the preceding family, and lately placed by Professor Marsh, in the new order, Tillodontia. Professor Cope has since described three other genera from the Eocene of New

Mexico: *Meniscotherium*, having resemblances to *Palæosyops*, *Hyopotamus*, and the Limnotheridæ; *Phenacodus*, the size of a hog, of doubtful position, but perhaps near *Elotherium;* and *Achænodon*, as large as a cow, but more hog-like than the preceding.   Another new genus from the Miocene of Colorado—*Pelonax*—is said by Professor Cope to come between *Elotherium* and *Hippopotamus*.

The Camelidæ are very abundant, and form one of the most striking features of the ancient fauna of America. *Procamelus*, *Homocamelus*, and *Megalomeryx*, are extinct genera found in the Pliocene formation; the first very closely allied to the Old World camel, the last smaller and more sheep-like.   In the Miocene two other genera occur, *Pœbrotherium* and *Protomeryx*, the former allied to both the camel and the llama.

Deer are represented by a single species of *Cervus* in the Pliocene, while two extinct genera, *Leptomeryx* and *Merycodus*, are found in the Miocene deposits, the latter indicating a transition between camels and deer.   Two other genera, *Hypisodus* and *Hypertragulus*, of very small size, are said by Professor Cope to be allied to the Tragulidæ and to *Leptomeryx*.

The Bovidæ, or hollow-horned ruminants, are only represented in the Newer Pliocene by a single species of an extinct genus, *Casoryx*, said to be intermediate between antelopes and deer.

We now come to an exclusively American family, the *Oreodontidæ*, which consisted of small animals termed by Dr. Leidy, "ruminating hogs," and which had some general structural resemblances to deer and camels.   They abounded in North America during the Pliocene, and especially during the Miocene epoch, no less than six genera and twenty species having been discovered.   *Merychus* contains the Pliocene forms; while *Oreodon*, *Eporeodon*, *Merychochœrus*, *Leptauchenia*, and *Agriochœrus* are Miocene.   The last genus extends back into the Eocene period, and shows affinity to the European Anoplotheridæ of the same epoch.

Proboscidea.—The Elephantidæ are only represented in America by one species of *Mastodon* and one of *Elephas*, in the Newer Pliocene deposits.   In the Older Pliocene, Miocene and Upper Eocene, no remains of this order have been found; and in 1869, Dr. Leidy remarked on the small average size of the extinct North American mammalia, which were almost all smaller than their living analogues. Since then, however, wonderful discoveries have been made in deposits of Middle Eocene age in Wyoming and Colorado, of a group of huge animals not only rivalling the elephants in size, but of so remarkable and peculiar a structure as to require the formation of a new order of mammals—Dinocerata—for their reception.

This order consists of animals with generalised Ungulate and Proboscidean affinities. The lower jaw resembles that of the hippopotamus; they had five toes on the anterior feet and four on the posterior; three pairs of horns, the first pair on the top of the head, large and perhaps palmated, the second pair above the eyes, while the third and smallest stood out sideways on the snout. They had enormous upper canines, of which the roots entered the middle horn cores, no upper incisors, and small molars. Professor Marsh believes that they had no trunk. The remains discovered indicate four genera, *Dinoceras* (3 sp.), *Tinoceras* (2 sp.), *Uintatherium* (1 sp.), and *Eobasileus* (2 sp.). Many other names have been given to fragments of these animals, and even those here given may not be all distinct.

Another new order, Tillodontia, recently established by Professor Marsh, is perhaps yet more remarkable in a zoological point of view, since it combines the characters of Carnivora, Ungulata, and Rodents. These animals have been formed into two families, Tillotheridæ and Stylinodontidæ; and three genera, *Tillotherium*, *Anchippodus*, and *Stylinodontia*. All are from the Eocene of Wyoming and New Jersey. Perhaps to these must be added *Elotherium* from the Miocene of Dakota, the other forms being all Eocene. They were mostly animals of small size, between that of the capybara and tapir. The skull resembled in form that of a bear; the molar teeth were of Ungulate type, and the incisors like those of a Rodent; but the skeleton was more that of the Ursidæ, the feet being plantigrade. Professor Cope has since described three new genera from the Eocene of New Mexico, *Ectoganus*, *Calamodon*, and *Esthonyx*, comprising seven species allied to *Tillotherium* and *Anchippodus*, and having also relations, as Professor Cope believes, with the South American Toxodontidæ.

Rodentia.—This order is represented in the Pliocene by a beaver, a porcupine, and an American mouse (*Hesperomys*), all extinct species of living genera, the *Hystrix* being an Old World type; and Professor Cope has recently described *Panolax*, a new genus of hares from the Pliocene of New Mexico. The Miocene deposits have furnished an extinct genus allied to the hares—*Palæolagus;* one of the squirrel family—*Ischyromys;* a small extinct form of beaver—*Palæocastor;* and an extinct mouse—*Eumys.* The Eocene strata of Wyoming have lately furnished two extinct forms of squirrel, *Paramys* and *Sciuravus;* and another of the Muridæ (or mouse family), *Mysops.*

Cetacea.—Numerous remains of dolphins and whales, belonging to no less than twelve genera, mostly extinct, have been found in the Miocene deposits of the Atlantic and Gulf States, from New Jersey to South Carolina and Louisiana; while seven genera of the extinct family, Zeuglodontidæ, have been found in Miocene and Eocene beds of the same districts. Some remains associated with these are doubtfully referred to the Seal family (Phocidæ) among the Carnivora.

Edentata.—Till quite recently no remains of this order have occurred in any North American deposits below the Post-Pliocene; but in 1874 Prof. Marsh described some remains allied to *Megalonyx* and *Mylodon*, from the Pliocene beds of California and Idaho, and forming a new genus, *Morotherium.* As these remains have only occurred to the west of the Rocky Mountains, and in Pliocene deposits whose exact age is not ascertained, they hardly affect the remarkable absence of this group from the whole of the exceedingly rich Tertiary deposits in all other parts of North America.

General Relations of the extinct Tertiary Mammalia of North America and Europe.—Having now given a sketch of the extinct Mammalia which inhabited Europe and North America during the Tertiary period, we are enabled by comparing them, to ascertain their relations to each other, and to see how far they elucidate the problem of the birth-place and subsequent migrations of the several families and genera. We have already pointed out the remarkable features of the Quaternary (or Post-Pliocene) fauna of North America, and now proceed to discuss that of the various Tertiary periods, which is closely connected with the extinct fauna of Europe.

The Tertiary Mammalia of North America at present described belong to from eighty to one hundred genera, while those of Europe are nearly double that number; yet only eighteen genera are common to the two faunas, and of these eight are living and belong chiefly to the Pliocene period. Taking first, the genera which in America do not go back beyond the Pliocene period (ten in number), we find that eight of them in Europe go back to the Upper Miocene.   These are *Felis*, *Pseudœlurus*, *Hipparion*, *Cervus*, *Mastodon*, *Elephas* (in India), *Castor* and *Hystrix;* while another, *Canis*, goes back to the Upper Eocene and the tenth, *Equus*, confined to the newer Pliocene or perhaps to the Post-Pliocene in America, extends back to the older Pliocene in Europe.   Of the seven European genera which are confined to the Miocene period in America, three, *Hyœnodon*, *Anchitherium*, and *Lophiodon* go back to the Eocene in Europe; three others, *Machairodus*, *Rhinoceros*, and *Aceratherium*, are also of Miocene age in Europe; *Amphicyon* goes back to the Lower Miocene of Europe. *Lophiotherium* belongs to the Eocene of both countries.

If we turn now to families instead of genera, we find that the same general rule prevails. Mustelidæ (weasels), Ursidæ (bears), true Equidæ (horses), and Bovidæ (oxen &c.), go no further back in America than the Pliocene, while they all go back to the Miocene in Europe. Suidæ (swine) and Anoplotheridæ (extinct) are found in the American Miocene and in the European Eocene. Anchitheridæ (extinct) reach the Upper Eocene in America, while in Europe they range through Upper, Middle, and Lower Eocene. Cervidæ (deer) alone are Miocene in both countries. There remain two families in which America has the pre-eminence. Camelidæ (camels) were wonderfully developed in the American Pliocene and Miocene periods, abounding in genera and species; whereas in Europe the group only exists in the Post-Pliocene or Lower Pliocene, with one Upper Miocene species of *Camelus* in N. India. The Anthracotheridæ (extinct), found only in the Upper Miocene of France and India, reach even the Lower Eocene in America.

These facts may be due, in part, to a want of strict co-ordination between the Tertiary deposits of Europe and North America, —in part to the imperfection of the record in the latter country. Yet it does not seem probable that they are altogether due to these causes, because the Miocene beds, which are by far the best known in America as in Europe, exhibit deficiencies of the same kind as the less known Eocene deposits. The fossil fauna of both countries is so rich, that we can hardly impute great and well marked differences to imperfect knowledge; yet we find such important families as the Civets, Hyænas, Giraffes, and Hippopotami absent from America, with the Weasels, and Antelopes almost so; while America possesses almost all the Camelidæ, two peculiar orders, Dinocerata and Tillodontia, and four remarkably peculiar families, Limnotheridæ, Lemuravidæ, Oreodontidæ and Brontotheridæ. If then the facts at present known represent approximately the real time-relations of the groups in question on the two continents, they render it probable that weasels, bears, true horses, swine, oxen, sheep and antelopes, originated on the Old World continent, and were transmitted to America during some part of the Miocene period; while camels originated in the New World, and somewhere about the same time passed over to Europe. Of the extinct families common to the two hemispheres, the Anthracotheridæ alone seem to have had an American origin. Of the genera common to the two countries, almost all seem to have had a European origin, the only genera of equal date being the two rhinoceroses and three Anchitheridæ; but if the Brontotheridæ are allied to the Rhinocerotidæ, these latter may have originated in America, although now an exclusively Old World type. These conclusions are not improbable when we consider the much greater size of the Old World continents, extending far into the tropics and probably always more or less united to the tropical areas; while the evidence of the extinct mammalia themselves shows, that South America has been for the most part isolated from the northern continent, and did not take part in the development of its characteristic Tertiary fauna.

Before speculating further on this subject, it will be well to lay before our readers a summary of South American palæontology, after which we shall be in a better position to draw correct inferences from the whole body of the evidence.

SOUTH AMERICA.

Unfortunately, our knowledge of the interesting fossil fauna of this continent, is almost wholly confined to the Post-Pliocene and Pliocene periods. A few remains have been discovered in deposits believed to be of Eocene age, but nothing whatever representing the vast intervening period, so rich in peculiar forms of animal life both in North America and Europe.

Fauna of the Brazilian caves.—What we know of the Post-Pliocene period is chiefly due to the long-continued researches of Dr. Lund in the caves of Central Brazil, mostly situated in a district near the head waters of the San Francisco river in the Province of Minas Geraes. The caves are formed in limestone rocks, and are so numerous that Dr. Lund visited thousands, but only sixty contained bones in any quantity. These caves have a floor of reddish earth, often crowded with bones. In one experiment, half a cubic foot of this earth contained jaws of 400 opossums, 2,000 mice, besides remains of bats, porcupines and small birds. In another trial, the whole of the earth in a cavern was carried out for examination, amounting to 6,552 firkins; and, from a calculation made by measured samples, it was estimated to contain nearly seven millions of jaw-bones of cavies, opossums, porcupines, and mice, besides small birds, lizards, and frogs. This immense accumulation is believed to have been formed from the bodies of animals brought into the cavern by owls; and, as these are unsocial birds, the quantity found implies an immense lapse of time, probably some thousands of years. More than 100 species of Mammalia, in all, were obtained in these caves. Some were living species or closely allied to such; but the majority were extinct, and a considerable number, about one-fourth, belonged to extinct genera, or genera not now inhabiting South America. Stone implements and human remains were found in several of the caves with extinct animals. The following enumeration of these remains is from the corrected list of M. Gervais.

Primates.—Extinct species of *Cebus, Callithrix,* and *Jacchus*—South American genera of monkeys; with an extinct genus, *Protopithecus*—an animal of large size but belonging to the American family Cebidæ.

Chiroptera.—Species belonging to the South American Phyllostomidæ, and to two South American genera of other families.

Carnivora.—Five species of *Felis,* some allied to living animals, others extinct; a species of the widespread extinct genus *Machairodus ;* and a small species referred to *Cynælurus,* the genus containing the hunting leopard now found only in Africa and India. Canidæ are represented by *Canis* and *Icticyon* (a living Brazilian species of the latter genus), and the extinct genus *Speothos.* Mustelidæ are represented by extinct species of the South American genera *Mephitis* and *Galictis.* Procyonidæ, by a species of *Nasua.* Ursidæ, by *Arctotherium,* a genus closely resembling, if not identical with, that containing the "spectacled bear" of Chili.

Ungulata.—*Equus, Tapirus, Dicotyles, Auchenia, Cervus, Leptotherium,* and *Antilope,* are the cave-genera of this order. *Equus* and *Antelope* are particularly interesting, as representing groups forming no part of existing South American zoology; while the presence also of *Leptotherium,* an extinct genus of antelopes, shows that the group was fairly represented in South America at this comparatively recent period.

Proboscidea.—A species of *Mastodon,* found also in the Pliocene of La Plata, represents this order.

Rodentia.—These abound. *Dasyprocta, Cœlogenys, Cavia, Kerodon,* all living genera of Caviidæ, are represented by extinct species. *Cercolabes*, the 'tree porcupine' (Cercolabidæ) has two species, one as large as a peccary; *Myopotamus, Loncheres, Carterodon*, are existing genera of spiny rats (Echimyidæ); and there are two extinct genera of the same family, *Loncho-phorus* and *Phyllomys*. *Lagostomus* (Chinchillidæ), the visca-cha of the Pampas, is represented by an extinct species. There is also an extinct species of *Lepus*; several species of *Hesperomys* and *Oxymycterus*; and a large *Arvicola*, a genus not living in South America.

Edentata.—These, which constitute the great feature of the existing South American fauna, were still more abundant and varied in the Cave period, and it is remarkable that most of them are extinct *genera*. The armadillos are alone represented by living forms, *Dasypus*, and *Xenurus*; *Eurydon* and *Hetero-don*, are extinct genera of the same family, as well as *Chlamydo-therium*—huge armadillos the size of a tapir or rhinoceros, and *Pachytherium*, which was nearly as large. The ant-eaters are represented only by *Glossotherium*, an extinct form allied to *Myrmecophaga* and *Manis*. The sloths were more numerous, being represented by the extinct genera *Cælodon*, *Sphenodon* and *Ochotherium*, the last of large size. The huge terrestrial sloths—Megatheridæ, also abounded; there being species of *Megatherium* and *Megalonyx*, as well as the allied *Scelidotherium*, supposed to have some affinity for the African *Orycteropus*.

Marsupials.—No new forms of these appear, but numerous species of *Didelphys*, all closely allied to opossums still living in South America.

The preceding sketch of the wonderful cave fauna of Central Brazil, is sufficient to show that it represents, in the main, a period of great antiquity. Not only are almost the whole of the species extinct, but there are twenty extinct genera, and three others not now inhabitants of South America. The fact that so few remains of the living animals of the country are found in these caves, indicates that some change of physical conditions has occurred since they were the receptacles of so many of the larger animals; and the presence of many extinct genera of

L

large size, especially among the Edentata and American families of Rodents, are additional proofs of a very high antiquity. Yet many of these cave animals are closely allied to those which are found in North America in the Post-Pliocene deposits only, so that we have no reason to suppose the cave-fauna to be of much earlier date. But the great amount of organic change it implies, must give us an enlarged idea of the vast periods of time, as measured by years, which are included in this, the most recent of all geological epochs.

Pliocene Period of Temperate South America.—We have now to consider the numerous remains of extinct animals found in various deposits in the Pampas, and in Patagonia, and a few in Bolivia. The age of these is uncertain; but as they are very similar to the cave-fauna, though containing a somewhat larger proportion of extinct genera and some very remarkable new forms, they cannot be *very* much older, and are perhaps best referred at present to the newer portion of the Pliocene formation.

Carnivora.—The genus *Machairodus* or sabre-toothed tigers, represents the Felidæ. There are several species of wolves (*Canis*); a weasel (*Mustela*); two bears of the Brazilian cave-genus *Arctotherium;* and the extinct European genus *Hyænarctos.*

Ungulata.—There are two species of *Equus,* found in the Pampas, Chili, and Bolivia; two of *Macrauchenia,* an extraordinary extinct group allied to the tapir and *Palæotherium,* but with the long neck, and general size of a camel. A second species found on the highlands of Bolivia is much smaller.

A more recent discovery, in Patagonia, is the almost perfect series of teeth of a large animal named *Homalodontotherium;* and which is believed by Professor Flower, who has described it, to have been allied to *Rhinoceros,* and still more to the Miocene *Hyracodon* from North America; and also to present some resemblances to *Macrauchenia,* and though much more remotely, to the curious genus *Nesodon* mentioned further on.

The Artiodactyla, or even-toed Ungulates, are represented by a species of *Dicotyles,* or peccary, found in the deposits of the

Pampas; by *Auchenia*, or llama, of which three extinct species inhabited Bolivia, in which country two allied but extinct genera, *Palæolama* and *Camelotherium*, have also been found. Three species of deer (*Cervus*), from the Pampas deposits, complete the list of Pliocene Ungulates.

Proboscidea.—The cave species of *Mastodon* is found also in the Pampas deposits, and another in the Andes of Chili and Bolivia.

Rodents.—These are not so numerous as in the caves. There are species of the existing genera, *Kerodon* and *Cavia* (Caviidæ); *Lagostomus* (Chinchillidæ); *Ctenomys* (Octodontidæ); *Lepus* (hare); *Hesperomys* and *Oxymycterus* (Muridæ); *Arvicola*, a genus not living in South America; and an extinct genus, *Cardiodus*. There is also a remarkable extinct form, *Typotherium*, larger than the capybara, and having affinities to Edentates and Ungulates. Three species have been found in the Pampas deposits.

Edentata.—These are as abundant and remarkable as in the cave deposits. *Scelidotherium*, *Megatherium*, *Megalonyx*, *Glossotherium* and *Dasypus*, have already been noticed as from the Brazilian caves. We have here, in addition, the huge *Mylodon* allied to the *Megatherium*, and the allied genera—*Gnathopsis* and *Lestodon*. We then come to the huge extinct armadillos, *Glyptodon* and *Schistopleurum*, the former consisting of numerous species, some of which were as large as an elephant. Another genus, *Eutatus*, is allied to the living three-handed armadillos; and a species of the existing genus *Euphractus* has been found in Bolivia.

Toxodontidæ.—There remain a number of huge animals rivalling the Megatherium in size, and forming the genera *Toxodon* and *Nesodon*, but whose position is doubtful. Several species have been found in the deposits of the Pampas and Patagonia. They are allied at once to Ungulates, Rodents, Edentates, and the aquatic Sirenia, in so puzzling a manner that it is impossible to determine to what order they belong, or whether they require a new order to be formed for their reception. Some are believed to date back to the Miocene period, and they indicate what strange forms may still be discovered, should any

L 2

productive deposits be found in South America of middle Tertiary age.

Pliocene Mammalia of the Antilles.—These may be noticed here, as they are of special interest, proving the connection of the larger West Indian Islands with the Continent some time in the later Tertiary period. They consist of remains of two large animals belonging to the South American Chinchillidæ, found in cave deposits in the island of Anguilla, and forming two new genera, *Amblyrhiza* and *Loxomylus;* and remain allied to *Megalonyx* from Cuba, which have been named *Megalocnus* and *Myomorphus.*

Eocene fauna of South America.—The few remains yet discovered in the Tertiary deposits of the Pampas which are believed to be of Eocene age, are exceedingly interesting, because they show us another change in the scenery of the great drama of life; there being apparently a considerable resemblance, at this epoch, between South America and Europe. They consist of a large extinct feline animal, *Eutemnodus;* of *Palæotherium* and *Anoplotherium*, the well-known extinct Ungulates of the European Tertiaries, and which have never been found in North America; and of three genera of Rodents,—*Theridromys*, allied to *Echimys*, and found also in the Eocene and Miocene of France; *Megamys*, allied to the living *Capromys* of the Antilles, and also to *Palæomys*, an extinct form of the French Miocene; and a very large animal referred to *Arvicola*, a genus found also in the Pliocene deposits of South America, and abundant in the northern hemisphere. No Edentates have been found.

The resemblances of this fauna to that of Europe rather than to any part of America, are so strong, that they can hardly be accidental. We greatly want, however, more information on this point, as well as some corresponding evidences as to the condition of West and South Africa about the same epoch, before we can venture to speculate on their bearing as regards the early migrations of organic forms.

General Remarks on the Extinct Mammalian Fauna of the Old

and New Worlds.—Leaving the more special applications of palæontological evidence to be made after discussing the relations of the existing fauna of the several regions, we propose here to indicate briefly, some of the more general deductions from the evidence which has now been laid before our readers.

The first, and perhaps the most startling fact brought out by our systematic review, is the very recent and almost universal change that has taken place in the character of the fauna, over all the areas we have been considering; a change which seems to be altogether unprecedented in the past history of the same countries as revealed by the geological record.   In Europe, in North America, and in South America, we have evidence that a very similar change occurred about the same time.   In all three we find, in the most recent deposits—cave-earths, peat-bogs, and gravels—the remains of a whole series of large animals, which have since become wholly extinct or only survive in far-distant lands.   In Europe, the great Irish elk, the *Machairodus* and cave-lion, the rhinoceros, hippopotamus, and elephant ;—in North America, equally large felines, horses and tapirs larger than any now living, a llama as large as a camel, great mastodons and elephants, and abundance of huge megatheroid animals of almost equal size ;—in South America these same megatheroids in greater variety, numerous huge armadillos, a mastodon, large horses and tapirs, large porcupines, two forms of antelope, numerous bears and felines, including a *Machairodus*, and a large monkey,—have all become extinct since the deposition of the most recent of the fossil-bearing strata.   This is certainly not a great while ago, geologically ; and it is *almost* certain that this great organic revolution, implying physical changes of such vast proportions that they must have been due to causes of adequate intensity and proportionate range, has taken place since man lived on the earth. This is proved to have been the case in Europe, and is supported by much evidence both as regards North and South America.

It is clear that so complete and sudden a change in the higher forms of life, does not represent the normal state of things. Species and genera have not, at all times, become so rapidly extinct.   The time occupied by the " Recent period," that is the

time *since* these changes took place is, geologically, minute.
The time of the whole of the Post-Pliocene period, as measured
by the amount of physical and *general* organic change known to
have taken place, is exceedingly small when compared with the
duration of the Pliocene period, and still smaller, probably, as
compared with the Miocene. Yet during these two periods we
meet with no such break in the continuity of the forms of life, no
such radical change in the *character* of the fauna (though the
number of specific and generic changes may be as great) as we
find in passing from the Post-Pliocene to recent times. For
example, in Central Europe numerous hyænas, rhinoceroses, and
antelopes, with the great *Machairodus*, continued from Miocene
all through Pliocene into Post-Pliocene times; while hippopotami and elephants continued to live through a good part of
the Pliocene and Post-Pliocene periods,—and then all suddenly
became extinct or left the country. In North America there has
been more movement of the fauna in all the periods; but we
have similar great felines, horses, mastodons, and elephants, in
the Pliocene and Post-Pliocene periods, while *Rhinoceros* is common to the Miocene and Pliocene, and camels range continuously
from Miocene, through Pliocene, to Post-Pliocene times;—when
all alike became extinct. Even in South America the evidence is,
as far as it goes, all the same way. We find *Machairodus*, *Equus*,
Mastodon, *Megatherium*, *Scelidotherium*, *Megalonyx*, and numerous
gigantic armadillos, alike in the caves and in the stratified
tertiary deposits of the Pampas;—yet all have since passed away.

It is clear, therefore, that we are now in an altogether
exceptional period of the earth's history. We live in a zoologically impoverished world, from which all the hugest, and fiercest,
and strangest forms have recently disappeared ; and it is, no,
doubt, a much better world for us now they have gone. Yet it
is surely a marvellous fact, and one that has hardly been sufficiently dwelt upon, this sudden dying out of so many large
mammalia, not in one place only but over half the land surface
of the globe. We cannot but believe that there must have been
some physical cause for this great change ; and it must have
been a cause capable of acting almost simultaneously over large

portions of the earth's surface, and one which, as far as the Tertiary period at least is concerned, was of an exceptional character. Such a cause exists in the great and recent physical change known as " the Glacial epoch." We have proof in both Europe and North America, that just about the time these large animals were disappearing, all the northern parts of these continents were wrapped in a mantle of ice; and we have every reason to believe that the presence of this large quantity of ice (known to have been thousands of feet if not some miles in thickness) must have acted in various ways to have produced alterations of level of the ocean as well as vast local floods, which would have combined with the excessive cold to destroy animal life.   There is great difference of opinion among geologists and physicists as to the extent, nature, and duration of the Glacial epoch.   Some believe it to have prevailed alternately in the northern and southern hemispheres; others that it was simultaneous in both. Some think there was a succession of cold periods, each lasting many thousands of years, but with intercalated warm periods of equal duration; others deny that there is any evidence of such changes, and maintain that the Glacial epoch was one continuous period of arctic conditions in the temperate zones, with some fluctuations perhaps but with no regular alternations of warm periods.   Some believe in a huge ice-cap covering the whole northern hemisphere from the pole to near 50° north latitude in the eastern, and 40° in the western hemisphere ; while others impute the observed effects either to glaciers from local centres, or to floating icebergs of vast size passing over the surface during a period of submersion.

Without venturing to decide which of these various theories will be ultimately proved to be correct, we may state, that there is an increasing belief among geologists in the long duration of this ice-period, and the vast extent and great thickness attained by the ice-sheet.   One of the most recent, and not the least able, of the writers on this question (Mr. Belt) shows strong reasons for adopting the view that the ice-period was simultaneous in both hemispheres ; and he calculates that the vast amount of water abstracted from the ocean and locked up

in mountains of ice around the two poles, would lower the general
level of the ocean about 2,000 feet. This would be equivalent
to a general elevation of the land to the same amount, and would
thus tend to intensify the cold; and this elevation may enable
us to understand the recent discoveries of signs of glacial
action at moderate elevations in Central America and Brazil, far
within the tropics. At the same time, the weight of ice piled up
in the north would cause the land surface to sink there, perhaps
unequally, according to the varying nature of the interior crust
of the earth; and since the weight has been removed land would
rise again, still somewhat irregularly; and thus the phenomena
of raised beds of arctic shells in temperate latitudes, are explained.

Now, it is evident, that the phenomena we have been considering—of the recent changes of the mammalian fauna in
Europe, North America, South Temperate America, and the
highlands of Brazil—are such as might be explained by the most
extreme views as to the extent and vastness of the ice-sheet,
and especially as to its simultaneous occurrence in the northern
and southern hemispheres; and where two such completely independent sets of facts are found to combine harmoniously, and
supplement each other on a particular hypothesis, the evidence
in favour of that hypothesis is greatly strengthened. An objection that will occur to zoologists, may here be noticed. If
the Glacial epoch extended over so much of the temperate and
even parts of the tropical zone, and led to the extinction of so
many forms of life even within the tropics, how is it that so
much of the purely tropical fauna of South America has maintained itself, and that there are still such a vast number of
forms, both of mammalia, birds, reptiles, and insects, that seem
organized for an exclusive existence in tropical forests? Now
Mr. Belt's theory, of the subsidence of the ocean to the extent of
about 2,000 feet, supplies an answer to this objection; for we
should thus have a tract of lowland of an average width of
some hundreds of miles, added to the whole east coast of Central
and South America. This tract would, no doubt, become covered
with forests as it was slowly formed, would enjoy a perfectly

tropical climate, and would thus afford an ample area for the continued existence and development of the typical South American fauna; even had glaciers descended in places so low as what is now the level of the sea, which, however, there is no reason to believe they ever did. It is probable too, that this low tract, which all round the Gulf of Mexico would be of considerable width, offered that passage for intermigration between North and South America, which led to the sudden appearance in the former country in Post-Pliocene times, of the huge Megatheroids from the latter; a migration which took place in opposite directions as we shall presently show.

The birth-place and migrations of some mammalian families and genera.—We have now to consider a few of those cases in which the evidence already at our command, is sufficiently definite and complete, to enable us to pronounce with some confidence as to the last movements of several important groups of mammalia.

Primates.—The occurrence in North America of numerous forms of Lemuroidea, forming two extinct families, which are believed by American palæontologists to present generalized features of both Lemuridæ and Hapalidæ, while in Europe only Lemurine forms allied to those of Africa have occurred in deposits of the same age (Eocene), renders it possible that the Primates may have originated in America, and sent one branch to South America to form the Hapalidæ and Cebidæ, and another to the Old World, giving rise to the lemurs and true apes. But the fact that apes of a high degree of organization occur in the European Miocene, while in the Eocene, a monkey believed to have relations to the Lemuroids and Cebidæ has also been discovered, make it more probable that the ancestral forms of this order originated in the Old World at a still earlier period. The absence of any early tertiary remains from the tropical parts of the two hemispheres, renders it impossible to arrive at any definite conclusions as to the origin of groups which were, no doubt, always best developed in tropical regions.

Carnivora.—This is a very ancient and wide-spread group, the families and genera of which had an extensive range in very early times. The true bears (*Ursus*) are almost the only important genus that seems to have recently migrated. In Europe it dates back to the Older Pliocene, while in North America it is Post-Pliocene only. Bears, therefore, seem to have passed into America from the Palæarctic region in the latter part of the Pliocene period. They probably came in on the north-west, and passed down the Andes into South America, where one isolated species still exists.

Ungulata.—Horses are very interesting. In Europe they date back under various forms to the Miocene period, and true *Equus* to the Older Pliocene. In North America they are chiefly Pliocene, true *Equus* being Post-Pliocene, with perhaps one or two species Newer Pliocene ; but numerous ancestral forms date back to the Miocene and Eocene, giving a more perfect " pedigree of the horse" than the European forms, and going back to a more primitive type—*Orohippus.* In South America, *Equus* is the only genus, and is Post-Pliocene or at most Newer Pliocene. While, therefore, the ancient progenitors of the Equidæ were common to North America and Europe, in Miocene and even Eocene times, true horses appear to have arisen in the Palæarctic region, to have passed into North America in the latter part of the Pliocene period, and thence to have spread over all suitable districts in South America. They were not, however, able to maintain themselves permanently in their new territory, and all became extinct; while in their birth-place, the Old World, they continue to exist under several varied forms.

True tapirs are an Old World group. They go back to the Lower Miocene in Europe, while in both North and South America they are exclusively Post-Pliocene. They occur in France down to the Newer Pliocene, and must, about that time, have entered America. The land connection by which this and so many other animals passed between the Old and New Worlds in late Tertiary times, was almost certainly in the North Pacific, south of Behring's Straits, where, as will be seen by our general map, there is a large expanse of shallow water, which a moderate elevation would convert into dry land, in a sufficiently temperate latitude.

The peccary (*Dicotyles*), now a characteristic South American genus, is a recent immigrant from North America, where it appears to have been developed from ancestral forms of swine dating back to the Miocene period.

Antelopes are an Old World type, but a few of them appear to have entered North, and reached South America in late Pliocene times. Camels, strange to say, are a special North American type, since they abounded in that continent under various ancient forms in the Miocene period. Towards the end of that period they appear to have entered eastern Asia, and developed into the Siberian *Merycotherium* and the North Indian *Camelus*, while in the Pliocene age the ancestral llamas entered South America.

Cervidæ are a wide-spread northern type in their generalized form, but true deer (*Cervus*) are Palæarctic. They abounded in Europe in Miocene times, but only appear in North and South America in the later Pliocene and Post-Pliocene periods.

True oxen (*Bovinæ*) seem to be an Oriental type (Miocene), while they appear in Europe only late in the Pliocene period, and in America are confined to the Post-Pliocene.

Elephants (*Elephantidæ*) are an Old World type, abounding in the Miocene period in Europe and India, and first appearing in America in Post-Pliocene or later Pliocene times. Ancestral forms, doubtfully Proboscidean (*Dinocerata*), existed in North America in the Eocene period, but these became extinct without leaving any direct descendants, unless the *Brontotheridæ* and rhinoceroses may be so considered.

Marsupials are almost certainly a recent introduction into South and North America from Asia. They existed in Europe in Eocene and Miocene times, and presumably over a considerable part of the Old World; but no trace of them appears in North or South America before the Post-Pliocene period.

Edentata.—These offer a most curious and difficult problem. In South America they abound, and were so much more numerous and varied in the Post-Pliocene and Pliocene, that we may be sure they lived also in the preceding Miocene period. A few living Edentates are scattered over Africa and Asia, and they flourished in Europe during the Miocene age—animals as large (in some species) as a rhinoceros, and most allied to living African forms. In North America no trace of Edentata has been found earlier than the Post-Pliocene period, or perhaps the Newer Pliocene on the west coast. Neither is there any trace of them in South America in the Eocene formations; but this may well be owing to our very imperfect knowledge of the forms of that epoch. Their absence from North America is, however, probably real; and we have to account for their presence in the Old World and in South America. Their antiquity is no doubt very great, and the point of divergence of the Old World and South American groups, may take us back to early Eocene, or even to Pre-Eocene times. The distribution of land and sea may then have been very different from what it is now; and to those who would create a continent to account for the migrations of a beetle, nothing would seem more probable than that a South Atlantic continent, then united parts of what are now Africa and South America. There is, however, so much evidence for the general permanence of what are now the great continents and deep oceans, that Professor Huxley's supposition of a considerable extension of land round the borders of the North Pacific Ocean in Mesozoic times, best indicates the probable area in which the Edentate type originated, and thence spread over much of the Old World and South America. But while in the latter country it flourished and increased with little check, in the other great continents it was soon overcome by the competition of higher forms, only leaving a few small-sized representatives in Africa and Asia.

CHAPTER VIII.

VARIOUS EXTINCT ANIMALS;—AND ON THE ANTIQUITY OF THE GENERA OF INSECTS AND LAND MOLLUSCA.

EXTINCT MAMMALIA OF AUSTRALIA.

THESE have all been obtained from caves and late Tertiary or Post-Tertiary deposits, and consist of a large number of extinct forms, some of gigantic size, but all marsupials and allied to the existing fauna. There are numerous forms of kangaroos, some larger than any living species; and among these are two genera, *Protemnodon* and *Sthenurus*, which Professor Garrod has lately shown to have been allied, not to any Australian forms, but to the *Dendrolagi* or tree-kangaroos of New Guinea. We have also remains of *Thylacinus* and *Dasyurus*, which now only exist in Tasmania; and extinct species of *Hypsiprymnus* and *Phascolomys*, the latter as large as a tapir. Among the more remarkable extinct genera are *Diprotodon*, a huge thick-limbed animal allied to the kangaroos, but nearly as large as an elephant; *Nototherium*, having characters of *Macropus* and *Phascolarctos* combined, and as large as a rhinoceros; and *Thylacoleo*, a phalanger-like marsupial nearly as large as a lion, and supposed by Professor Owen to have been of carnivorous habits, though this opinion is not held by other naturalists.

Here then we find the same phenomena as in the other countries we have already discussed,—the very recent disappearance of a large number of peculiar forms, many of them far surpassing in size any that continue to exist. It hardly seems probable that in this case their disappearance can have been due to the direct effects of the Glacial epoch, since no very extensive glaciation could have occurred in a country like Australia; but if the ocean sank 2,000 feet, the great eastern mountain range might have given rise to local glaciers. It is, however, almost certain that during late Tertiary times Australia must have been much more extensive than it is now. This is necessary to allow of the development of its peculiar and extensive fauna, especially as we see that that fauna comprised animals rivalling in bulk those of the great continents. It is further indicated by the relations with New Guinea, already alluded to, and by the general character of the various faunas which compose the Australian region, details of which will be found in the succeeding part of this work. The lowering of the ocean during the Glacial period would be favourable to the still further development of the fauna of such a country; and it is to the unfavourable conditions produced by its subsequent rising—equivalent to a depression of the land to the amount of two thousand feet—that we must impute the extinction of so many remarkable groups of animals. It is not improbable, that the disappearance of the ice and the consequent (apparent) subsidence of the land, might have been rapid as compared with the rate at which large animals can become modified to meet new conditions. Extensive tracts of fertile land might have been submerged, and the consequent crowding of large numbers of species and individuals on limited areas would have led to a struggle for existence in which the less adapted and less easily modifiable, not the physically weaker, would succumb.

There is, however, another cause for the extinction of large rather than small animals whenever an important change of conditions occurs, which has been suggested to me by a correspondent,[1] but which has not, I believe, been adduced by Mr. Darwin or by any other writer on the subject. It is dependent on the fact, that large animals as compared with small ones are almost invariably slow breeders, and as they also necessarily exist in much smaller numbers in a given area, they offer far less materials for favourable variations than do smaller animals. In such an extreme case as that of the rabbit and elephant, the

[1] Mr. John Hickman of Desborough.

young born each year in the world are probably as some millions to one; and it is very easily conceivable that in a thousand years the former might, under pressure of rapidly changing conditions, become modified into a distinct species, while the latter, not offering enough favourable variations to effect a suitable adaptation, would become extinct. We must also remember the extreme specialization of many of the large animals that have become extinct—a specialization which would necessarily render modification in any new direction difficult, since the inherited tendency of variation would probably be to increase the specialization in the same directions which had heretofore been beneficial. If to these two causes we add the difficulty of obtaining sufficient food for such large animals, and perhaps the injurious effects of changes of climate, we shall not find it difficult to understand how such a vast physical revolution as the Glacial epoch, with its attendant phenomena of elevations and subsidences, icy winds, and sudden floods by the bursting of lake barriers, might have led to the total extinction of a vast number of the most bulky forms of mammalia, while the less bulky were able to survive, either by greater hardiness of constitution or by becoming more or less modified. The result is apparent in the comparatively small or moderate size of the species constituting the temperate fauna, in all parts of the globe.

It is much to be regretted that no mammalian remains of earlier date have been found in Australia, as we should then see if it is really the case that marsupials have always formed its highest type of mammalian life. At present its fossil fauna is chiefly interesting to the zoologist, but throws little light on the past relations of this isolated country with other parts of the globe.

MAMMALIAN REMAINS IN THE SECONDARY FORMATIONS.

In the oldest Tertiary beds of Europe and North America, we have (even with our present imperfect record) a rich and varied mammalian fauna. As compared with our living or recent highly specialized forms, it may be said to consist of generalised types; but as compared with any primeval mammalian type, it must be pronounced highly specialised. Not only are such diversified groups as Carnivora, Perrissodactyle and Artiodactyle Ungulates, Primates, Chiroptera, Rodents, and Marsupials already well marked, but in many of these there is a differentiation into numerous families and genera of diverse character. It is impossible therefore to doubt, that many peculiar forms of mammalia must have lived long anterior to the Eocene period; but there is unfortunately a great gap in the record between the Eocene and Cretaceous beds, and these latter being for the most part marine continue the gap as regards mammals over an enormous lapse of time. Yet far beyond both these chasms in the Upper Oolitic strata, remains of small mammalia have been found; again, in the Stonesfield slate, a member of the Lower Oolite, other forms appear. Then comes the marine Lias formation with another huge gap; but beyond this again in the Upper Trias, the oldest of the secondary formations, mammalian teeth have been discovered in both England and Germany, and these are, as nearly as can be ascertained, of the same age as the *Dromatherium* already noticed, from North America. They have been named *Microlestes*, and show some resemblance to those of the West Australian *Myrmecobius*. In the Oolitic strata numerous small jawbones have been found, which have served to characterise eight genera, all of which are believed to have been Marsupials, and in some of them a resemblance can be traced to some of the smaller living Australian species. These, however, are mere indications of the number of mammalia that must have lived in the secondary period, so long thought to be exclusively "the age of reptiles;" and the fact that the few yet found are at all comparable with such specialised forms as still exist, must convince us, that we shall have to seek far beyond even the earliest of these remains, for the first appearance of the mammalian type of vertebrata.

EXTINCT BIRDS.

Compared with those of mammalia, the remains of birds are exceedingly scarce in Europe and America; and from the wandering habits of so many of this class, they are of much less value as indications of past changes in physical geography. A large proportion of the remains belong to aquatic or wading types, and as these have now often a world-wide range, the occurrence of extinct forms can have little bearing on our present inquiry. There are, however, a few interesting cases of extinct land-birds belonging to groups now quite strangers to the country in which they are found; and others scarcely less interesting, in which groups now peculiar to certain areas are shown to have been preceded by allied species or genera of gigantic size.

Palæarctic Region and N. India.—In the caves and other Post-Pliocene deposits of these countries, the remains of birds almost all belong to genera now inhabiting the same districts. Almost the only exceptions are, the great auk and the capercailzie, already mentioned as being found in the Danish mounds; the latter bird, with *Tetrao albus*, in Italian caverns; and a species of pheasant (*Phasianus*) said to have occurred in the Post-Pliocene of France, considerably west of the existing range of the genus in a wild state.

In the preceding Pliocene deposits, but few remains have been found, and all of existing genera but one, a gallinaceous bird (*Gallus bravardi*) allied to the domestic fowl and peacock.

The Miocene beds of France and Central Europe have produced many more remains of birds, but these, too, are mostly of existing European genera, though there are some notable exceptions. Along with forms undistinguishable from crows (*Corvus*), shrikes (*Lanius*), wagtails (*Motacilla*), and woodpeckers (*Picus*), are found remains allied to the Oriental edible-nest swift (*Collocalia*) and *Trogon;* a parrot resembling the African genus *Psittacus;* an extinct form *Necrornis*, perhaps allied to the plantain-eaters (*Musophaga*); *Homalophus*, doubtfully allied to woodpeckers, and *Limnatornis* to the hoopoes. The gallinaceous birds are represented by three species of pheasants, some very close to the domesticated species; *Palæoperdix* allied to the partridges; and *Palæortyx*, small birds allied to the American genus *Ortyx*, but with larger wings. There are also species of *Pterocles* allied to living birds, and a small pigeon. There are numerous living genera of Accipitres; such as eagle (*Aquila*),

M

kite (*Milvus*), eagle-owl (*Bubo*), and screech-owl (*Strix*); with
the African secretary-bird (*Serpentarius*), and some extinct forms,
as *Palæocercus*, *Palæohierix* and *Palæetus*.

Aquatic and wading birds were abundant, including numerous
rails, bustards, herons, sandpipers, gulls, divers, and pelicans.
There were also many ducks, some allied to the genus *Dendro-
cygna*; the Oriental genus of storks, *Leptoptilus*; *Ibidipodia*, a
remarkable form allied to *Ibis* and *Ciconia*; *Elornis*, near
Limosa; *Pelagornis*, a large bird allied to gannets and pelicans;
Hydrornis, allied to the ducks and petrels; *Dolichopterus*, allied
to plovers. Perhaps the most interesting of these extinct birds
are, however, the flamingoes, represented by forms hardly distinguishable from living species, and by one extinct genus *Palæ-
lodus*, which had very long toes, and probably walked on aquatic
plants like the tropical jacanas.

The Miocene beds of North India have furnished few birds;
the only one of geographical interest being an extinct species
of ostrich, not very different from that now inhabiting Arabia.

On the whole, the birds of Europe at this period were very
like those now living, with the addition of a few tropical forms.
These latter were, however, perhaps more numerous and important than they appear to be, as they belong to inland and forest-
haunting types, which would not be so frequently preserved as
the marsh and lake-dwelling species. Taking this into consideration, the assemblage of Miocene birds accords well with
what we know of the mammalian fauna. We have the same
indications of a luxuriant vegetation and subtropical climate,
and the same appearance of Oriental and especially of African
types. *Trogon* is perhaps the most interesting of all the forms
yet discovered, since it furnishes us with a central point whence
the living trogons of Asia, Africa, and South America might
have diverged.

In the Eocene we find ourselves almost wholly among extinct
forms of birds. The earliest known Passerine bird is here
met with, in *Protornis*, somewhat similar to a lark, found in
the Lower Eocene of Switzerland; while another Passerine form,
Palægithalus, and one allied to the nuthatch (*Sitta*), have been

discovered in the Upper Eocene of Paris. Picariæ of equal antiquity are found. *Cryptornis*, from the Paris Eocene, and *Hal-
cyornis* from the Lower Eocene of the Isle of Sheppey, were
both allied to kingfishers; while a form allied to *Centropus* a
genus of cuckoos, or, as Milne-Edwards thinks, to the Madagascar *Leptosomus*, has been found in the Upper Eocene of France.
Several *Accipitres* of somewhat doubtful affinities have been
found in the same country; while *Lithornis*, from the Lower
Eocene of the Isle of Sheppey, was a small vulturine bird supposed to be allied to the American group, *Cathartes*. Among
the waders, some extinct forms of plovers have been found, and a
genus (*Agnopterus*), allied to the flamingoes; while there are
many swimming birds, such as pelicans, divers, and several
extinct types of doubtful affinities. Most intersting of all is a
portion of a cranium discovered in the Lower Eocene of Sheppey, and lately pronounced by Professor Owen to belong to a
large Struthious bird, allied to the New Zealand *Dinornis* and
also perhaps to the ostrich. Another gigantic bird is the *Gas-
tornis*, from the Lower Eocene of Paris, which was as large as an
ostrich, but which is believed to have been a generalised type,
allied to wading and swimming birds as well as to the Struthiones.
     Beyond this epoch we have no remains of birds in European
strata till we come to the wonderful *Archæopteryx* from the
Upper Oolite of Bavaria; a bird of a totally new type, with a
bony tail, longer than the body, each vertebra of which carried
a pair of diverging feathers.
     North America.—A number of bird-remains have lately been
found in the rich Tertiary and Cretaceous deposits of the United
States; but here, too, comparatively few are terrestrial forms.
No Passerine bird has yet been found. The Picariæ are represented by *Uintornis*, an extinct form allied to woodpeckers, from
the Eocene of Wyoming. Species of turkey (*Meleagris*) occur
in the Post-Pliocene and as far back as the Miocene strata,
showing that this interesting type is a true denizen of temperate
North America. The other birds are, *Accipitres;* waders and
aquatics of existing genera; and a number of extinct forms of
the two latter orders—such as, *Aletornis* an Eocene wader;

Palæotringa, allied to the sandpipers, and *Telmatobius* to the rails, both Cretaceous; with *Graculavus*, allied to *Graculus;* *Laornis* allied to the swans; *Hesperornis* a gigantic diver; and *Icthyornis* a very low form, with biconcave vertebra, such as are only found in fishes and some reptiles—also from Cretaceous deposits.

South America.—The caverns of Brazil produced thirty-four species of birds, most of them referable to Brazilian genera, and many to still existing species. The most interesting were two species of American ostrich (*Rhea*), one larger than either of the living species; a large turkey-buzzard (*Cathartes*); a new species of the very isolated South American genus *Opisthocomus;* and a *Cariama*, or allied new genus.

Madagascar and the Mascarene Islands.—We have here only evidence of birds that have become extinct in the historical period or very little earlier. First we have a group of birds incapable of flight, allied to pigeons, but forming a separate family, *Dididæ*; and which, so far as we yet know, inhabited Mauritius, Rodriguez, and probably Bourbon. *Aphanapteryx*, an extinct genus of rails, inhabited Mauritius; and another genus, (*Erythromachus*), Rodriguez. A large parrot, said by Prof. Milne Edwards to be allied to *Ara* and *Microglossus*, also inhabited Mauritius; and another allied to *Eclectus*, the island of Rodriguez. None of these have been found in Madagascar; but a gigantic Struthious bird, *Æpyornis*, forming a peculiar family distinct both from the ostriches of Africa and the *Dinornis* of New Zealand inhabited that island; and there is reason to believe that this may have lived less than 200 years ago.

New Zealand.—A number of extinct Struthious birds, forming two families, *Dinornithidæ* and *Palapterygidæ*, have been found in New Zealand. Some were of gigantic size. They seem allied both to the living *Apteryx* of New Zealand and the emu of Australia. They are quite recent, and some of them have probably lived within the last few centuries. Remains of *Dinornis* have also been found in a Post-Pliocene deposit in Queensland, N. E. Australia [1]—a very important discovery, as it

---

[1] *Trans. Zool. Soc. of London*, vol. viii. p. 381.

gives support to the theory of a great eastward extension of Australia in Tertiary times.

EXTINCT TERTIARY REPTILES.

These will not occupy us long, as no very great number are known, and most of them belong to a few principal forms of comparatively little geographical interest.

Tortoises are perhaps the most abundant of the Tertiary reptiles. They are numerous in the Eocene and Miocene formations both in Europe and North America. The genera *Emys* and *Trionyx* abound in both countries, as well as in the Miocene of India. Land tortoises occur in the Eocene of North America and in the Miocene of Europe and India, where the huge *Colossochelys*, twelve feet long, has been found. In the Pliocene deposits of Switzerland the living American genus *Chelydra* has been met with. These facts, together with the occurrence of a living *species* in the Miocene of India, show that this order of reptiles is of great antiquity, and that most of the genera once had a wider range than now.

Crocodiles, allied to the three forms now characteristic of India, Africa, and America, have been found in the Eocene of our own country, and several species of *Crocodilus* have occurred in beds of the same age in North America.

Lizards are very ancient, many small terrestrial forms occurring in all the Tertiary deposits. A species of the genus *Chamæleo* is recorded from the Eocene of North America, together with several extinct genera.

Snakes were well developed in the Eocene period, where remains of several have been found which must have been from twelve to twenty feet long. An extinct species of true viper has occurred in the Miocene of France, and one of the Pythonidæ in the Miocene brown coal of Germany.

Batrachia occur but sparingly in a fossil state in the Tertiary deposits. The most remarkable is the large Salamander *Andreas*) from the Upper Miocene of Switzerland, which is allied to the *Menopoma* living in North America. Species of frog (*Rana*), and *Palæophryus* an extinct genus of toads, have been found in the Miocene deposits of Germany and Switzerland.

Fresh water fish are almost unknown in the Tertiary deposits of Europe, although most of the families and some genera of living marine fish are represented from the Eocene downwards.

ANTIQUITY OF THE GENERA OF INSECTS.

Fossil insects are far too rarely found, to aid us in our determination of difficult questions of geographical distribution; but in discussing these questions it will be important to know, whether we are to look upon the existing generic forms of insects as of great or small antiquity, compared with the higher vertebrates; and to decide this question the materials at our command are ample.

The conditions requisite for the preservation of insects in a fossil state are no doubt very local and peculiar; the result being, that it is only at long intervals in the geological record that we meet with remains of insects in a recognisable condition. None appear to have been found in the Pliocene formation; but in the Upper Miocene of Œninghen in Switzerland, associated with the wonderfully rich fossil flora, are found immense quantities of insects. Prof. Heer examined more than 5,000 specimens belonging to over 800 species, and many have been found in other localities in Switzerland; so that more than 1,300 species of Miocene insects have now been determined. Most of the orders are represented, but the beetles (Coleoptera) are far the most abundant. Almost all belong to existing genera, and the majority of these genera now inhabit Europe, only three or four being exclusively Indian, African, or American.

In the Lower Miocene of Croatia there is another rich deposit of insects, somewhat more tropical in character, comprising large white-ants and dragon-flies differently marked from any now inhabiting Europe. A butterfly is also well preserved, with all the markings of the wings; and it seems to be a *Junonia*, a tropical genus, though it may be a *Vanessa*, which is European, but the fossil most resembles Indian species of *Junonia*.

The Eocene formations seem to have produced no insect remains; but they occur again in the Upper Cretaceous at Aix-la-Chapelle, where two butterflies have been found, *Cyllo sepulta* and *Satyrites Reynesii*, both belonging to the Satyridæ, and the former to a genus now spread over Africa, India, and Australia.

A little earlier, in the Wealden formation of our own country, numerous insects have been found, principally dragon flies (*Libellula, Æshna*); aquatic Hemiptera (*Velia Hydrometra*); crickets cockroaches, and cicadas, of familiar types.

Further back in the Upper Oolite of Bavaria—which produced the wonderful long-tailed bird, *Archæopteryx*—insects of all orders have been found, including a moth referred to the existing genus *Sphinx*.

In the Lower Oolite of Oxfordshire many fossil beetles have been found whose affinities are shown by their names :—*Buprestidium, Curculionidium, Blapsidium, Melolonthidium*, and *Prionidium ;* a wing of a butterfly has also been found, allied to the Brassolidæ now confined to tropical America, and named *Palæontina oolitica*.

Still more remote are the insects of the Lias of Gloucestershire, yet they too can be referred to well-known family types—Carabidæ, Melolonthidæ, Telephoridæ, Elateridæ, and Curculionidæ, among beetles; Gryllidæ and Blattidæ among Orthoptera; with *Libellula, Agrion, Æshna, Ephemera*, and some extinct genera. When we consider that almost the only vertebrata of this period were huge Saurian reptiles like the *Icthyosaurus, Plesiosaurus*, and *Dinosaurus*, with the flying Pterodactyles; and that the great mass of our existing genera, and even families, of fish and reptiles had almost certainly not come into existence, we see at once that types of insect-form are, proportionately, far more ancient. At this remote epoch we find the chief family types (the *genera* of the time of Linnæus) perfectly differentiated

and recognisable. It is only when we go further back still, into
the Palæozoic formations, that the insect forms begin to show that
generalization of type which renders it impossible to classify
them in any existing groups. Yet even in the coal formation of
Nova Scotia and Durham, the fossil insects are said by competent
entomologists to be "allied to *Ephemera*," "near *Blatta*," "near
Phasmidæ;" and in deposits of the same age at Saarbrück near
Trèves, a well-preserved wing of a grasshopper or locust has been
found, as well as a beetle referred to the Scarabeidæ. More
remarkable, however, is the recent discovery in the carboniferous
shales of Belgium, of the clearly-defined wing of a large moth
(*Breyeria borinensis*), closely resembling some of the Saturniidæ;
so that we have now all the chief orders of Insects—including
those supposed to be the most highly developed and the most
recent—well represented at this very remote epoch. Even the
oldest insects, from the Devonian rocks of North America, can
mostly be classed as Neuroptera or Myriapoda, but appear to
form new families.

We may consider it, therefore, as proved, that many of the
larger and more important genera of insects date back to the
beginning of the Tertiary period, or perhaps beyond it; but the
family types are far older, and must have been differentiated very
early in the Secondary period, while some of them perhaps go
back to Palæozoic times. The great comparative antiquity of
the *genera* is however the important fact for us, and we shall
have occasion often to refer to it, in endeavouring to ascertain
the true bearing of the facts of insect distribution, as elucidating
or invalidating the conclusions arrived at from a study of the
distribution of the higher animals.

ANTIQUITY OF THE GENERA OF LAND AND FRESH-WATER SHELLS.

The remains of land and fresh-water shells are not much more
frequent than those of insects. Like them, too, their forms are
very stable, continuing unchanged through several geological

periods. In the Pliocene and Miocene formations, most of the shells are very similar to living species, and some are quite identical. In the Eocene we meet with ordinary forms of the genera *Helix, Clausilia, Pupa, Bulimus, Glandina, Cyclostoma, Megalostoma, Planorbis, Paludina* and *Limnæa*, some resembling European species, others more like tropical forms. A British Eocene species of *Helix* is still living in Texas ; and in the South of France are found species of the Brazilian sub-genera *Megaspira* and *Anastoma*. In the secondary formation no true land shells have been found, but fresh water shells are tolerably abundant, and almost all are still of living forms. In the Wealden (Lower Cretaceous) and Purbeck (Upper Oolite) are found *Unio, Melania, Paludina, Planorbis*, and *Limnæa ;* while the last named genus occurs even in the Lias.

The notion that land shells were really not in existence during the secondary period is, however, proved to be erroneous by the startling discovery, in the Palæozoic coal measures of Nova Scotia, of two species of Helicidæ, both of living genera—*Pupa vetusta*, and *Zonites priscus*. They have been found in the hollow trunk of a *Sigillaria*, and in great quantities in a bed full of Stigmarian rootlets. The most minute examination detects no important differences of form or of microscopic structure, between these shells and living species of the same genera ! These mollusca were the contemporaries of Labyrinthodonts and strange Ganoid fishes, which formed almost the whole vertebrate fauna. This unexpected discovery renders it almost certain, that numbers of other existing genera, of which we have found no traces, lived with these two through the whole secondary period; and we are thus obliged to assume as a probability, that any particular genus has lived through a long succession of geological ages. In estimating the importance of any peculiarities or anomalies in the geographical distribution of land shells as compared with the higher vertebrates, we shall, therefore, have to keep this possible, and even probable high antiquity, constantly in mind.

We have now concluded our sketch of Tertiary Palæontology as a preparation for the intelligent study of the Geographical

Distribution of Land Animals; and however imperfectly the task
has been performed, the reader will at all events have been convinced that some such preliminary investigation is an essential
and most important part of our work.   So much of palæontology
is at present tentative and conjectural, that in combining the
information derived from numerous writers, many errors of detail
must have been made.   The main conclusions have, however, been
drawn from as large a basis of facts as possible; and although
fresh discoveries may show that our views as to the past history
of some of the less important genera or families are erroneous,
they can hardly invalidate our results to any important degree,
either as regards the intercommunications between separate
regions in the various geological epochs, or as to the centres
from which some of the more important groups have been dispersed.

PART III.

ZOOLOGICAL GEOGRAPHY:

*A REVIEW OF THE CHIEF FORMS OF ANIMAL LIFE IN THE
SEVERAL REGIONS AND SUB-REGIONS, WITH THE INDICA-
TIONS THEY AFFORD OF GEOGRAPHICAL MUTATIONS.*

CHAPTER IX.

HAVING discussed, in our First Part, such general and preliminary matters as are necessary to a proper comprehension of our subject; and having made ourselves acquainted, in our Second Part, with the most important results of Palæontology, we now come to our more immediate subject, which we propose to treat first under its geographical aspect. Taking each of our six regions in succession, we shall point out in some detail the chief zoological features they present, as influenced by climate, vegetation, and other physical features. We shall then treat each of the sub-regions by itself, as well as such of the islands or other sub-divisions as present features of special interest; endeavouring to ascertain their true relations to each other, and the more important changes of physical geography that seem necessary to account for their present zoological condition.

Order of Succession of the Regions.—We may here explain the reason for taking the several regions in a different succession from that in which they appear in the tabular or diagrammatic headings to each family, in the Fourth, and concluding part of this work. It will have been seen, by our examination of extinct animals (and it will be made still clearer during our study of the several regions) that all the chief types of animal life appear to have originated in the great north temperate or northern continents; while the southern continents—now represented by

South America, Australia, and South Africa with Madagascar—
have been more or less completely isolated, during long periods,
both from the northern continent and from each other. These
latter countries have, however, been subject to more or less immigration from the north during rare epochs of approximation
to, or partial union with it. In the northern, more extensive, and
probably more ancient land, the process of development has
been more rapid, and has resulted in more varied and higher
types; while the southern lands, for the most part, seem to have
produced numerous diverging modifications of the lower grades
of organization, the original types of which they derived either
from the north, or from some of the ancient continents in Mesozoic or Palæozoic times. Hence those curious resemblances in
the fauna of South America, Australia, and, to a less extent,
Madagascar, which have led to a somewhat general belief that
these distant countries must at one time or other have been
united ; a belief which, after a careful examination of all the
facts, does not seem to the author of this work to be well
founded. On the other hand, there is the most satisfactory
evidence that each southern region has been more or less
closely united (during the tertiary or later secondary epoch)
with the great northern continents, leading to numerous resemblances and affinities in their productions.

In endeavouring to present at a glance in the most convenient
manner, the distribution of the families in the several regions
and sub-regions, it was necessary to arrange them, so that those
whose relations to each other were closest should stand side by
side ; the first and last being those between which the relations
were least numerous and least important. Influenced by the
usual opinions as to the relations between Australia and South
America, the series was at first begun with the Nearctic, and
terminated with the Australian and Neotropical regions; and it
was not till the whole of the vertebrate families had been gone
through, and their distribution carefully studied, that these last
two regions were seen to be really wider apart than any others
of the series. It was therefore decided to alter the arrangement,
beginning with the Neotropical, and ending with the Australian

regions; and a careful inspection of the diagrams themselves, taken in their entirety, will, it is believed, show that this is the most natural plan, and most truly exhibits the relations of the several regions.

In the portion of our work now commencing, we are not, however, by any means bound to begin at either end of this series. Each region is studied by itself, but reference will often have to be made to all the other regions; and wherever we begin, we must occasionally refer to facts which will be given further on. As, however, the great northern continents form the central mass from which the southern regions, as it were, diverge, and as the Palæarctic region is both more extensive and much better known than any other, it undoubtedly forms the most convenient starting-point for our proposed survey of the zoological history of the earth. We thus pass from the better known to the less known—from Europe to Africa and tropical Asia, and thence to Australia, completing the series of regions of the Eastern Hemisphere. Beginning again with the Neotropical region, we pass to the Nearctic, which has such striking relations with the preceding and with the Palæarctic region, that it can only be properly understood by constant reference to both. We thus keep separate the Eastern and Western hemispheres, which form, from our point of view, the most radical and most suggestive division of terrestrial faunas; and as we are able to make this also the dividing point of our two volumes, reference to the work will be thereby facilitated.

Cosmopolitan Groups.—Before proceeding to sketch the zoological features of the several Regions it will be well to notice those family groups which belong to the earth as a whole, and which are so widely and universally distributed over it that it will be unnecessary, in some cases, to do more than refer to them under the separate geographical divisions.

The only absolutely cosmopolitan families of Mammalia are those which are aerial or marine; and this is one of the striking proofs that their distribution has been effected by natural causes, and that the permanence of barriers is one of the chief

agencies in the limitation of their range. Even among the
aerial bats, however, only one family—the Vespertilionidæ—is
truly cosmopolitan, the others having a more or less restricted
range. Neither are the Cetacea necessarily cosmopolitan, most of
the families being restricted either to warm or to cold seas ; but
one family, the dolphins (Delphinidæ), is truly so. This order
however will not require further notice, as, being exclusively
marine the groups do not enter into any of our terrestrial
regions. The only other family of mammals that may be considered to be cosmopolitan, is the Muridæ (rats and mice) ; yet
these are not entirely so, since none are known to be truly
indigenous in any part of the Australian region except Australia
itself.

In the class of Birds, a number of families are cosmopolites,
if we reckon as such all which are found in each region and
sub-region ; but several of these are so abundant in some parts,
while they are so sparingly represented in others, that they
cannot fairly be considered so. We shall confine that term
therefore, to such as, there is reason to believe, inhabit every
important sub-division of each region. Such are, among the
Passerine birds the crows (Corvidæ), and swallows (Hirundinidæ) ; among the Picariæ the kingfishers (Alcedinidæ) ; among
other Land birds the pigeons (Columbidæ), grouse and partridges
(Tetraonidæ), hawks (Falconidæ), and owls (Strigidæ) ; among
the Waders the rails (Rallidæ), snipes (Scolopacidæ), plovers
(Charadriadæ), and herons (Ardeidæ) ; and among the Swimmers
the ducks (Anatidæ), gulls (Laridæ), petrels (Procellariidæ),
pelicans (Pelecanidæ), and grebes (Podicipidæ).

In the class of Reptiles there are few absolutely cosmopolitan
families, owing to the scarcity of members of this group in some
insular sub-regions, such as New Zealand and the Pacific Islands.
Those which are most nearly so are the Colubridæ among snakes.
and the Scincidæ among lizards.

There is no cosmopolitan family of Amphibia, the true frogs
(Ranidæ) being the most widely distributed.

Neither is any family of Freshwater Fishes cosmopolitan,
the Siluridæ, which have the widest range, being confined

to warm regions, and becoming very scarce in the temperate zones.

Among the Diurnal and Crepuscular Lepidoptera (butterflies and sphinges) the following families are cosmopolitan :—Satyridæ, Nymphalidæ, Lycænidæ, Pieridæ, Papilionidæ, Hesperidæ, Lycænidæ, and Sphingidæ.

Of the Coleoptera almost all, except some of the small and obscure families, are cosmopolitan.

Of the terrestrial Mollusca, the Helicidæ alone are true cosmopolites.

Tables of Distribution of Families and Genera.—Having been obliged to construct numerous tables of the distribution of the various groups for the purposes of the descriptive part of the work, I have thought it well to append the most important of them, in a convenient form, to the chapter on each region ; as much information will thereby be given, which can only be obtained from existing works at the cost of great labour. All these tables are drawn up on a uniform plan, the same generic and family names being used in each ; and all are arranged in the same systematic order, so as to be readily comparable with each other. This, although it seems a simple and natural thing to do, has involved a very great amount of labour, because hardly two authors use the same names or follow the same arrangement. Hence comparison between them is impossible, till all their work has been picked to pieces, their synonymy unravelled, their differences accounted for, and the materials recast ; and this has to be done, not for two or three authors only, but for the majority of those whose works have been consulted on the zoology of any part of the globe.

Except in the two higher orders—Mammalia and Birds— materials do not exist for complete tables of the genera brought down to the present time. We have given therefore, first, a complete table of all the families of Vertebrata and Diurnal Lepidoptera found in each region, showing the sub-regions in which they occur, and their range beyond the limits of the region. Families which are wholly peculiar to the region, or

N

very characteristic and almost exclusively confined to it, are in *italics*. The number prefixed to each family corresponds to that of the series of families in the Fourth Part of this work, so that if further information is required it can be readily referred to without consulting the index. Names inclosed in parentheses—(. . .) thus—indicate families which only just enter a region from an adjacent one, to which they properly belong. The eye is thus directed to the more, and the less important families; and a considerable amount of information as to the general features of the zoology of the region, is conveyed in the easiest manner.

The tables of genera of Mammalia and Birds, are arranged on a somewhat different plan. Each genus is given under its Family and Order, and they follow in the same succession in all the tables. The number of species of each genus, inhabiting the region, is given as nearly as can be ascertained; but in many cases this can only be a general approximation. The distribution of the genera within the region, is then given with some detail; and, lastly, the range of the genus beyond the region is given in general terms, the words " Oriental," "Ethiopian," &c., being used for brevity, to indicate that the genus occurs over a considerable part of such regions. Genera which are restricted to the region (or which are very characteristic of it though just transgressing its limits) are given in *italics ;* while those which only just enter the region from another to which they really belong, are enclosed in parentheses—(. . .) thus. The genera are here numbered consecutively, in order that the number of genera in each family or each order, in the region, may be readily ascertained (by one process of subtraction), and thus comparisons made with other regions or with any other area. As the tables of birds would be swelled to an inconvenient length by the insertion in each region of all the genera of Waders and Aquatics, most of which have a very wide range and would have to be repeated in several or all the regions, these have been omitted; but a list has been given of such of the genera as are peculiar to, or highly characteristic of each region.

As this is the first time that any such extensive tables of

distribution have been constructed for the whole of the Mammalia and Birds, they must necessarily contain many errors of detail; but with all their imperfections it is believed they will prove very useful to naturalists, to teachers, and to all who take an intelligent interest in the wider problems of geography and natural history.

CHAPTER X.

THIS region is of immense extent, comprising all the temperate portions of the great eastern continents. It thus extends from the Azores and Canary Islands on the west to Japan on the east, a distance not far short of half the circumference of the globe. Yet so great is the zoological unity of this vast tract, that the majority of the genera of animals in countries so far removed as Great Britain and Northern Japan are identical. Throughout its northern half the animal productions of the Palæarctic region are very uniform, except that the vast elevated desert-regions of Central Asia possess some characteristic forms; but in its southern portion, we find a warm district at each extremity with somewhat contrasted features. On the west we have the rich and luxuriant Mediterranean sub-region, possessing many peculiar forms of life, as well as a few which are more especially characteristic of the Ethiopian region. On the east we have the fertile plains of Northern China and the rich and varied islands of Japan, possessing a very distinct set of peculiar forms, with others belonging to the Oriental region, into which this part of the Palæarctic region merges gradually as we approach the Tropic of Cancer. Thus, the countries roughly indicated by the names—Northern Europe, the Mediterranean district, Central and Northern Asia, and China with Japan—have each well-marked minor characteristics which entitle them to the rank of sub-regions. Their boundaries are often indefinable; and those here adopted have been fixed upon to some extent by considera-

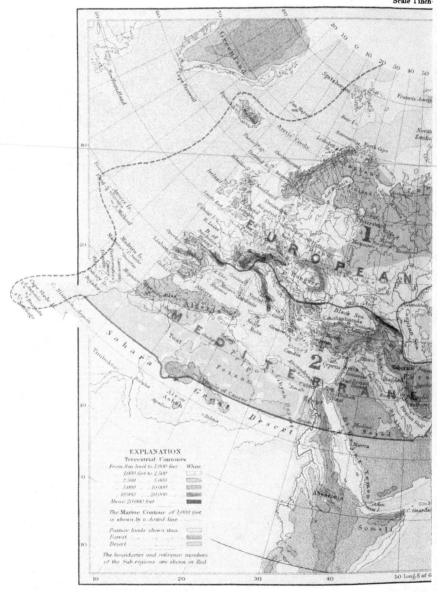

EXPLANATION

Terrestrial Contours

From Sea level to 1,000 feet .... White

1,000 feet to 2,500	
2,500 " 5,000	
5,000 " 10,000	
10,000 " 20,000	
Above 20,000 feet	

The Marine Contour of 1,000 feet
is shewn by a dotted line ....

Pasture lands shewn thus ....
Forest ....
Desert ....

The boundaries and reference numbers
of the Sub-regions are shewn in Red.

London ; Mac

millan & C?

tions of convenience, dependent on custom and on the more or less perfect knowledge we possess of some of the intervening countries.

Zoological Characteristics of the Palæarctic Region.—The Palæarctic region has representatives of thirty-five families of mammalia, fifty-five of birds, twenty-five of reptiles, nine of amphibia, and thirteen of freshwater fishes. Comparing it with the only other wholly temperate region, the Nearctic, we find a much greater variety of types of mammalia and birds. This may be due in part to its greater area, but more, probably, to its southern boundary being conterminous for an enormous distance with two tropical regions, the Ethiopean and Oriental ; whereas the Nearctic has a comparatively short southern boundary conterminous with the Neotropical region only. This is so very important a difference, that it is rather a matter of surprise that the two north temperate regions should not be more unequal in the number of their higher vertebrate forms, than they actually are.

It is also to the interblending of the Palæarctic with the two adjacent tropical regions, that we must attribute its possession of so few peculiar family groups. These are only three ; two of reptiles, *Trogonophidæ* and *Ophiomoridæ*, and one of fishes, *Comephoridæ*. The number of peculiar genera is, however, considerable, as the following enumeration will show.

Mammalia.—The monkey of Gibraltar and North Africa, and an allied species found in Japan, are now considered to belong to the extensive eastern genus *Macacus*. The former, however, is peculiar in the entire absence of the tail, and has by many naturalists, been held to form a distinct genus, *Inuus*, confined to the Palæarctic region.

Of bats there are one or two genera (*Barbastellus*, *Plecotus*) which seem to be mainly or wholly Palæarctic, but the classification of these animals is in such an unsettled state that the distribution of the genera is of little importance.

In the next order, Insectivora, we have almost the entire family of the Moles confined to the region. *Talpa* just enters Northern India ; and *Urotrichus* is common to Japan and North-

Western America, but the remaining genera, six in number, are all exclusively Palæarctic.

Among Carnivora we have *Nyctereutes*, the curious racoon-dog of Japan and North-Eastern Asia; *Lutronectes*, an otter peculiar to Japan; and the badger (*Meles*), which ranges over the whole region, and just enters the Oriental region as far as Hongkong; *Æluropus*, a curious form of the Himalayan panda, inhabiting the high mountains of Eastern Thibet; and *Pelagius*, a genus of seals, ranging from the shores of Madeira to the Black Sea.

The Ungulata, or hoofed animals, are still more productive of forms peculiar to this region. First we have the Camels, whose native home is the desert region of Central and Western Asia and Northern Africa, and which, even in their domesticated condition, are confined almost wholly within the limits of the Palæarctic region. Of Deer we have six peculiar genera, *Dama* and *Capreolus* found in Europe, with *Elaphodus, Lophotragus, Hydropotes*, and *Moschus*, confined to Northern China and Mongolia. The great family Bovidæ—comprising the oxen, sheep, goats and antelopes —furnishes no less than seven peculiar Palæarctic genera. These are *Poephagus*, the yak of Thibet; *Addax*, a well-known antelope of Northern Africa and Syria; *Procapra, Pantholops* and *Budorcas*, antelopine genera peculiar to Thibet and Mongolia; with *Rupicapra* (the chamois), and the extraordinary large-nosed antelope *Saiga*, confined to Europe and Western Asia. Besides these we have *Capra* (the wild sheep and goats), all the numerous species of which, except two, are exclusively Palæarctic.

Coming to the Rodents, we have again many peculiar forms. Of Muridæ (the mouse and rat tribe), we have six peculiar genera, the more important being *Cricetus, Rhombomys Sminthus*, and *Myospalax*. Of Spalacidæ (mole-rats) both the Palæarctic genera, *Ellobius* and *Spalax*, are peculiar. *Ctenodactylus*, a genus of the South American family Octodontidæ, is found only in North Africa. To these we may add *Myoxus* (the dormice) and *Lagomys* (the pikas or tail-less hares) as essentially Palæarctic, since but one species of each genus is found beyond the limits of the region.

Birds.—It appears to have been the opinion of many natural-

ists that the Palæarctic region could not be well characterised by
its peculiar genera of birds.   In Mr. Sclater's celebrated paper
already referred to, he remarks, "It cannot be denied that the
ornithology of the Palæarctic region is more easily characterised
by what it has not than by what it has," and this has been quite
recently quoted by Mr. Allen, in his essay on the distribution
of North American birds, as if it represented our present know
ledge of the subject.   But, thanks to the labours of Dr.
Jerdon, Mr. Swinhoe, Père David and others, we have now
learnt that a large number of birds included in the Indian
list, are either mere winter emigrants from Central Asia, or only
inhabit the higher ranges of the Himalayas, and thus really
belong to the Palæarctic region.   The result is, that a host of
genera are now seen to be either exclusively or characteristically
Palæarctic, and we have no further difficulty in giving positive
ornithological characters to the region.   In the tables appended
to this chapter, all these truly Palæarctic genera will be found
printed in *italics*, with an indication of their distribution, which
will sometimes be found more fully given under the respective
families in the third part of this work.   Referring to this table
for details we shall here summarise the results.

Of the Sylviidæ or warblers, no less than fourteen genera are
either exclusively or characteristically Palæarctic, of which
Locustella, *Sylvia*, *Curruca* and *Erithacus* are good examples.
Of the oriental family Timaliidæ, the genus *Pterorhinus* is Palæarctic. Of Panuridæ, or reedlings, there are four peculiar genera
(comprising almost the whole family) ; of Certhiidæ, or creepers,
one—*Tichodroma*—which extends southward to the Abyssinian
highlands. Of Paridæ, or tits, one—*Acredula* ; of Corvidæ, or
crows, four—*Pica* (containing our magpie) being a good example; of
Fringillidæ, or finches and buntings, twelve, among which *Acanthis*,
Pyrrhula and *Emberiza* are good illustrations ; of Alaudidæ, or
larks, there are two peculiar genera. Leaving the Passeres we next
come to peculiar forms among the gallinaceous birds : *Syrrhaptes*
among the Pteroclidæ or sand grouse; four genera of Tetraonidæ or
grouse and partridges, and five of Phasianidæ or pheasants, comprising some of the most magnificent birds in the world. Lastly

among the far-wandering aquatic birds we have no less than five
genera which are more especially Palæarctic,—*Ortygometra,* the
corn-crake, and *Otis,* the great bustard, being typical examples.
We may add to these, several genera almost confined to this
region, such as *Garrulus* (jays), *Fringilla* (true finches), *Yunx*
(wrynecks) and some others ; so that in proportion to its total
generic forms a very large number are found to be peculiar or
characteristic.

This view, of the high degree of speciality of the Palæarctic
region, will no doubt be objected to by some naturalists, on the
ground that many of the genera reckoned as exclusively
Palæarctic are not so, but extend more or less into other regions.
It is well, therefore, to consider what principles should guide us
in a matter of this kind, especially as we shall have to apply
the same rules to each of the other regions. We may remark
first, that the limits of the regions themselves are, when not
formed by the ocean, somewhat arbitrary, depending on the
average distribution of a number of characteristic forms ; and
that slight local peculiarities of soil, elevation, or climate, may
cause the species of one region to penetrate more or less deeply
into another. The land boundary between two regions will be,
not a defined line but a neutral territory of greater or less
width, within which the forms of both regions will intermingle ;
and this neutral territory itself will merge imperceptibly into
both regions. So long therefore as a species or genus does not
permanently reside considerably beyond the possible limits of
this neutral territory, we should not claim it as an inhabitant
of the adjacent region. A consideration of perhaps more importance arises, from the varying extent of the range of a genus,
over the area occupied by the region. Some genera are represented by single species existing only in a very limited area ;
others by numerous species which occupy, entirely or very
nearly, the whole extent of the region ; and there is every intermediate grade between these extremes. Now, the small local-
ised genera, are always reckoned as among the best examples
of types peculiar to a region ; while the more wide-spread groups
are often denied that character if they extend a little beyond

the supposed regional limits, or send one or two, out of a large number of species, into adjacent regions; yet there is some reason to believe that the latter are really more important as characterising a zoological region than the former. In the case of a single isolated species or genus we have a dying-out group; and we have so many cases of discontinuous species of such groups (of which *Urotrichus* in Japan and British Columbia, *Eupetes* in Sumatra and New Guinea are examples), that it is quite as probable as not, that any such isolated species has only become peculiar to the region by the recent extinction of an allied form or forms in some other region. On the other hand, a genus consisting of numerous species ranging over an entire region or the greater part of one, is a dominant group, which has most likely been for some time extending its range, and whose origin dates back to a remote period. The slight extension of such a group beyond the limits of the region to which it mainly belongs, is probably a recent phenomenon, and in that case cannot be held in any degree to detract from its value as one of the peculiar forms of that region.

The most numerous examples of this class, are those birds of the temperate regions which in winter migrate, either wholly or partially, into adjacent warmer countries. This migration most likely began subsequent to the Miocene period, during that gradual refrigeration of the temperate zones which culminated in the glacial epoch, and which still continues in a mitigated form. Most of the genera, and many even of the species of birds which migrate southwards in winter, have therefore, most likely, always been inhabitants of our present Palæarctic and Nearctic regions; permanent residents during warm epochs, but only able now to maintain their existence by migration in winter. Such groups belong truly to the temperate zones, and the test of this is the fact of their not having any, or very few, representatives, which are permanent residents in the adjacent tropical regions. When there are such representative species, we do not claim them as peculiar to the Northern regions. Bearing in mind these various considerations, it will be found that we have been very moderate in our estimate of the number of genera that may fairly be considered as exclusively or characteristically Palæarctic.

Reptiles and Amphibia.—The Palæarctic region possesses, in proportion to its limited reptilian fauna, a full proportion of peculiar types. We have for instance two genera of snakes, *Rhinechis* and *Halys*; seven of lizards, *Trigonophis, Psammodromus, Hyalosaurus, Scincus, Ophiomorus, Megalochilus,* and *Phrynocephalus*; eight of tailed batrachians, *Proteus, Salamandra, Seiranota, Chioglossa, Hynobius, Onychodactylus, Geotriton,* and *Sieboldia;* and eight of tail-less batrachians, *Bombinator, Pelobates, Didocus, Alytes, Pelodytes, Discoglossus, Laprissa,* and *Latonia.* The distribution of these and other Palæarctic genera will be found in our second vol. chap. xix.

Freshwater Fish.—About twenty genera of freshwater fishes are wholly confined to this region, and constitute a feature which ought not to be overlooked in estimating its claim to the rank of a separate primary division of the earth. They belong to the following families:—Percidæ (three genera), *Acerina, Percarina, Aspro;* Comephoridæ (one genus), *Comephorus,* found only in Lake Baikal; Salmonidæ (three genera), *Brachymystax, Luciotrutta,* and *Plecoglossus;* Cyprinodontidæ (one genus), *Tellia,* found only in Alpine pools on the Atlas Mountains; Cyprinidæ (thirteen genera), *Cyprinus, Carassus, Paraphoxinus, Tinca, Achilognathus, Rhodeus, Chondrostoma, Pseudoperilampus, Ochetebius, Aspius, Alburnus, Misgurnus,* and *Nemachilus.*

Summary of Palæarctic Vertebrata.—Summarising these details, we find that the Palæarctic region possesses thirty-five peculiar genera of mammalia, fifty-seven of birds, nine of reptiles, sixteen of amphibia, and twenty-one of freshwater fishes; or a total of 138 peculiar generic types of vertebrata. Of these, 87 are mammalia and land-birds out of a total of 274 genera of these groups; or rather less than one-third peculiar, a number which will serve usefully to compare with the results obtained in other regions.

In our chapter on Zoological Regions we have already pointed out the main features which distinguish the Palæarctic from the Oriental and Ethiopian regions. The details now given will strengthen our view of their radical distinctness, by showing to how considerable an extent the former is inhabited by peculiar, and often very remarkable generic types.

Insects : Lepidoptera.—The Diurnal Lepidoptera, or butterflies, are not very abundant in species, their number being probably somewhat over 500,and these belong to not more than fifty genera. But no less than fifteen of these genera are wholly confined to the region. Nine of the families are represented, as follows :—
1. *Danaidæ;* having only a single species in South Europe.
2. *Satyridæ;* well represented, there being more than 100 species in Europe, and three peculiar genera. 3. *Nymphalidæ;* rather poorly represented, Europe having only about sixty species, but there is one peculiar genus. 4. *Libytheidæ;* a very small family, represented by a single species occurring in South Europe.
5. *Nemeobiidæ;* a rather small family, also having only one species in Europe, but which constitutes a peculiar genera.
6. *Lycænidæ;* an extensive family, fairly represented, having about eighty European species ; there are two peculiar genera in the Palæarctic region. 7. *Pieridæ;* rather poorly represented with thirty-two European species; two of the genera are, how-ever, peculiar. 8. *Papilionidæ;* very poorly represented in Europe with only twelve species, but there are many more in Siberia and Japan. No less than five of the small number of genera in this family are wholly confined to the region, a fact of much importance, and which to a great extent redeems the character of the Palæarctic region as regard this order of insects. Their names are *Mesapia, Hypermnestra, Doritis, Sericinus,* and *Thais;* and besides these we have *Parnassius*—the "Apollo" butterflies—highly characteristic, and only found elsewhere in the mountains of the Nearctic region. 9. *Hesperidæ;* poorly represented with about thirty European species, and one peculiar genus.

Four families of *Sphingina* occur in the Palæarctic region, and there are several peculiar genera.

In the *Zygænidæ* there are two exclusively European genera, and the extensive genus *Zygæna* is itself mainly Palæarctic. The small family *Stygiidæ* has two out of its three genera confined to the Palæarctic region. In the *Ægeriidæ* the genus *Ægeria* is mainly Palæarctic. The *Sphingidæ* have a wider general range, and none of the larger genera are peculiar to any one region.

Coleoptera.—The Palæarctic region is the richest portion of the globe in the great family of *Carabidæ*, or predacious ground-beetles, about 50 of the genera being confined to it, while many others, including the magnificent genus *Carabus,* have here their highest development. While several of the smaller genera are confined to the eastern or western sub-regions, most of the larger ones extend over the whole area, and give it an unmistakable aspect; while in passing from east to west or *vice-versâ*, allied species and genera replace each other with considerable regularity, except in the extreme south-east, where, in China and Japan, some Oriental forms appear, as do a few Ethiopian types in the south-west.

Cicindelidæ, or tiger-beetles, are but poorly represented by about 70 species of the genus *Cicindela,* and a single *Tetracha* in South Europe.

Lucanidæ, or stag-beetles, are also poor, there being representatives of 8 genera. One of these, *Æsalus* (a single species), is peculiar to South Europe, and two others, *Cladognathus* and *Cyclopthalmus,* are only represented in Japan, China, and Thibet.

Cetoniidæ, or rose-chafers, are represented by 13 genera, two of which are peculiar to South Europe (*Tropinota* and *Heterocnemis*), while *Stalagmosoma,* ranging from Persia to Nubia, and the fine *Dicranocephalus* inhabiting North China, Corea, and Nipal, may also be considered to belong to it. The genera *Trichius, Gnorimus,* and *Osmoderma* are confined to the two north temperate regions.

Buprestidæ, or metallic beetles, are rather abundant in the warmer parts of the region, 27 genera being represented, nine of which are peculiar. By far the larger portion of these are confined to the Mediterranean sub-region. A considerable number also inhabit Japan and China.

The Longicorns, or long-horned beetles, are represented by no less than 196 genera, 51 of which are peculiar. They are much more abundant in the southern than the northern half of the region. Several Oriental genera extend to Japan and North China, and a few Ethiopian genera to North Africa. Thirteen genera are confined to the two north temperate regions. Several large genera, such as *Dorcadion* (154 species), *Phytœcia* (85 species), *Pogonochœrus* (22 species), *Agapanthia* (22 species), and *Vesperus* (7 species), are altogether peculiar to the Palæarctic region; and with a preponderance of *Leptura*, *Grammoptera*, *Stenocorus*, and several others, strongly characterise it as distinct from the Nearctic and Oriental regions.

The other families which are well developed in the Palæarctic regions, are, the Staphylinidæ or rove-beetles, Silphidæ or burying-beetles, Histeridæ or mimic-beetles, Nitidulidæ, Aphodiidæ, Copridæ (especially in South Europe), Geotrupidæ or dung-beetles, Melolonthidæ or chafers, Elateridæ or click-beetles, the various families of Malacoderms and Heteromera, especially Pimeliidæ in the Mediterranean sub-region, Curculionidæ or weevils, the Phytophaga or leaf-eaters, and Cocinellidæ or lady-birds.

The number of species of Coleoptera in the western part of the Palæarctic region is about 15,000, and there are probably not more than 2,000 to add to this number from Siberia, Japan, and North China; but were these countries as well explored as Europe, we may expect that they would add at least 5,000 to the number above given, raising the Palæarctic Coleopterous fauna to 20,000 species. As the total number of species at present known to exist in collections is estimated (and perhaps somewhat over-estimated) at 70,000 species, we may be sure that were the whole earth as thoroughly investigated as Europe, the number would be at least doubled, since we cannot suppose that Europe, with the Mediterranean basin, can contain more than one-fifth of the whole of the Coleoptera of the globe.

Of the other orders of insects we here say nothing, because in their case much more than in that of the Coleoptera and Lepidoptera, is the disproportion enormous between our knowledge of the European fauna and that of almost all the rest of the globe.

They are, therefore, at present of comparatively little use for purposes of geographical distribution, even were it advisable to enter into the subject in a work which will, perhaps, be too much overburdened with details only of interest to specialists.

Land Shells.—These are very numerous in the warmer parts of the region, but comparatively scarce towards the North. South Europe alone possesses over 600 species, whereas there are only 200 in all Northern Europe and Asia. The total number of species in the whole region is probably about 1,250, of which the great majority are Helicidæ; the Operculated families being very poorly represented. Several small genera or sub-genera are peculiar to the region, as *Testacella* (West Europe and Canaries); *Leucochroa* (Mediterranean district); *Acicula* (Europe); *Craspedopoma* (Atlantic Islands); *Leonia* (Algeria and Spain); *Pomatias* (Europe and Canaries); *Cecina* (Mongolia). The largest genera are *Helix* and *Clausilia*, which together comprise more than half the species; *Pupa*, very numerous; *Bulimus* and *Achatina* in moderate numbers, and all the rest small *Helix* is the only genus which contains large and handsome species; *Bulimus* and *Achatina*, so magnificent in tropical countries, being here represented by small and obscure forms only. *Daudebardia* is confined to Central and South Europe and New Zealand; *Glandina* is chiefly South American; *Hyalina* is only American and European; *Buliminus* ranges over all the world except America; and the other European genera of Helicidæ are widely distributed. Of the Operculata, *Cyclotus*, *Cyclophorus*, and *Pupina* extend from the Oriental region into Japan and North China; *Tudoria* is found in Algeria and the West Indies; *Hydrocena* is widely scattered, and occurs in South Europe and Japan. The genera of freshwater shells are all widely distributed.

THE PALÆARCTIC SUB-REGIONS.

The four sub-regions which are here adopted, have been fixed upon as those which are, in the present state of our knowledge, at once the most natural and the only practicable ones.

No doubt all of them could be advantageously again sub-divided, in a detailed study of the geographical distribution of *species*. But in a general work, which aims at treating all parts of the world with equal fulness, and which therefore is confined almost wholly to the distribution of families and genera, such further subdivision would be out of place. It is even difficult, in some of the classes of animals, to find peculiar or even characteristic genera for the present sub-regions; but they all have well marked climatic and physical differences, and this leads to an assemblage of species and of groups which are suffi-ciently distinctive.

I. Central and Northern Europe.

This sub-region, which may perhaps be termed the "European," is zoologically and botanically the best known on the globe. It can be pretty accurately defined, as bounded on the south by the Pyrenees, the Alps, the Balkans, the Black Sea, and the Caucasus range; and by the Ural Mountains, or perhaps more correctly the valley of the Irtish and Caspian Sea, on the east; while Ireland and Iceland are its furthest outliers in the west. To the north, it merges so gradually into the Arctic zone that no demarcation is possible. The great extent to which this sub-region is interpenetrated by the sea, and the prevalence of westerly winds bringing warmth and moisture from an ocean influenced by the gulf-stream, give it a climate for the most part genial, and free from extremes of heat and cold. It is thus broadly distinguished from Siberia and Northern Asia generally, where a more extreme and rigorous climate prevails.

The whole of this sub-region is well watered, being pene-trated by rivers in every direction; and it consists mainly of plains and undulating country of moderate elevation, the chief mountain ranges being those of Scandinavia in the north-west, and the extensive alpine system of Central Europe. But these are both of moderate height, and a very small portion of their surface is occupied either by permanent snow-fields, or by barren uplands inimical to vegetable and animal life. It is, in fact, to these, and the numerous lesser mountains and hills which everywhere diversify the surface of Europe, that the variety and abundance of its animal life is greatly due. They afford the perennial supplies to rivers, and furnish in their valleys and ever varying slopes, stations suited to every form of existence. A considerable area of Central Europe is occupied by uplands of moderate elevation, a comparatively small portion being flat and marshy plains.

Most of the northern and much of the central portions of Europe are covered with vast forests of coniferous trees; and these, occupying as they do those tracts where the winter is most severe, supply food and shelter to many animals who could not otherwise maintain their existence. It is probable that the original condition of the greater part, if not the whole, of temperate Europe, except the flat marshes of the river valleys and the sandy downs of the coast, was that of woodland and forest, mostly of deciduous trees, but with a plentiful admixture of such hardy evergreens as holly, ivy, privet, and yew. A sufficient proportion of these primeval woods, and of artificial plantations which have replaced them, fortunately remain, to preserve for us most of the interesting forms of life, which were developed before man had so greatly modified the surface of the earth, and so nearly exterminated many of its original tenants. Almost exactly in proportion to the amount of woodland that still remains in any part of Europe, do we find (other things being equal) the abundance and variety of wild animals; a pretty clear indication that the original condition of the country was essentially that of a forest, a condition which only now exists in the thinly inhabited regions of the north.

Although the sub-region we are considering is, for its extent and latitude, richly peopled with animal life, the number of genera altogether peculiar to it is not great. There are, however, several which are very characteristic, and many species, both of the smaller mammalia and of birds, are wholly restricted to it.

Mammalia.—The genera wholly confined to this sub-region are only two. *Myogale*, the desman, is a curious long-snouted Insectivorous animal somewhat resembling the water-rat in its habits. There are two species, one found only on the banks of streams in the French Pyrenees, the other on the great rivers of Southern Russia. The other peculiar genus, *Rupicapra* (the chamois of the Alps), is found on all the high mountains of Central Europe. Almost peculiar are *Spalax* (the mole-rat) found only in Eastern Europe and Western Siberia; and *Saiga*, an extraordinary large-nosed antelope which has a nearly similar distribution. Highly characteristic forms, which inhabit nearly every part of the sub-region, are, *Talpa* (the mole), *Erinaceus*, (the hedgehog), *Sorex* (the shrew), *Meles* (the badger), *Ursus* (the bear), *Canis* (the wolf and fox), *Mustela* (the weasel), *Lutra* (the otter), *Arvicola* (the vole), *Myoxus* (the dormouse), and *Lepus* (the hare and rabbit) ; while *Bos* (the wild bull) was, until exterminated by man, no doubt equally characteristic. Other genera inhabiting the sub-region will be found in the list given at the end of this chapter.

Birds.—It is difficult to name the birds that are most characteristic of this sub-region, because so many of the most familiar and abundant are emigrants from the south, and belong to groups that have a different range. There is perhaps not a single genus wholly confined to it, and very few that have not equal claims to be placed elsewhere. Among the more characteristic we may name *Turdus* (the thrushes), *Sylvia* (the warblers), *Panurus* (the reedling) *Parus* (the tits), *Anthus* (the pipits), *Motacilla* (the wagtails), which are perhaps more abundant here than in any other part of the world, *Emberiza* (the buntings), *Plectrophanes* (the snow buntings), *Passer* (the house sparrows), *Loxia* (the crossbills), *Linota* (the linnets), *Pica* (the magpies), *Tetrao* (grouse), *Lagopus* (ptarmigan) and many others.

I am indebted to Mr. H. E. Dresser, who is personally acquainted with the ornithology of much of the North of Europe, for some valuable notes on the northern range of many European birds. Those which are characteristic of the extreme Arctic zone, extending beyond 70° north latitude, and tolerably abundant, are two falcons (*Falco gyrfalco* and *F. peregrinus*) ;

O

the rough-legged buzzard (*Archibuteo lagopus*); the snowy owl
(*Nyctea scandiaca*); the raven (*Corvus corax*); three buntings
(*Emberiza schœniculus, Plectrophanes nivalis* and *P. calcarata*); a
lark (*Otocorys alpestris*); several pipits, the most northern being
Anthus cervinus; a wagtail (*Budytes cinereocapilla*); a dipper
(*Cinclus melanogaster*); a warbler (*Cyanecula suecica*); the
wheatear (*Saxicola œnanthe*); and two ptarmigans (*Lagopus
albus* and *L. salicetus*). Most of these birds are, of course, only
summer visitors to the Arctic regions, the only species noted as
a permanent resident in East Finmark (north of latitude 70°)
being the snow-bunting (*Plectrophanes nivalis*).

The birds that are characteristic of the zone of pine forests,
or from about 61° to 70° north latitude, are very numerous, and
it will be sufficient to note the genera and the number of
species (where more than one) to give an idea of the ornithology of this part of Europe. The birds of prey are, *Falco* (three
species), *Astur* (two species), *Buteo, Pandion, Surnia, Bubo,
Syrnium, Asio, Nyctala*. The chief Passerine birds are, *Corvus*
(two species), *Pica, Garrulus* (two species), *Nucifraga, Bomby-
cilla, Hirundo* (two species), *Muscicapa* (two species), *Lanius,
Sturnus, Passer* (two species), *Pyrrhula, Carpodacus, Loxia* (two
species), *Pinicola, Fringilla* (eight species), *Emberiza* (five
species), *Alauda, Anthus, Turdus* (five species), *Ruticilla, Pratincola, Accentor, Sylvia* (four species), *Hypolais, Regulus, Phylloscopus*
(two species), *Acrocephalus, Troglodytes*, and *Parus* (six species).
Woodpeckers are abundant, *Picus* (four species), *Gecinus*, and
Yunx. The kingfisher (*Alcedo*), goatsucker (*Caprimulgus*), and
swift (*Cypselus*) are also common. The wood-pigeon (*Columba*)
is plentiful. The gallinaceous birds are three grouse, *Tetrao* (two
species) and *Bonasa*, and the common quail (*Coturnix*).

The remaining genera and species of temperate or north-
European birds, do not usually range beyond the region of
deciduous trees, roughly indicated by the parallel of 60° north
latitude.

Plate I.— Illustrating the Zoology of Central Europe.—
Before considering the distribution of the other classes of
vertebrata, it will be convenient to introduce our first illustra-

PLATE I.

THE ALPS OF CENTRAL EUROPE, WITH CHARACTERISTIC ANIMALS.

tion, which represents a scene in the Alps of Central Europe, with figures of some of the most characteristic Mammalia and Birds of this sub-region. On the left is the badger (*Meles Taxus*) one of the weasel family, and belonging to a genus which is strictly Palæarctic. It abounds in Central and Northern Europe and also extends into North Asia, but is represented by another species in Thibet and by a third in Japan. The elegantly-formed creatures on the right are chamois (*Rupicapra tragus*), almost the only European antelopes, and wholly confined to the higher mountains, from the Pyrenees to the Carpathians and the Caucasus. The chamois is the only species of the genus, and is thus perhaps the most characteristic European mammal. The bird on the left, above the badgers, is the Alpine chough, (*Fregilus pyrrhocorax*). It is found in the high mountains from the Alps to the Himalayas, and is allied to the Cornish chough, which is still found on our south-western coasts, and which ranges to Abyssinia and North China. The Alpine chough differs in having a shorter bill of an orange colour, and vermilion red feet as in the other species. In the foreground are a pair of ruffs (*Machetes pugnax*) belonging to the Scolopacidæ or snipe family, and most nearly allied to the genus *Tringa* or sandpiper. This bird is remarkable for the fine collar of plumes which adorns the males in the breeding season, when they are excessively pugnacious. It is the only species of its genus, and ranges over all Europe and much of Northern Asia, migrating in the winter to the plains of India, and even down the east coast of Africa as far as the Cape of Good Hope; but it only breeds in the Palæarctic region, over the greater part of which it ranges.

Reptiles and Amphibia.—There are no genera of reptiles peculiar to this sub-region. Both snakes and lizards are comparatively scarce, there being about fourteen species of the former and twelve of the latter. Our common snake (*Tropidonotus natrix*) extends into Sweden and North Russia, but the viper (*Viperus berus*) goes further north, as far as Archangel (64° N.), and in Scandinavia (67° N.), and is the most Arctic of all known

snakes. Of the lizards, *Lacerta stirpium* (the sand lizard) has
the most northerly range, extending into Poland and Northern
Russia; and *Anguis fragilis* (the blind or slow-worm) has almost
an equal range.

Amphibia, being more adapted to a northern climate, have
acquired a more special development, and thus several forms
are peculiar to the North European sub-region. Most remarkable
is *Proteus*, a singular eel-like aquatic creature with small legs,
found only in the subterranean lakes in Carniola and Carinthia;
Alytes, a curious toad, the male of which carries about the eggs
till they are hatched, found only in Central Europe from
France to the east of Hungary; and *Pelodytes*, a frog found only
in France. Frogs and toads are very abundant all over Europe, the
common frog (*Rana temporaria*) extending to the extreme north.
The newts (*Triton*) are also very abundant and widely spread,
though not ranging. so far north as the frogs. The genera *Bombinator* (a toad-like frog), and *Hyla* (the tree frog) are also common in Central Europe.

Freshwater Fish.—Two genera of the perch family (Percidæ) are
peculiar to this sub-region,—*Percarina*, a fish found only in the
river Dniester, and *Aspro*, confined to the rivers of Central
Europe. Of the very characteristic forms are, *Gasterosteus*
(stickle-back), which alone forms a peculiar family—Gasteros-
teidæ; *Perca*, *Acerina* and *Lucioperca*, genera of the perch family;
Silurus, a large fish found in the rivers of Cenrtal Europe, of
the family Siluridæ; *Esox* (the pike), of the family Esocidæ;
Cyprinus (carp), *Gobio* (gudgeon), *Leuciscus* (roach, chub, dace,
&c.), *Tinca* (tench), *Abramus* (bream), *Alburnus* (bleak), *Cobitis*
(loach), all genera of the family Cyprinidæ.

Insects—Lepidoptera.—No genera of butterflies are actually
confined to this sub-region, but many are characteristic of it.
Parnassius, *Aporia*, *Leucophasia*, *Colias*, *Melitœa*, *Argynnis*,
Vanessa, *Limenitis*, and *Chionobas*, are all very abundant and
widespread, and give a feature to the entomology of most of the
countries included in it.

Coleoptera.—This sub-region is very rich in Carabidæ; the
genera *Elaphrus*, *Nebria*, *Carabus*, *Cychrus*, *Pterostichus*, *Amara*,

Trechus and *Peryphus* being especially characteristic. Staphylinidæ abound. Among Lamellicorns the genus *Aphodius* is most characteristic. Buprestidæ are scarce ; Elateridæ more abundant. Among Malacoderms *Telephorus* and *Malachius* are characteristic. Curculionidæ abound : *Otiorhyuchus, Omias, Erirhinus, Bagous, Rhynchites* and *Ceutorhynchus* being very characteristic genera. Of Longicorns *Callidium, Dorcadion, Pogonochœrus, Pachyta* and *Leptura* are perhaps the best representatives. *Donacia, Crioceris, Chrysomela,* and *Altica,* are typical Phytophaga; while *Coccinella* is the best representative of the Securipalpes.

North European Islands.—The British Islands are known to have been recently connected with the Continent, and their animal productions are so uniformly identical with continental species as to require no special note. The only general fact of importance is, that the number of species in all groups is much less than in continental districts of equal extent, and that this number is still farther diminished in Ireland. This may be accounted for by the smaller area and less varied surface of the latter island ; and it may also be partly due to the great extent of low land, so that a very small depression would reduce it to the condition of a cluster of small islands capable of supporting a very limited amount of animal life. Yet further, if after such a submergence had destroyed much of the higher forms of life in Great Britain and Ireland, both were elevated so as to again form part of the Continent, a migration would commence by which they would be stocked afresh ; but this migration would be a work of time, and it is to be expected that many species would never reach Ireland or would find its excessively moist climate unsuited to them.

Some few British species differ slightly from their continental allies, and are considered by many naturalists to be distinct. This is the case with the red grouse (*Lagopus scoticus*) among birds ; and a few of the smaller Passeres have also been found to vary somewhat from the allied forms on the Continent, showing that the comparatively short interval since the glacial period, and the slightly different physical conditions dependent on insularity, have sufficed to commence the work of specific modification.   There are also a few small land-shells and several insects not yet found elsewhere than in Britain ; and even one of the smaller Mammalia—a shrew (*Sorex rusticus*).   These facts are all readily explained by the former union of these islands with the Continent, and the alternate depressions and elevations which are proved by geological evidence to have occurred, by which they have been more than once separated and united again in recent times.   For the evidence of this elevation and depression, the reader may consult Sir Charles Lyell's *Antiquity of Man.*

Iceland is the only other island of importance belonging to this sub-region, and it contrasts strongly with Great Britain, both in its Arctic climate and oceanic position.   It is situated just south of the Arctic circle and considerably nearer Greenland than Europe, yet its productions are almost wholly European. The only indigenous land mammalia are the Arctic fox (*Canis lagopus*), and the polar bear as an occasional visitant, with a mouse (*Mus islandicus*), said to be of a peculiar species.   Four species of seals visit its shores.   The birds are more interesting. According to Professor Newton, ninety-five species have been observed ; but many of these are mere stragglers.   There are twenty-three land, and seventy-two aquatic birds and waders. Four or five are peculiar species, though very closely related to others inhabiting Scandinavia or Greenland.   Only two or three species are more nearly related to Greenland birds than to those of Northern Europe, so that the Palæarctic character of the fauna is unmistakable.   The following lists, compiled from a paper by Professor Newton, may be interesting as showing more exactly the character of Icelandic ornithology.

1. Peculiar species. —*Troglodytes borealis* (closely allied to the common wren, found also in the Faroe Islands); *Falco islandicus* (closely allied to *F. gyrfalco*); *Lagopus islandorum* (closely allied to *L. rupestris* of Greenland).

2. European species resident in Iceland.—*Emberiza nivalis, Corvus corax, Haliæetus albicilla, Rallus aquaticus, Hæmatopus ostralegus, Cygnus ferus, Mergus* (two species), *Phalacocorax* (two

species), *Sula bassana*, *Larus* (two species), *Stercorarius catarractes*, *Puffinus anglorum*, *Mergulus alle*, *Uria* (three species),
Alca torda.

3. American species resident in Iceland.—*Clangula islandica,
Histrionicus torquatus*.

4. Annual visitants from Europe.—*Turdus iliacus*, *Ruticilla
tithys*, *Saxicola œnanthe*, *Motacilla alba*, *Anthus pratensis*, *Linota
linaria*, *Chelidon urbica*, *Hirundo rustica*, *Falco œsalon*, *Surnia
nyctea*, *Otus brachyotus*, *Charadrius pluvialis*, *Ægialites hiaticula*,
Strepsilas interpres, *Phalaropus fulicarius*, *Totanus calidris*,
Limosa (species), *Tringa* (three species), *Calidris arenaria*,
Gallinago media, *Numenius phœopus*, *Ardea cinerea*, *Anser* (two
species), *Bernicla* (two species), *Anas* (four species), *Fuligula
marila*, *Harelda glacialis*, *Somateria mollissima*, *Œdemia nigra*,
Sterna macrura, *Rissa tridactyla*, *Larus luecopterus*, *Stercorarius*
(two species), *Fratercula artica*, *Colymbus* (two species), *Podi-
ceps cornutus*.

5. Annual visitant from Greenland.—*Falco candicans*.

6.—Former resident, now extinct.—*Alca impennis* (the
great auk).

II.—*Mediterranean Sub-region.*

This is by far the richest portion of the Palæarctic region,
for although of moderate extent much of it enjoys a climate in
which the rigours of winter are almost unknown. It includes
all the countries south of the Pyrenees, Alps, Balkans, and
Caucasus mountains; all the southern shores of the Mediterranean to the Atlas range, and even beyond it to include the
extra-tropical portion of the Sahara; and in the Nile valley
as far as the second cataract. Further east it includes the
northern half of Arabia and the whole of Persia, as well as
Beluchistan, and perhaps Affghanistan up to the banks of the
Indus. This extensive district is almost wholly a region of
mountains and elevated plateaus. On the west, Spain is
mainly a table-land of more than 2000 feet elevation, deeply
penetrated by extensive valleys and rising into lofty mountain chains. Italy, Corsica, Sardinia, and Sicily, are all very

mountainous, and much of their surface considerably elevated. Further east we have all European Turkey and Greece, a mountain region with a comparatively small extent of level plain. In Asia the whole country, from Smyrna through Armenia and Persia to the further borders of Affghanistan, is a vast mountainous plateau, almost all above 2000, and extensive districts above 5000 feet in elevation. The only large tract of low-land is the valley of the Euphrates. There is also some low-land south of the Caucasus, and in Syria the valley of the Jordan. In North Africa the valley of the Nile and the coast plains of Tripoli and Algiers are almost the only exceptions to the more or less mountainous and plateau-like character of the country. Much of this extensive area is now bare and arid, and often even of a desert character; a fact no doubt due, in great part, to the destruction of aboriginal forests. This loss is rendered permanent by the absence of irrigation, and, it is also thought, by the abundance of camels and goats, animals which are exceedingly injurious to woody vegetation, and are able to keep down the natural growth of forests. Mr. Marsh (whose valuable work *Man and Nature* gives much information on this subject) believes that even large portions of the African and Asiatic deserts would become covered with woods, and the climate thereby greatly improved, were they protected from these destructive domestic animals, which are probably not indigenous to the country. Spain, in proportion to its extent, is very barren; Italy and European Turkey are more woody and luxuriant; but it is perhaps in Asia Minor, on the range of the Taurus, along the shores of the Black Sea, and to the south of the Caucasus range, that this sub-region attains its maximum of luxuriance in vegetation and in animal life. From the Caspian eastward extends a region of arid plains and barren deserts, diversified by a few more fertile valleys, in which the characteristic flora and fauna of this portion of the Palæarctic region abounds. Further east we come to the forests of the Hindoo Koosh, which probably form the limit of the sub-region. Beyond these we enter on the Siberian sub-region to the north, and on the outlying portion of the Oriental region on the south.

In addition to the territories now indicated as forming part of the Mediterranean sub-region, we must add the group of Canary Islands off the west coast of Africa which seem to be an extension of the Atlas mountains, and the oceanic groups of Madeira and the Azores; the latter about 1,000 miles from the continent of Europe, yet still unmistakably allied to it both in their vegetable and animal productions. The peculiarities of the faunas of these islands will be subsequently referred to.

It seems at first sight very extraordinary, that so large and wide a sea as the Mediterranean should not separate distinct faunas, and this is the more remarkable when we find how very deep the Mediterranean is, and therefore how ancient we may well suppose it to be. Its eastern portion reaches a depth of 2,100 fathoms or 12,600 feet, while its western basin is about 1,600 fathoms or 9,600 feet in greatest depth, and a considerable area of both basins is more than 1,000 fathoms deep. But a further examination shows, that a comparatively shallow sea or submerged bank incloses Malta and Sicily, and that on the opposite coast a similar bank stretches out from the coast of Tripoli leaving a narrow channel the greatest depth of which is 240 fathoms. Here therefore is a broad plateau, which an elevation of about 1,500 feet would convert into a wide extent of land connecting Italy with Africa; while the same elevation would also connect Morocco with Spain, leaving two extensive lakes to represent what is now the Mediterranean Sea, and affording free communication for land animals between Europe and North Africa. That such a state of things existed at a comparatively recent period, is almost certain; not only because a considerable number of identical *species* of mammalia inhabit the opposite shores of the Mediterranean, but also because numerous remains of three species of elephants have been found in caves in Malta,—now a small rocky island in which it would be impossible for such animals to live even if they could reach it. Remains of hippopotami are also found at Gibraltar, and many other animals of African types in Greece; all indicating means of communication between South Europe and North Africa which no longer exist. (See Chapter VI. pp. 113—115.)

Mammalia.—There are a few groups of Palæarctic Mammalia that are peculiar to this sub-region. Such are, *Dama*, the fallow deer, which is now found only in South Europe and North Africa; *Psammomys*, a peculiar genus of Muridæ, found only in Egypt and Palestine; while *Ctenodactylus*, a rat-like animal classed in the South American family Octodontidæ, inhabits Tripoli. Among characteristic genera not found in other sub-regions, are, *Dysopes*, a bat of the family Noctilionidæ; *Macroscelides*, the elephant shrew, in North Africa; *Genetta*, the civet, in South Europe; *Herpestes*, the ichneumon, in North Africa and (?) Spain; *Hyæna*, in South Europe; *Gazella, Oryx, Alcephalus*, and *Addax*, genera of antelopes in North Africa and Palestine; *Hyrax*, in Syria: and *Hystrix*, the porcupine, in South Europe. Besides these, the camel and the horse were perhaps once indigenous in the eastern parts of the sub-region; and a wild sheep (*Ovis musmon*) still inhabits Sardinia, Corsica, and the mountains of the south-east of Spain. The presence of the large feline animals—such as the lion, the leopard, the serval, and the hunting leopard—in North Africa, together with several other quadrupeds not found in Europe, have been thought by some naturalists to prove, that this district should not form part of the Palæarctic region. No doubt several Ethiopian groups and species have entered it from the south, but the bulk of its Mammalia still remains Palæarctic, although several of the species have Asiatic rather than European affinities. The *Macacus innuus* is allied to an Asiatic rather than an African group of monkeys, and thus denotes an Oriental affinity. Ethiopian affinity is apparently shown by the three genera of antelopes, by *Herpestes*, and by *Macroscelides;* but our examination of the Miocene fauna has shown that these were probably derived from Europe originally, and do not form any part of the truly indigenous or ancient Ethiopian fauna. Against these, however, we have the occurrence in North Africa of such purely Palæarctic and non-Ethiopian genera as *Ursus, Meles, Putorius, Sus, Cervus, Dama, Capra, Alactaga;* together with actual European or West Asiatic species of *Canis, Genetta, Felis, Putorius, Lutra,* many bats, *Sorex, Crocidura, Crossopus, Hystrix,*

Dipus, Lepus, and *Mus.* It is admitted that, as regards every other group of animals, North Africa is Palæarctic, and the above enumeration shows that even in Mammalia, the intermixture of what are now true Ethiopian types is altogether insignificant. It must be remembered, also, that the lion inhabited Greece even in historic times, while large carnivora were contemporary with man all over Central Europe.

Birds.—So many of the European birds migrate over large portions of the region, and so many others have a wide permanent range, that we cannot expect to find more than a few genera, consisting of one or two species, each, confined to a sub-region; and such appear to be, *Lusciniola* and *Pyrophthalma,* genera of Sylviidæ. But many are characteristic of this, as compared with other Palæarctic sub-regions; such as, *Bradyptetus, Aedon, Dromolœa,* and *Cercomela,* among Sylviidæ; *Crateropus* and *Malacocercus,* among Timaliidæ; *Telophonus* among Laniidæ; *Certhilauda* and *Mirafra* among larks; *Pastor* among starlings; *Upupa,* the hoopoe; *Halycon* and *Ceryle* among kingfishers; *Turnix* and *Caccabis* among Gallinæ, and the pheasant as an indigenous bird; together with *Gyps, Vultur* and *Neophron,* genera of vultures. In addition to these, almost all our summer migrants spend their winter in some part of this favoured land, mostly in North Africa, together with many species of Central Europe that rarely or never visit us. It follows, that a large proportion of all the birds of Europe and Western Asia are to be found in this sub-region, as will be seen by referring to the list of the genera of the region. Palestine is one of the remote portions of this region which has been well explored by Canon Tristram, and it may be interesting to give his summary of the range of the birds. We must bear in mind that the great depression of the Dead Sea has a tropical climate, which accounts for the presence here only, of such a tropical form as the sun-bird (*Nectarinea osea*).

The total number of the birds of Palestine is 322, and of these no less than 260 are European, at once settling the question of the general affinities of the fauna. Of the remainder eleven belong to North and East Asia, four to the Red Sea, and thirty-

one to East Africa, while twenty-seven are peculiar to Palestine.
It is evident therefore that an unusual number of East African
birds have extended their range to this congenial district, but
most of these are desert species and hardly true Ethiopians,
and do not much interfere with the general Palæarctic character
of the whole assemblage. As an illustration of how wide-spread
are many of the Palæarctic forms, we may add, that seventy-
nine species of land birds and fifty-five of water birds, are common to Palestine and Britain. The Oriental and Ethiopian
genera *Pycnonotus* and *Nectarinea* are found here, while *Bessornis*
and *Dromolœa* are characteristically Ethiopian. Almost all the
other genera are Palæarctic.

Persia is another remote region generally associated with the
idea of Oriental and almost tropical forms, but which yet undoubtedly belongs to the Palæarctic region. Mr. Blanford's recent
collections in this country, with other interesting information, is
summarised in Mr. Elwes's paper on the " Geographical Distribution of Asiatic Birds " (*Proc. Zool. Soc.* 1873, p. 647). No less
than 127 species are found also in Europe, and thirty-seven
others belong to European genera; seven are allied to birds of
Central Asia or Siberia, and fifteen to those of North-East Africa,
while only three are purely of Indian affinities. This shows a
preponderance of nearly nine-tenths of Palæarctic forms, which
is fully as much as can be expected in any country near the
limits of a great region.

Reptiles and Amphibia.—The climatal conditions being here
more favourable to these groups, and the genera being often of
limited range, we find some peculiar, and several very interesting
forms. *Rhinechis,* a genus of Colubrine snakes, is found only in
South Europe; *Trogonophis,* one of the Amphisbænians—
curious snake-like lizards—is known only from North Africa;
Psammosaurus, belonging to the water lizards (Varanidæ) is
found in North Africa and North-West India; *Psammodromus,*
a genus of Lacertidæ, is peculiar to South Europe; *Hyalosaurus,*
belonging to the family Zonuridæ, is a lizard of especial interest, as it inhabits North Africa while its nearest ally is the
Ophisaurus or " glass snake " of North America; the family of

the sciuks is represented by *Scincus* found in North Africa and Arabia. Besides these *Seps*, a genus of sand lizards (Sepidæ) and *Agama*, a genus of Agamidæ, are abundant and characteristic.

Of Amphibia we have *Seiranota*, a genus of salamanders found only in Italy and Dalmatia; *Chioglossa*, in Portugal, and *Geotriton*, in Italy, belonging to the same family, are equally peculiar to the sub-region.

Freshwater Fish.—One of the most interesting is *Tellia*, a genus of Cyprinodontidæ found only in alpine pools in the Atlas mountains. *Paraphoxinius*, found in South-East Europe, and *Chondrostoma*, in Europe and Western Asia, genera of Cyprimidæ, seem almost peculiar to this sub-region.

Insects—Lepidoptera.—Two genera of butterflies, *Thais* and *Doritis*, are wholly confined to this sub-region, the former ranging over all Southern Europe, the latter confined to Eastern Europe and Asia Minor. *Anthocharis* and *Zegris* are very characteristic of it, the latter only extending into South Russia, while *Danais*, *Charaxes*, and *Libythea* are tropical genera unknown in other parts of Europe.

Coleoptera.—This sub-region is very rich in many groups of Coleoptera, of which a few only can be noticed here. Among Carabidæ it possesses *Procerus* and *Procrustes*, almost exclusively, while *Brachinus*, *Cymindis*, *Lebia*, *Graphipterus*, *Scarites*, *Chlœnius*, *Calathus*, and many others, are abundant and characteristic. Among Lamellicorns—Copridæ, Glaphyridæ, Melolonthidæ, and Cetoniidæ abound. Buprestidæ are plentiful, the genera *Julodis*, *Acmœodera*, *Buprestis*, and *Sphenoptera* being characteristic. Among Malacoderms—Cebrionidæ, Lampyridæ, and Malachiidæ abound. The Tenebrioid Heteromera are very varied and abundant, and give a character to the sub-region. The Mylabridæ, Cantharidæ, and Œdemeridæ are also characteristic. Of the immense number of Curculionidæ—*Thylacites*, *Brachycerus*, *Lixus*, and *Acalles* may be mentioned as among the most prominent. Of Longicorns there are few genera especially characteristic, but perhaps *Prinobius*, *Purpuricenus*, *Hesperophanes*, and *Parmena* are most so. Of the remaining families, we may mention Clythridæ, Hispidæ, and Cassididæ as being abundant.

The Mediterranean and Atlantic Islands.—The various islands
of the Mediterranean are interesting to the student of geographical distribution as affording a few examples of local species
of very restricted range, but as a rule they present us with
exactly the same forms as those of the adjacent mainland.[1]
Their peculiarities do not, therefore, properly come within the
scope of this work. The islands of the Atlantic Ocean belonging to this sub-region are, from their isolated position and the
various problems they suggest, of much more interest, and their
natural history has been carefully studied. We shall therefore
give a short account of their peculiar features.

Of the three groups of Atlantic islands belonging to this sub-
region, the Canaries are nearest to the Continent, some of the
islands being only about fifty miles from the coast of Africa.
They are, however, separated from the mainland by a very deep
channel (more than 5,000 feet), as shown on our general map.
The islands extend over a length of 300 miles ; they are very
mountainous and wholly volcanic, and the celebrated peak of
Teneriffe rises to a height of more than 12,000 feet. The small
Madeira group is about 400 miles from the coast of Morocco
and 600 from the southern extremity of Portugal; and there is
a depth of more than 12,000 feet between it and the continent.
The Azores are nearly 1,000 miles west of Lisbon. They are
quite alone in mid-Atlantic, the most westerly islands being
nearer Newfoundland than Europe, and are surrounded by ocean
depths of from 12,000 to 18,000 feet. It will be convenient to
take these islands first in order.

Azores.—Considering the remoteness of this group from every
other land, it is surprising to find as many as fifty-three species
of birds inhabiting or visiting the Azores; and still more to

---

[1] Malta is interesting as forming a resting-place for migratory birds, while
crossing the Mediterranean. It has only eight land and three aquatic birds
which are permanent residents ; yet no less than 278 species have been
recorded by Mr. E. A. Wright as visiting or passing over it, comprising a
large proportion of the European migratory birds. The following are the
permanent residents : *Cerchneis tinnunculus, Strix flammea, Passer salicicola,
Emberiza miliaria, Corvus monedula, Monticola cyanea, Sylvia conspicillata,
Columba livia, Puffinus cinereus, P. anglorum, Thalassidroma pelagica.*

find that they are of Palæarctic genera and, with one exception, all of species found either in Europe, North Africa, Madeira, or the Canaries. The exception is a bullfinch peculiar to the islands, but closely allied to a European species. Of land birds there are twenty-two, belonging to twenty-one genera, all European. These genera are *Cerchneis, Buteo, Asio, Strix, Turdus, Oriolus, Erithacus, Sylvia, Regulus, Saxicola, Motacilla, Plectrophanes, Fringilla, Pyrrhula, Serinus, Sturnus, Picus, Upupa, Columba, Caccabis,* and *Coturnix.* Besides the bullfinch (*Pyrrhula*) other species show slight differences from their European allies, but not such as to render them more than varieties. The only truly indigenous mammal is a bat of a European species. Nine butterflies inhabit the Azores; eight of them are European species, one North American. Of beetles 212 have been collected, of which no less than 175 are European species; of the remainder, nineteen are found in the Canaries or Madeira, three in South America, while fourteen are peculiar to the islands.

Now these facts (for which we are indebted to Mr. Godman's *Natural History of the Azores*) are both unexpected and exceedingly instructive. In most other cases of remote Oceanic islands, a much larger proportion of the fauna is endemic, or consists of peculiar species and often of peculiar genera; as is well shown by the case of the Galapagos and Juan Fernandez, both much nearer to a continent and both containing peculiar genera and species of birds. Now we know that the cause and meaning of this difference is, that in the one case the original immigration is very remote and has never or very rarely been repeated, so that under the unchecked influence of new conditions of life the species have become modified; in the other case, either the original immigration has been recent, or if remote has been so frequently repeated that the new comers have kept up the purity of the stock, and have not allowed time for the new conditions to produce the effect we are sure they would in time produce if not counteracted. For Mr. Godman tells us that many of the birds are modified—instancing the gold-crested wren, blackcap, and rock dove—and he adds, that the modification all tends in one direction—to produce a more sombre plumage, a greater strength of feet and legs, and a more robust bill. We further find, that four of the land-birds, including the oriole, snow-bunting, and hoopoe, are not resident birds, but straggle accidentally to the islands by stress of weather; and we are told that every year some fresh birds are seen after violent storms. Add to this the fact, that the number of species diminishes in the group as we go from east to west, and that the islands are subject to fierce ánd frequent storms blowing from every point of the compass,—and we have all the facts requisite to enable us to understand how this remote archipelago has become stocked with animal life without ever probably being much nearer to Europe than it is now. For the islands are all volcanic, the only stratified rock that occurs being believed to be of Miocene date.

Madeira and the Canaries.—Coming next to Madeira, we find the number of genera of land birds has increased to twenty-eight, of which seventeen are identical with those of the Azores. Some of the commonest European birds—swallows, larks, sparrows, linnets, goldfinches, ravens, and partridges, are among the additions. A gold-crested warbler, *Regulus Maderensis,* and a pigeon, *Columba Trocaz,* are peculiar to Madeira.

In the Canaries we find that the birds have again very much increased, there being more than fifty genera of land birds; but the additions are wholly European in character, and almost all common European species. We find a few more peculiar species (five), while some others, including the wild canary, are common to all the Atlantic Islands or to the Canaries and Madeira. Here, too, the only indigenous mammalia are two European species of bats.

Land Shells.—The land shells of Madeira offer us an instructive contrast to the birds of the Atlantic Islands. About fifty-six species have been found in Madeira, and forty-two in the small adjacent island of Porto Santo, but only twelve are common to both, and all or almost all are distinct from their nearest allies in Europe and North Africa. Great numbers of fossil shells are also found in deposits of the Newer Pliocene period; and although these comprise many fresh species, the two faunas and that of the continent still remain almost as distinct from each other as before. It has been already stated (p. 31) that the means by which land mollusca have been carried across arms of the sea are unknown, although several modes may be suggested; but it is evidently a rare event, requiring some concurrence of favourable conditions not always present. The diversity and specialization of the forms of these animals is, therefore, easily explained by the fact, that, once introduced they have been left to multiply under the influence of a variety of local conditions, which inevitably lead, in the course of ages, to the formation of new varieties and new species.

Coleoptera.—The beetles of Madeira and the Canaries have been so carefully collected and examined by Mr. T. V. Wollaston, and those of the Azores described and compared by Mr. Crotch, and they illustrate so many curious points in geographical distribution, that it is necessary to give some account of them. No less than 1,480 species of beetles have been obtained from the Canaries and Madeira, only 360 of which are European, the remainder being peculiar to the islands. The Canaries are inhabited by a little over 1,000 species, Madeira by about 700, while 240 are common to both; but it is believed that many of these have been introduced by man. In the Azores, 212 species have been obtained, of which 175 are European; showing, as in the birds, a closer resemblance to the European fauna than in the other islands which, although nearer to the continent, offer more shelter and are situated in a less tempestuous zone. Of the non-European species in the Azores, 19 are found also in the other groups of islands, 14 are peculiar, while 3 are American. Of the European species, 132 are found also in the other Atlantic islands, while 43 have reached the Azores only. This is interesting as showing to how great an extent the same insects reach all the islands, notwithstanding the difference of latitude and position; and it becomes of great theoretical importance, when we find how many extensive families and genera are altogether absent.

The Madeira group has been more thoroughly explored than

P

any other, and its comparatively remote situation, combined with its luxuriant vegetation, have been favourable to the development and increase of the peculiar forms which characterize all the Atlantic islands in a more or less marked degree. A consideration of some of its peculiarities will, therefore, best serve to show the bearing of the facts presented by the insect fauna of the Atlantic islands, on the general laws of distribution. The 711 species of beetles now known from the Madeira group, belong to 236 genera; and no less than 44 of these genera are not European but are peculiar to the Atlantic islands. Most of them are, however, closely allied to European genera, of which they are evidently modifications. A most curious general feature presented by the Madeiran beetles, is the total absence of many whole families and large genera abundant in South Europe. Such are the Cicindelidæ, or tiger beetles; the Melolonthidæ, or chafers; the Cetoniidæ, or rose-chafers; the Eumolpidæ and Galerucidæ, large families of Phytophagous, or leaf-eating beetles; and also the extensive groups of Elateridæ and Buprestidæ, which are each represented by but one minute species. Of extensive genera abundant in South Europe, but wholly absent in Madeira, are *Carabus, Rhizotrogus, Lampyris*, and other genera of Malacoderms; *Otiorhynchus, Brachycerus*, and 20 other genera of Curculionidæ, comprising more than 300 South European and North African species; *Pimelia, Tentyra, Blaps*, and 18 other genera of Heteromera, comprising about 550 species in South Europe and North Africa; and *Timarcha*, containing 44 South European and North African species.

Another most remarkable feature of the Madeiran Coleoptera is the unusual prevalence of apterous or wingless insects. This is especially the case with groups which are confined to the Atlantic islands, many of which consist wholly of wingless species; but it also affects the others, no less than twenty-two genera which are usually or sometimes winged in Europe, having only wingless species in Madeira; and even the same species which is winged in Europe becomes, in at least three cases, wingless in Madeira, without any other perceptible change having taken place. But there is another most curious fact noticed by

Mr. Wollaston; that those species which possess wings in Madeira, often have them rather larger than their allies in Europe. These two facts were connected by Mr. Darwin, who suggested that flying insects are much more exposed to be blown out to sea and lost, than those which do not fly (and Mr. Wollaston had himself supposed that the "stormy atmosphere" of Madeira had something to do with the matter); so that the most frequent fliers would be continually weeded out, while the more sluggish individuals, who either could not, or would not fly, remained to continue the race; and this process going on from generation to generation, would, on the well-ascertained principles of selection and abortion by disuse, in time lead to the entire loss of wings by those insects to whom wings were *not a necessity*. But those whose wings were essential to their existence would be acted upon in another way. All these must fly to obtain their food or provide for their offspring, and those that flew best would be best able to battle with the storms, and keep themselves safe, and thus those with the longest and most powerful wings would be preserved. If however all the individuals of the species were too weak on the wing to resist the storms, they would soon become extinct.[1]

Now this explanation of the facts is not only simple and probable in itself, but it also serves to explain in a remarkable manner some of the peculiarities and deficiencies of the Madeiran insect fauna, in harmony with the view (supported by the distribution of the birds and land shells, and in particular by the immigrant birds and insects of the Azores) that all the insects have been derived from the continent or from other islands, by

---

[1] A remarkable confirmation of this theory, is furnished in the Report to the Royal Society of the naturalist to the Kerguelen Island, "Transit Expedition"—the Rev. A. E. Eaton. Insects were assiduously collected, and it was found that almost all were either completely apterous, or had greatly abbreviated wings. The only moth found, several flies, and numerous beetles, were alike incapable of flight. As this island is subject to violent, and almost perpetual gales, even in the finest season, the meaning of the extraordinary loss of wings in almost all the insects, can, in this case, hardly be misunderstood.

immigration across the ocean, in various ways and during a long period. These deficiencies are, on the other hand, quite inconsistent with the theory (still held by some entomologists) that a land communication is absolutely necessary to account for the origin of the Madeiran fauna.

First, then, we can understand how the tiger-beetles (Cicindelidæ) are absent; since they are insects which have a short weak flight, but yet to whom flight is necessary. If a few had been blown over to Madeira, they would soon have become exterminated. The same thing applies to the Melolonthidæ, Cetoniidæ, Eumolpidæ, and Galerucidæ,—all flower and foliage-haunting insects, yet bulky and of comparatively feeble powers of flight. Again, all the large genera abundant in South Europe, which have been mentioned above as absent from Madeira, are wholly apterous (or without wings), and thus their absence is a most significant fact; for it proves that in the case of all insects of moderate size, flight was essential to their reaching the island, which could not have been the case had there been a land connection. There are, however, one or two curious exceptions to the absence of these wholly apterous European genera in Madeira, and as in each case the reason of their being exceptions can be pointed out, they are eminently exceptions that prove the rule. Two of the apterous species common to Europe and Madeira are found always in ants' nests; and as ants, when winged, fly in great swarms and are carried by the wind to great distances, they may have conveyed the minute eggs of these very small beetles. Two European species of *Blaps* occur in Madeira, but these are house beetles, and are admitted to have been introduced by man. There are also three species of *Meloe*, of which two are European and one peculiar. These are large, sluggish, wingless insects, but they have a most extraordinary and exceptional metamorphosis, the larvæ in the first state being minute active insects parasitic on bees, and thus easily conveyed across the ocean. This case is most suggestive, as it accounts for what would be otherwise a difficult anomaly. Another case, not quite so easily explained, is that of the genus *Acalles*, which is very abundant in all the Atlantic islands and also occurs in South Europe, but is always apterous. It is however closely allied to another genus, *Cryptorhynchus*, which is apterous in some species, winged in others. We may therefore well suppose that the ancestors of *Acalles* were once in the same condition, and that some of the winged forms reached Madeira, the genus having since become wholly apterous.

We may look at this curious subject in another way. The Coleoptera of Madeira may be divided into those which are found also in Europe or the other islands, and those which are peculiar to it. On the theory of introduction by accidental immigration across the sea, the latter must be the more ancient, since they have had time to become modified ; while the former are comparatively recent, and their introduction may be supposed to be now going on. The peculiar influence of Madeira in aborting the wings should, therefore, have acted on the ancient and changed forms much more powerfully than on the recent and unchanged forms. On carefully comparing the two sets of insects (omitting those which have almost certainly been introduced by man) we find, that out of 263 species which have a wide range, only 14 are apterous ; while the other class, consisting of 393 species, has no less than 178 apterous ; or about 5 per cent in the one case, and 45 per cent in the other.[1] On the theory of a land connection as the main agent in introducing the fauna, both groups must have been introduced at or about the same time, and why one set should have lost their wings and the other not, is quite inexplicable.

Taking all these singular facts, in connection with the total absence of all truly indigenous terrestrial mammalia and reptiles from these islands—even from the extensive group of the Canaries so comparatively near to the continent, we are forced to reject the theory of a land connection as quite untenable ; and this view becomes almost demonstrated by the case of the Azores, which being so much further off, and surrounded by such a vast expanse of deep ocean, could only have been con-

---

[1] The facts on which these statements rest, will be found more fully detailed in the Author's Presidential Address to the Entomological Society of London for the year 1871.

nected with Europe at a far remoter epoch, and ought therefore
to exhibit to us a fauna composed almost entirely of peculiar
forms both of birds and insects. Yet, so far from this being the
case, the facts are exactly the reverse. Far more of the birds
and insects are identical with those of Europe than in the
other islands, and this difference is clearly traced to the more
tempestuous atmosphere, which is shown to be even now
annually bringing fresh inmigrants (both birds and insects) to
its shores. We here see nature actually at work; and if the
case of Madeira rendered her mode of action probable, that of
the Azores may be said to demonstrate it.

Mr. Wollaston has objected to this view that "storms and
hurricanes" are somewhat rare in the latitude of Madeira and
the Canaries; but this little affects the question, since the *time*
allowed for such operations is so ample. If but one very
violent storm happened in a century, and ten such storms
recurred before a single species of insect was introduced into
Madeira, that would be more than sufficient to people it, as we
now find it, with a varied fauna. But he also adds the important information that the ordinary winds blow almost uninterruptedly from the north-east, so that there would be always a
chance of a little stronger wind than usual bringing insect, or
larva, or egg, attached to leaves or twigs. Neither Mr. Wollaston,
Mr. Crotch, Mr. A. Murray, nor any other naturalist who
upholds the land-connection theory, has attempted to account
for the fact of the absence of so many extensive groups of
insects that ought to be present, as well as of all small
mammalia and reptiles.

Cape Verd Islands.—There is yet another group of Atlantic
islands which is very little known, and which is usually considered to be altogether African—the Cape Verd Islands, situated
between 300 and 400 miles west of Senegal, and a little to the
south of the termination of the Sahara. The evidence that we
possess as to the productions of these islands, shows that, like
the preceding groups, they are truly oceanic, and have probably
derived their fauna from the desert and the Canaries to the
north-east of them rather than from the fertile and more truly

Ethiopian districts of Senegal and Gambia to the east. There is a mingling of the two faunas, but the preponderance seems to be undoubtedly with the Palæarctic rather than with the Ethiopian. I owe to Mr. R. B. Sharpe of the British Museum, a MS. list of the birds of these islands, twenty-three species in all. Of these eight are of wide distribution and may be neglected. Seven are undoubted Palæarctic species, viz.:—*Milvus ictinus, Sylvia atricapilla, S. conspicillata, Corvus corone, Passer salicarius, Certhilauda desertorum, Columba livia.* Three are peculiar species, but of Palæarctic genera and affinities, viz.:—*Calamoherpe brevipennis, Ammomanes cinctura,* and *Passer jagoensis.* Against this we have to set two West African species, *Estrilda cinerea* and *Numida meleagris,* both of which were probably introduced by man; and three which are of Ethiopian genera and affinities, viz.:—*Halcyon erythrorhyncha,* closely allied to *H. semicærulea* of Arabia and North-east Africa, and therefore almost Palæarctic, *Accipiter melanoleucus;* and *Pyrrhulauda nigriceps,* an Ethiopian form; but the same species occurs in the Canaries.

The Coleoptera of these islands have been also collected by Mr. Wollaston, and he finds that they have generally the same European character as those of the Canaries and Madeira, several of the peculiar Atlantic genera, such as *Acalles* and *Hegeter,* occurring, while others are represented by new but closely allied genera. Out of 275 species 91 were found also in the Canaries and 81 in the Madeiran group; a wonderful amount of similarity when we consider the distance and isolation of these islands and their great diversity of climate and vegetation.

This connection of the four groups of Atlantic islands now referred to, receives further support from the occurrence of landshells of the subgenus *Leptaxis* in all the groups, as well as in Majorca; and by another subgenus, *Hemicycla,* being common to the Canaries and Cape Verd islands. Combining these several classes of facts, we seem justified in extending the Mediterranean sub-region to include the Cape Verd Islands.

III.—The Siberian Sub-region, or Northern Asia.

This large and comparatively little-known subdivision of the Palæarctic region, extends from the Caspian Sea to Kamschatka and Behring's Straits, a distance of about 4,000 miles; and from the shores of the Arctic Ocean to the high Himalayas of Sikhim in North Latitude 29°, on the same parallel as Delhi. To the east of the Caspian Sea and the Ural Mountains is a great extent of lowland which is continued round the northern coast, becoming narrower as it approaches the East Cape. Beyond this, in a general E.N.E. direction, rise hills and uplands, soon becoming lofty mountains, which extend in an unbroken line from the Hindu Koosh, through the Thian Shan, Altai and Yablonoi Mountains, to the Stanovoi range in the north-eastern extremity of Asia. South of this region is a great central basin, which is almost wholly desert; beyond which again is the vast plateau of Thibet, with the Kuenlun, Karakorum, and Himalayan snow-capped ranges, forming the most extensive elevated district on the globe.

The superficial aspects of this vast territory, as determined by its vegetable covering, are very striking and well contrasted. A broad tract on the northern coast, varying from 150 to 300 and even 500 miles wide, is occupied by the Tundras or barrens, where nothing grows but mosses and the dwarfest Arctic plants, and where the ground is permanently frozen to a great depth. This tract has its greatest southern extension between the rivers Obi and Yenesi, where it reaches the parallel of 60° north latitude. Next to this comes a vast extent of northern forests; mostly of conifers in the more northern and lofty situations, while deciduous trees preponderate in the southern portions and in the more sheltered valleys. The greatest extension of this forest region is north of Lake Baikal, where it is more than 1,200 miles wide. These forests extend along the mountain ranges to join those of the Hindu Koosh. South of the forests the remainder of the sub-region consists of open pasture-lands and vast intervening deserts, of which the Gobi, and those of Turkestan between the Aral and Balkash lakes, are the most extensive. The former is nearly 1,000 miles long, with a width of from 200 to 350 miles, and is almost as complete a desert as the Sahara.

With very few exceptions, this vast territory is exposed to an extreme climate, inimical to animal life. All the lower parts being situated to the north, have an excessively cold winter, so that the limit of constantly frozen ground descends below the parallel of 60° north latitude. To the south, the land is greatly elevated, and the climate extremely dry. In summer the heat is excessive, while the winter is almost as severe as further north. The whole country, too, is subject to violent storms, both in summer and winter; and the rich vegetation that clothes the steppes in spring, is soon parched up and replaced by dusty plains. Under these adverse influences we cannot expect animal life to be so abundant as in those sub-regions subject to more favourable physical conditions; yet the country is so extensive and so varied, that it does actually, as we shall see, possess a very considerable and interesting fauna.

Mammalia.—Four genera seem to be absolutely confined to this sub-region, *Nectogale*, a peculiar form of the mole family (Talpidæ); *Poephagus*, the yak, or hairy bison of Thibet; with *Procapra* and *Pantholops*, Thibetan antelopes. Some others more especially belong here, although they just enter Europe, as *Saiga*, the Tartarian antelope; *Sminthus*, a desert rat; and *Ellobius*, a burrowing mole-rat; while *Myospalax*, a curious rodent allied to the voles, is found only in the Altai mountains and North China; and *Moschus*, the musk-deer, is almost confined to this sub-region. Among the characteristic animals of the extreme north, are *Mustela*, and *Martes*, including the ermine and sable; *Gulo*, the glutton; *Tarandus*, the reindeer; *Myodes*, the lemming; with the lynx, arctic fox, and polar bear; and here, in the Post-pliocene epoch, ranged the hairy rhinoceros and Siberian mammoth, whose entire bodies still remain preserved in the ice-cliffs near the mouths of the great rivers. Farther south, species of wild cat, bear, wolf, deer, and pika (*Lagomys*) abound; while in the mountains we find wild goats and sheep of several species, and in the plains and deserts wild horses and asses, gazelles, two species of antelopes, flying squirrels (*Pteromys*), ground squirrels (*Tamias*), marmots, of the genus *Spermophilus*, with camels and dromedaries, probably natives of the south-western part of this sub-region. The most abundant and conspicuous of the mammalia are the great herds of reindeer in the north, the wolves of the steppes, with the wild horses, goats, sheep, and antelopes of the plateaus and mountains.

Among the curiosities of this sub-region we must notice the seal, found in the inland and freshwater lake Baikal, at an elevation of about 2,000 feet above the sea. It is a species of *Callocephalus*, closely allied to, if not identical with, one inhabiting northern seas as well as the Caspian and Lake Aral. This would indicate that almost all northern Asia was depressed beneath the sea very recently; and Mr. Belt's view, of the ice during the glacial epoch having dammed up the rivers and converted much of Siberia into a vast freshwater or brackish lake, perhaps offers the best solution of the difficulty.[1]

Plate II.—Characteristic Mammalia of Western Tartary.— Several of the most remarkable animals of the Palæarctic region inhabit Western Tartary, and are common to the European and Siberian sub-regions. We therefore choose this district for one of our illustrative plates. The large animals in the centre are the remarkable saiga antelopes (*Saiga Tartarica*), distinguished from all others by a large and fleshy proboscis-like nose, which gives them a singular appearance. They differ so much from all other antelopes that they have been formed into a distinct family by some naturalists, but are here referred to the great family Bovidæ. They inhabit the open plains from Poland to the Irtish River On the left is the mole-rat, or sand-rat (*Spalax murinus*). This animal burrows under ground like a mole, feeding on bulbous roots. It inhabits the same country as the saiga, but extends farther south in Europe. On the right is a still more curious animal, the desman (*Myogale Muscovitica*), a long-snouted water-mole. This creature is fifteen inches long, including the tail; it burrows in the banks of streams, feeding on insects,

---

[1] *Quarterly Journal of the Geological Society*, 1874, p. 494.

PLATE II.

CHARACTERISTIC MAMMALIA OF WESTERN TARTARY.

worms, and leeches; it swims well, and remains long under water, raising the tip of the snout, where the nostrils are situated, to the surface when it wants to breathe. It is thus well concealed; and this may be one use of the development of the long snout, as well as serving to follow worms into their holes in the soft earth. This species is confined to the rivers Volga and Don in Southern Russia, and the only other species known inhabits some of the valleys on the north side of the Pyrenees. In the distance are wolves, a characteristic feature of these wastes.

Birds.—But few genera of birds are absolutely restricted to this sub-region. *Podoces*, a curious form of starling, is the most decidedly so; *Mycerobas* and *Pyrrhospiza* are genera of finches confined to Thibet and the snowy Himalayas; *Leucosticte*, another genus of finches, is confined to the eastern half of the sub-region and North America; *Tetraogallus*, a large kind of partridge, ranges west to the Caucasus; *Syrrhaptes*, a form of sand-grouse, and *Lerwa* (snow-partridge), are almost confined here, only extending into the next sub-region; as do *Grandala* and *Calliope*, genera of warblers, *Uragus*, a finch allied to the North American cardinals, and *Crossoptilon*, a remarkable group of pheasants.

Almost all the genera of central and northern Europe are found here, and give quite a European character to the ornithology, though a considerable number of the species are different. There are a few Oriental forms, such as *Abrornis* and *Larvivora* (warblers); with *Ceriornis* and *Ithaginis*, genera of pheasants, which reach the snow-line in the Himalayas and thus just enter this sub-region, but as they do not penetrate farther north, they hardly serve to modify the exclusively Palæarctic character of its ornithology.

According to Middendorf, the extreme northern Asiatic birds are the Alpine ptarmigan (*Lagopus mutus*); the snow-bunting (*Plectrophanes nivalis*); the raven, the gyrfalcon and the snowy-owl. Those which are characteristic of the barren "tundras," but which do not range so far north as the preceding are,—the willow-grouse (*Lagopus albus*); the Lapland-bunting (*Plectrophanes lapponica*) ; the shore-lark (*Otocorys alpestris*) ; the sand-martin (*Cotyle riparia*), and the sea-eagle (*Haliæetus albicilla*).

Those which are more characteristic of the northern forests, and which do not pass beyond them, are—the linnet ; two crossbills (*Loxia Leucoptera* and *L. Curvirostra*) ; the pine grosbeak (*Pinicola enucleator*) ; the waxwing ; the common magpie ; the common swallow ; the peregrine falcon ; the rough-legged buzzard ; and three species of owls.

Fully one-half of the land-birds of Siberia are identical with those of Europe, the remainder being mostly representative species peculiar to Northern Asia, with a few stragglers and immigrants from China and Japan or the Himalayas. A much larger proportion of the wading and aquatic families are European or Arctic, these groups having always a wider range than land birds.

Reptiles and Amphibia.—From the nature of the country and climate these are comparatively few, but in the more temperate districts snakes and lizards seem to be not uncommon. *Halys*, a genus of Crotaline snakes, and *Phrynocephalus*, lizards of the family Agamidæ, are characteristic of these parts. *Simotes*, a snake of the family Oligodontidæ, reaches an elevation of 16,000 feet in the Himalayas, and therefore enters this sub-region.

Insects.—*Mesapia* and *Hypermnestra*, genera of Papilionidæ, are butterflies peculiar to this sub-region ; and *Parnassius* is as characteristic as it is of our European mountains. Carabidæ are also abundant, as will be seen by referring to the Chapter on the Distribution of Insects in the succeeding part of this work. The insects, on the whole, have a strictly European character, although a large proportion of the species are peculiar, and several new genera appear.

IV.—*Japan and North China, or the Manchurian Sub-region.*

This is an interesting and very productive district, corresponding in the east to the Mediterranean sub-region in the west, or rather perhaps to all western temperate Europe. Its limits are not very well defined, but it probably includes all Japan ; the Corea and Manchuria to the Amour river and to the lower slopes of the Khingan and Peling mountains ; and China to the Nanlin mountains south of the Yang-tse-kiang. On the coast of China the dividing line between it and the Oriental region seems to be somewhere about Foo-chow, but as there is here no natural barrier, a great intermingling of northern and southern forms takes place.

Japan is volcanic and mountainous, with a fine climate and a most luxuriant and varied vegetation. Manchuria is hilly, with a high range of mountains on the coast, and some desert tracts in the interior, but fairly wooded in many parts. Much of northern China is a vast alluvial plain, backed by hills and mountains with belts of forest, above which are the dry and barren uplands of Mongolia. We have a tolerable knowledge of China, of Japan, and of the Amoor valley, but very little of Corea and Manchuria. The recent researches of Père David in Moupin, in east Thibet, said to be between 31° and 32° north latitude, show, that the fauna of the Oriental region here advances northward along the flanks of the Yun-ling mountains (a continuation of the Himalayas); since he found at different altitudes representatives of the Indo-Chinese, Manchurian, and Siberian faunas. On the higher slopes of the Himalayas, there must be a narrow strip from about 8,000 to 11,000 feet elevation intervening between the tropical fauna of the Indo-Chinese sub-region and the almost arctic fauna of Thibet ; and the animals of this zone will for the most part belong to the fauna of temperate China and Manchuria, except in the extreme west towards Cashmere, where the Mediterranean fauna will in like manner intervene. On a map of sufficiently large scale, therefore, it would be necessary to extend our present sub-region westward along the Himalayas, in a narrow strip just below the upper limits of forests. It is evident that the large number of Fringillidæ, Corvidæ, Troglodytidæ, and Paridæ, often of south Palæarctic forms, that abound in the higher Himalayas, are somewhat out of place as members of the Oriental fauna, and are equally so in that of Thibet and Siberia ; but they form a natural portion of that of North China on the one side, or of South Europe on the other.

Mammalia.—This sub-region contains a number of peculiar and very interesting forms, most of which have been recently discovered by Père David in North and West China and East Thibet. The following are the peculiar genera :—*Rhinopithecus*, a sub-genus of monkeys, here classed under *Semnopithecus ; Anurosorex, Scaptochirus, Uropsilus* and *Scaptonyx*, new forms of Talpidæ or moles; *Æluropus* (Æluridæ); *Nyctereutes* (Canidæ); *Lutronectes* (Mustelidæ); *Cricetulus* (Muridæ); *Hydropotes, Moschus*, and *Elaphodus* (Cervidæ). The *Rhinopithecus* appears to be a permanent inhabitant of the highest forests of Moupin, in a cold climate. It has a very thick fur, as has also a new species of *Macacus* found in the same district. North China and East Thibet seem to be very rich in Insectivora. *Scaptochirus* is like a mole; *Uropsilus* between the Japanese *Urotrichus* and *Sorex ; Scaptonyx* between *Urotrichus* and *Talpa. Æluropus* seems to be the most remarkable mammal discovered by Père David. It is allied to the singular panda (*Ælurus fulgens*) of Nepal, but is as large as a bear, the body wholly white, with the feet, ears, and tip of the tail black. It inhabits the highest forests, and is therefore a true Palæarctic animal, as most likely is the *Ælurus. Nyctereutes*, a curious racoon-like dog, ranges from Canton to North China, the Amoor and Japan, and therefore seems to come best in this sub-region; *Hydropotes* and *Lophotragus* are small hornless deer confined to North China ; *Elaphodus*, from East Thibet, is another peculiar form of deer; while the musk deer (*Moschus*) is confined to this sub-region and the last. Besides the above, the following Palæarctic genera were found by Père David in this sub-region : *Macacus ;* five genera or sub-genera of bats (*Vespertilio, Vesperus, Vesperugo, Rhinolophus*, and *Murina*) ; *Erinaceus, Nectogale, Talpa, Crocidura* and *Sorex*, among Insectivora; *Mustela, Putorius, Martes, Lutra, Viverra, Meles, Ælurus, Ursus, Felis*, and *Canis*, among Carnivora ; *Hystrix, Arctomys, Myospalax, Spermophilus, Gerbillus, Dipus, Lagomys, Lepus, Sciurus, Ptcromys, Arvicola*, and *Mus*, among Rodentia ; *Budorcas, Nemorhedus, Antilope, Ovis, Moschus, Cervulus* and *Cervus* among Ruminants; and the widespread *Sus* or wild boar. The following Oriental genera are also included in Père David's list, but no doubt occur only in the lowlands and warm valleys, and can hardly be considered to belong to the Palæarctic region : *Paguma, Helictis, Arctonyx, Rhizomys, Manis.* The *Rhizomys* from Moupin is a peculiar species of this tropical genus, but all the others inhabit Southern China.

A few additional forms occur in Japan : *Urotrichus*, a peculiar Mole, which is found also in north-west America ; *Enhydra*, the sea otter of California; and the dormouse (*Myoxus*). Japan also possesses peculiar species of *Macacus, Talpa, Meles, Canis,* and *Sciuropterus.*

It will be seen that this sub-region is remarkably rich in Insectivora, of which it possesses ten genera; and that it has also several peculiar forms of Carnivora, Rodentia, and Ruminants.

Birds.—To give an accurate idea of the ornithology of this sub-region is very difficult, both on account of its extreme richness and the impossibility of defining the limits between it and the Oriental region. A considerable number of genera which are well developed in the high Himalayas, and some which are peculiar to that district, have hitherto always been classed as Indian, and therefore Oriental groups ; but they more properly belong to this sub-region. Many of them frequent the highest forests, or descend into the Himalayan temperate zone only in winter ; and others are so intimately connected with Palæarctic species, that they can only be considered as stragglers into the border land of the Oriental region. On these principles we consider the following genera to be confined to this sub-region :—

Grandala, Nemura (Sylviidæ) ; *Pterorhinus* (Timaliidæ) ; *Cholornis, Conostoma, Heteromorpha* (Panuridæ); *Cyanoptila* (Muscicapidæ); *Eophona* (Fringillidæ) ; *Dendrotreron* (Columbidæ) ; *Lophophorus, Tetraophasis, Crossoptilon, Pucrasia, Thaumalea,* and *Ithaginis* (Phasianidæ). This may be called the sub-region of Pheasants ; for the above six genera, comprising sixteen species of the most magnificent birds in the world, are all confined to the temperate or cold mountainous regions of the Himalayas, Thibet, and China ; and in addition we have most of the species of tragopan (*Ceriornis*), and some of the true pheasants (*Phasianus*).

The most abundant and characteristic of the smaller birds are warblers, tits, and finches, of Palæarctic types ; but there are also a considerable number of Oriental forms which penetrate far into the country, and mingling with the northern birds give a character to the Ornithology of this sub-region very different from that of the Mediterranean district at the western end of the region. Leaving out a large number of wide-ranging groups, this mixture of types may be best exhibited by giving lists of the more striking Palæarctic and Oriental genera which are here found intermingled.

PALÆARCTIC GENERA.

SYLVIIDÆ.
 Erithacus.
 Ruticilla.
 Locustella.
 Cyanecula.
 Sylvia.
 Potamodus.
 Reguloides.
 Regulus.
 Accentor.
CINCLIDÆ.
 Cinclus.
TROGLODYTIDÆ.
 Troglodytidæ.
CERTHIIDÆ.
 Certhia.
 Sitta.
 Tichodroma.
PARIDÆ.
 Parus.
 Lophophanes.
 Acredula.

CORVIDÆ.
 Fregilus.
 Nucifraga.
 Pica.
 Cyanopica.
 Garrulus.
AMPELIDÆ.
 Ampelis.
FRINGILLIDÆ.
 Fringilla.
 Chrysomitris.
 Chlorospiza.
 Passer.
 Coccothraustes.
 Pyrrhula.
 Carpodacus.
 Uragus.
 Loxia.
 Linota.
 Emberiza.
STURNIDÆ.
 Sturnus.

ALAUDIDÆ.
 Otocorys.
PICIDÆ.
 Picoides.
 Picus.
 Hyopicus.
 Dryocopus.
YUNGIDÆ.
 Yunx.
PTEROCLIDÆ.
 Syrrhaptes.
TETRAONIDÆ.
 Tetrao.
 Tetraogallus.
 Lerwa.
 Lagopus.
VULTURIDÆ.
 Gypaëtus.
 Vultur.
FALCONIDÆ.
 Archibuteo.

ORIENTAL GENERA.

SYLVIIDÆ.
 Suya.
 Calliope.
 Larvivora.
 Tribura.
 Horites.

SYLVIIDÆ—(*continued*).
 Abrornis.
 Copsychus.
TURDIDÆ.
 Oreocincla.

TIMALIIDÆ.
 Alcippe.
 Timalia.
 Pterocyclus.
 Garrulax.
 Trochalopteron.

ORIENTAL GENERA—*continued.*

TIMALIIDÆ—(*continued*).
Pomatorhinus.
Suthora.
PANURIDÆ.
Paradoxornis.
CINCLIDÆ.
Enicurus.
Myiophonus.
TROGLODYTIDÆ.
Pnœpyga.
LIOTRICHIDÆ.
Liothrix.
Yuhina.
Pteruthius.
PYCNONOTIDÆ.
Microscelis.
Pycnonotus.
Hypsipetes.
CAMPEPHAGIDÆ.
Pericrocotus.
DICRURIDÆ.
Dicrurus.
Chibia.
Buchanga.

MUSCICAPIDÆ.
Xanthopygia.
Niltava.
Tchitrea.
CORVIDÆ.
Urocissa.
NECTARINEIDÆ.
Æthopyga.
MOTACILLIDÆ.
Nemoricola.
DICÆIDÆ.
Zosterops.
FRINGILLIDÆ.
Melophus.
Pyrgilauda.
PLOCEIDÆ.
Munia.
STURNIDÆ.
Acridotheres.
Sturnia.
PITTIDÆ.
Pitta.

PICIDÆ.
Vivia.
Yungipicus.
Gecinus.
CORACIIDÆ.
Eurystomus.
ALCEDINIDÆ.
Halcyon.
Ceryle.
UPUPIDÆ.
Upupa.
PSITTACIDÆ.
Palæornis.
COLUMBIDÆ.
Treron.
Ianthænas.
Macropygia.
PHASIANIDÆ.
Phasianus.
Ceriornis.
STRIGIDÆ.
Scops.

In the above lists there are rather more Oriental than Palæarctic genera; but it must be remembered that most of the former are summer migrants only, or stragglers just entering the sub-region; whereas the great majority of the latter are permanent residents, and a large proportion of them range over the greater part of the Manchurian district. Many of those in the Oriental column should perhaps be omitted, as we have no exact determination of their range, and the limits of the regions are very uncertain. It must be remembered, too, that the Palæarctic genera of Sylviidæ, Paridæ, and Fringillidæ, are often represented by numerous species, whereas the corresponding Oriental genera have for the most part only single species; and we shall then find that, except towards the borders of the Oriental region the Palæarctic element is strongly predominant. Four of the more especially Oriental groups are confined to Japan, the southern

Q

extremity of which should perhaps come in the Oriental region. The great richness of this sub-region compared with that of Siberia is well shown by the fact, that a list of all the known land-birds of East Siberia, including Dahuria and the comparatively fertile Amoor Valley, contains only 190 species; whereas Père David's catalogue of the birds of Northern China with adjacent parts of East Thibet and Mongolia (a very much smaller area) contains for the same families 366 species. Of the Siberian birds more than 50 per cent. are European species, while those of the Manchurian sub-region comprise about half that proportion of land-birds which are identical with those of Europe.

Japan is no doubt very imperfectly known, as only 134 land-birds are recorded from it. Of these twenty-two are peculiar species, a number that would probably be diminished were the Corea to be explored. Of the genera, only nine are Indo-Malayan, while forty-three are Palæarctic.

Plate III.—Scene on the Borders of North-West China and Mongolia with Characteristic Mammalia and Birds. — The mountainous districts of Northern China, with the adjacent portions of Thibet and Mongolia, are the head-quarters of the pheasant tribe, many of the most beautiful and remarkable species being found there only. In the north-western provinces of China and the southern parts of Mongolia may be found the species figured. That in the foreground is the superb golden pheasant (*Thaumalea picta*), a bird that can hardly be surpassed for splendour of plumage by any denizen of the tropics. The large bird perched above is the eared pheasant (*Crossoptilon auritum*), a species of comparatively sober plumage but of remarkable and elegant form. In the middle distance is Pallas's sand grouse (*Syrrhaptes paradoxus*), a curious bird, whose native country seems to be the high plains of Northern Asia, but which often abounds near Pekin, and in 1863 astonished European ornithologists by appearing in considerable numbers in Central and Western Europe, in every part of Great Britain, and even in Ireland.

The quadruped figured is the curious racoon dog (*Nyctereutes*

PLATE III.

CHARACTERISTIC ANIMALS OF NORTH CHINA.

procyonoides), an animal confined to North China, Japan, and the Amoor Valley, and having no close allies in any other part of the globe. In the distance are some deer, a group of animals very abundant and varied in this part of the Palæarctic region.

Reptiles and Amphibia.—Reptiles are scarce in North China, only four or five species of snakes, a lizard and one of the Geckotidæ occurring in the country round Pekin. The genus *Halys* is the most characteristic form of snake, while *Callophis*, an oriental genus, extends to Japan. Among lizards, *Plestiodon, Maybouya, Tachydromus,* and *Gecko* reach Japan, the two latter being very characteristic of the Oriental region.

Amphibia are more abundant and interesting; *Hynobius, Onychodactylus,* and *Sieboldtia* (Salamandridæ) being peculiar to it, while most of the European genera are also represented.

Fresh-water Fish.—Of these there are a few peculiar genera; as *Plecoglossus* (Salmonidæ) from Japan; *Achilognathus, Pseudoperilampus, Ochetobius,* and *Opsariichthys* (Cyprinidæ); and there are many other Chinese Cyprinidæ belonging to the border land of the Palæarctic and Oriental regions.

Insects—The butterflies of this sub-region exhibit the same mixture of tropical and temperate forms as the birds. Most of the common European genera are represented, and there are species of *Parnassius* in Japan and the Amoor. *Isodema,* a peculiar genus of Nymphalidæ is found near Ningpo, just within our limits; and *Sericinus,* one of the most beautiful genera of Papilionidæ is peculiar to North China, where four species occur, thus balancing the *Thais* and *Doritis* of Europe. The genus *Zephyrus* (Lycænidæ) is well represented by six species in Japan and the Amoor, against two in Europe. *Papilio paris* and *P. bianor,* magnificent insects of wholly tropical appearance, abound near Pekin, and allied forms inhabit Japan and the Amoor, as well as *P. demetrius* and *P. alcinous* belonging to the "Protenor" group of the Himalayas. Other tropical genera occurring in Japan, the Amoor, or North China are, *Debis, Neope, Mycalesis, Ypthimia* (Satyridæ); *Thaumantis* (Morphidæ), at Shanghae; *Euripus, Neptis, Athyma* (Nymphalidæ); *Terias* (Pieridæ); and the above-mentioned Papilionidæ.

Q 2

Coleoptera.—The beetles of Japan decidedly exhibit a mixture
of tropical forms with others truly Palæarctic, and it has been
with some naturalists a matter of doubt whether the southern and
best known portion of the islands should not be joined to the
Oriental region. An important addition to our knowledge of
the insects of this country has recently been made by Mr. George
Lewis, and a portion of his collections have been described by
various entomologists in the *Transactions of the Entomological
Society of London.* As the question is one of considerable interest we shall give a summary of the results fairly deducible
from what is now known of the entomology of Japan; and it
must be remembered that almost all our collections come from
the southern districts, in what is almost a sub-tropical climate;
so that if we find a considerable proportion of Palæarctic forms,
we may be pretty sure that the preponderance will be much
greater a little further north.

Of Carabidæ Mr. Bates enumerates 244 species belonging to
84 genera, and by comparing these with the Coleoptera of a
tract of about equal extent in western Europe, he concludes that
there is little similarity, and that the cases of affinity to the forms
of eastern tropical Asia preponderate. By comparing his genera
with the distributions as given in *Gemminger and Harold's
Catalogue,* a somewhat different result is arrived at. Leaving
out the generic types altogether peculiar to Japan, and also those
genera of such world-wide distribution that they afford no clear
indications for our purpose, it appears that no less than twenty-
two genera, containing seventy-four of the Japanese species, are
either exclusively Palæarctic, Palæarctic and Nearctic, or highly
characteristic of the Palæarctic region; then come thirteen genera
containing eighty-seven of the species which have a very wide
distribution, but are also Palæarctic : we next have seventeen
genera containing twenty-four of the Japanese species which are
decidedly Oriental and tropical. Here then the fair comparison
is between the twenty-two genera and seventy-four species whose
affinities are clearly Palæarctic or at least north temperate, and
seventeen genera with twenty-four species which are Asiatic
and tropical; and this seems to prove that, although South

Japan (like North China) has a considerable infusion of tropical forms, there is a preponderating substratum of Palæarctic forms, which clearly indicate the true position of the islands in zoological geography. There are also a few cases of what may be called eccentric distribution; which show that Japan, like many other island-groups, has served as a kind of refuge in which dying-out forms continue to maintain themselves. These, which are worthy of notice, are as follows : *Orthotrichus* (1 sp.) has the only other species in Egypt; *Trechichus* (1 sp.) has two other species, of which one inhabits Madeira, the other the Southern United States ; *Perileptus* (1 sp.) has two other species, of which one inhabits Bourbon, the other West Europe ; and lastly, *Crepidogaster* (1 sp.) has the other known species in South Africa. These cases diminish the value of the indications afforded by some of the Japanese forms, whose only allies are single species in various remote parts of the Oriental region.

The Staphylinidæ have been described by Dr. Sharp, and his list exhibits a great preponderance of north temperate, or cosmopolitan forms, with a few which are decidedly tropical. The Pselaphidæ and Scydmenidæ, also described by Dr. Sharp, exhibit, according to that gentleman, " even a greater resemblance to those of North America than to those of Europe," but he says nothing of any tropical affinities. The water-beetles are all either Palæarctic or of wide distribution.

The Lucanidæ (*Gemm. and Har. Cat.*, 1868) exhibit an intermingling of Palæarctic and Oriental genera.

The Cetoniidæ (*Gemm. and Har. Cat.* 1869) show, for North China and Japan, three Oriental to two Palæarctic genera.

The Buprestidæ collected by Mr. Lewis have been described by Mr. Edward Saunders in the *Journal of the Linnæan Society*, vol. xi. p. 509. The collection consisted of thirty-six species belonging to fourteen genera. No less than thirteen of these are known also from India and the Malay Islands; nine from Europe; seven from Africa ; six from America, and four from China. In six of the genera the Japanese species are said to be allied to those of the Oriental region ; while in three they are allied to European forms, and in two to American. Considering the southern latitude and warm climate in which these insects were mostly collected, and the proximity to Formosa and the Malay Islands compared with the enormous distance from Europe, this shows as much Palæarctic affinity as can be expected. In the Palæarctic region the group is only plentiful in the southern parts of Europe, which is cut off by the cold plateau of Thibet from all direct communication with Japan; while in the Oriental region it everywhere abounds and is, in fact, one of the most conspicuous and dominant families of Coleoptera.

The Longicorns collected by Mr. Lewis have been described by Mr. Bates in the *Annals of Natural History for* 1873. The number of species now known from Japan is 107, belonging to sixty-four genera. The most important genera are *Leptura, Clytanthus, Monohammus, Praonetha, Exocentrus, Glenea,* and *Oberea.* There are twenty-one tropical genera, and seven peculiar to Japan, leaving thirty-six either Palæarctic or of very wide range. A number of the genera are Oriental and Malayan, and many characteristic European genera seem to be absent; but it is certain that not half the Japanese Longicorns are yet known, and many of these gaps will doubtless be filled up when the more northern islands are explored.

The Phytophaga, described by Mr. Baly, appear to have a considerable preponderance of tropical Oriental forms.

A considerable collection of Hymenoptera formed by Mr. Lewis have been described by Mr. Frederick Smith; and exhibit the interesting result, that while the bees and wasps are decidedly of tropical and Oriental forms, the Tenthredinidæ and Ichneumonidæ are as decidedly Palæarctic, "the general aspect of the collection being that of a European one, only a single exotic form being found among them."

Remarks on the General Character of the Fauna of Japan — From a general view of the phenomena of distribution we feel justified in placing Japan in the Palæarctic region; although some tropical groups, especially of reptiles and insects, have largely occupied its southern portions; and these same groups have in many cases spread into Northern China, beyond the usual dividing line of the Palæarctic and Oriental regions.   The causes of such a phenomenon are not difficult to conceive.   Even now, that portion of the Palæarctic region between Western Asia and Japan is, for the most part, a bleak and inhospitable region, abounding in desert plateaus, and with a rigorous climate even in its most favoured districts, and can, therefore, support but a scanty population of snakes, and of such groups of insects as require flowers, forests, or a considerable period of warm summer weather; and it is precisely these which are represented in Japan and North China by tropical forms.   We must also consider, that during the Glacial epoch this whole region would have become still less productive, and that, as the southern limit of the ice retired northward, it would be followed up by many tropical forms along with such as had been driven south by its advance, and had survived to return to their northern homes.

It is also evident that Japan has a more equable and probably moister climate than the opposite shores of China, and has also a very different geological character, being rocky and broken, often volcanic, and supporting a rich, varied, and peculiar vegetation.   It would thus be well adapted to support all the more hardy denizens of the tropics which might at various times reach it, while it might not be so well adapted for the more boreal forms from Mongolia or Siberia.   The fact that a mixture of such forms occurs there, is then, little to be wondered at, but we may rather marvel that they are not more predominant, and that even in the extreme south, the most abundant forms of mammal, bird, and insect, are modifications of familiar Palæarctic types.   The fact clearly indicates that the former land connections of Japan with the continent have been in a northerly rather than in a southerly direction, and that the tropical immigrants have had difficulties to contend with, and have found the land already fairly stocked with northern aborigines in almost every class and order of animals.

General Conclusions as to the Fauna of the Palæarctic Region.—From the account that has now been given of the fauna of the Palæarctic region, it is evident that it owes many of its deficiencies and some of its peculiarities to the influence of the Glacial epoch, combined with those important changes of physical geography which accompanied or preceded it. The elevation of the old Sarahan sea and the complete formation of the Mediterranean, are the most important of these changes in the western portion of the region.   In the centre, a wide arm of the Arctic Ocean extended southward from the Gulf of Obi to the Aral and the Caspian, dividing northern Europe and Asia.   At this time our European and Siberian sub-regions were probably more distinct than they are now, their complete fusion having been effected since the Glacial epoch. As we know that the Himalayas have greatly increased in altitude during the Tertiary period, it is not impossible that during the Miocene and Pliocene epochs the vast plateau of Central Asia was much less elevated and less completely cut off from the influence of rain-bearing winds.   It might then have been far more fertile, and have supported a rich and varied animal population, a few relics of which we see in the Thibetan antelopes, yaks, and wild horses.   The influence of yet earlier changes of physical geography, and the relations of the Palæarctic to the tropical regions immediately south of it, will be better understood when we have examined and discussed the faunas of the Ethiopian and Oriental regions.

TABLES OF DISTRIBUTION.

In constructing these tables showing the distribution of various classes of animals in the Palæarctic region, the following sources of information have been chiefly relied on, in addition to the general treatises, monographs, and catalogues used in compiling the fourth part of this work.

Mammalia.—Lord Clement's Mammalia and Reptiles of Europe ; Siebold's Fauna Japonica ; Père David's List of Mammalia of North China and Thibet ; Swinhoe's Chinese Mammalia ; Radde's List of Mammalia of South-Eastern Siberia ; Canon Tristram's, Lists for Sahara and Palestine ; Papers by Professor Milne-Edwards, Mr. Blanford, Mr. Sclater, and the local lists given by Mr. A. Murray in the Appendix to his Geographical Distribution of Mammalia.

Birds.—Blasius' List of Birds of Europe ; Godman, On Birds of Azores, Madeira, and Canaries ; Middendorf, for Siberia ; Père David and Mr. Swinhoe, for China and Mongolia ; Homeyer, for East Siberia ; Mr. Blanford, for Persia and the high Himalayas ; Mr. Elwes's paper on the Distribution of Asiatic Birds ; Canon Tristram, for the Sahara and Palestine ; Professor Newton, for Iceland and Greenland ; Mr. Dresser, for Scandinavia ; and numerous papers and notes in the Ibis ; Journal für Ornithologie ; Annals and Mag. of Nat. History ; and Proceedings of the Zoological Society.

Reptiles and Amphibia.—Schreiber's European Herpetology.

TABLE I.

FAMILIES OF ANIMALS INHABITING THE PALÆARCTIC REGION.

EXPLANATION.

Names in *italics* show families peculiar to the region.
Names inclosed thus (......) barely enter the region, and are not considered proper to belong to it.
Numbers are not consecutive, but correspond to those in Part IV.

Order and Family.	Europe.	Mediterranean.	Siberia.	Japan.	Range beyond the Region.
MAMMALIA.					
PRIMATES.					
3. Cynopithecidæ		—		—	Ethiopian, Oriental
CHIROPTERA.					
9. (Pteropidæ) ...				—	Tropics of E. Hemisphere
11. Rhinolophidæ	—	—	—	—	Warmer parts of E. Hemis.
12. Vespertilionidæ	—	—	—	—	Cosmopolite
13. Noctilionidæ...				—	Tropical regions
INSECTIVORA.					
15. Macroscelididæ		—			Ethiopian
17. Erinaceidæ ...	—	—	—	—	Oriental, S. Africa
21. Talpidæ... ...	—	—	—	—	Nearctic, Oriental
22. Soricidæ... ...	—	—	—	—	Cosmopolite, excl. Australia and S. America
CARNIVORA.					
23. Felidæ	—	—	—	—	All regions but Australian
25. Viverridæ ...		—			Ethiopian, Oriental
27. Hyænidæ ..		—			Ethiopian, Oriental
28. Canidæ	—	—	—	—	All regions but Australian
29. Mustelidæ ...	—	—	—	—	All regions but Australian
31. Æluridæ ...				—	Oriental
32. Ursidæ	—	—	—	—	Nearctic, Oriental, Andes
33. Otariidæ... ...				—	N. and S. temperate zones
34. Trichechidæ ...	—		—		Arctic regions
35. Phocidæ ...	—	—	—	—	N. and S. temperate zones
CETACEA.					
36 to 41.					Oceanic
SIRENIA.					
42. Manatidæ ...	—		—		Tropics, from Brazil to N. Australia
UNGULATA.					
43. Equidæ		—	—		Ethiopian
47. Suidæ	—	—	—	—	Cosmopolite, excl. Nearctic reg. and Australia
48. Camelidæ ...		—	—		Andes
50. Cervidæ... ...	—	—	—	—	All regions but Ethiopian and Australian
52. Bovidæ	—	—	—	—	All regions but Neotropical and Australian

Order and Family.	Europe.	Mediterranean.	Siberia.	Japan.	Range beyond the Region.
HYRACOIDAE.					
54. (Hyracidæ) ..		—			Ethiopian family
RODENTIA.					
55. Muridæ	—	—	—	—	Almost Cosmopolite
56. Spalacidæ ...	—	—	—		Ethiopian, Oriental
57. Dipodidæ ...		—	—	—	Ethiopian, Nearctic
58. Myoxidæ ..	—	—	—		Ethiopian
60. Castoridæ ...	—		—		Nearctic
61. Sciuridæ... ...	—	—	—	—	All regions but Australian
64. Octodontidæ ...		—			Abyssinia, Neotropical
67. Hystricidæ ...	—				Ethiopian, Oriental
69. Lagomyidæ ...			—		Nearctic
70. Leporidæ ...	—	—	—	—	All regions but Australian
BIRDS.					
PASSERES.	—	—	—	—	
1. Turdidæ... ...	—	—	—	—	Cosmopolite
2. Sylviidæ... ...	—	—		—	Cosmopolite
3. Timaliidæ ...	—	—	—	—	Ethiopian, Oriental, Australian
4. Panuridæ ...	—	—	—	—	Nearctic, Oriental
5. Cinclidæ ...	—	—	—	—	Oriental
6. Troglodytidæ...	—	—	—	—	American, Oriental
8. Certhiidæ ...	—	—	—	—	Oriental, Nearctic
9. Sittidæ	—	—	—	—	Nearctic, Oriental, Australian, Madagascar
10. Paridæ	—	—	—	—	Nearctic, Oriental, Australian [?]
13. Pycnonotidæ...		—		—	Oriental, Ethiopian
14. Oriolidæ... ...	—	—	—	—	Ethiopian, Oriental, Australian
17. Muscicapidæ...				—	Eastern Hemisphere
19. Laniidæ	—	—	—	—	Eastern Hemisphere and N. America
20. Corvidæ	—	—	—	—	Cosmopolite
23. (Nectariniidæ)		—			Ethiopian, Oriental, Australian
24. (Dicæidæ) ...				—	Ethiopian, Oriental, Australian
29. Ampelidæ ...	—	—	—	—	Nearctic
30. Hirundinidæ...	—	—	—	—	Cosmopolite
33. Fringillidæ ...	—	—	—	—	All regions but Australian
35. Sturnidæ ...	—	—	—	—	Eastern Hemisphere
37. Alaudidæ ...	—	—	—	—	All regions but Neotropical
38. Motacillidæ ...	—	—	—	—	Cosmopolite
47. (Pittidæ) ...				—	Oriental, Australian, Ethiopian
PICARIÆ.					
51. Picidæ	—	—	—	—	All regions but Australian
52. Yungidæ ...	—	—	—	—	N. W. India, N. E. Africa, S. Africa
58. Cuculidæ ...	—	—	—	—	Almost Cosmopolite
62. Coraciidæ ...	—	—	—	—	Ethiopian, Oriental, Australian
63. Meropidæ ...	—	—	—	—	Ethiopian, Oriental, Australian
67. Alcedinidæ ...	—	—	—	—	Cosmopolite
69. Upupidæ ...		—	—	—	Ethiopian, Oriental
73. Caprimulgidæ	—	—	—	—	Cosmopolite
74. Cypselidæ ...	—	—	—	—	Almost Cosmopolite

Order and Family.	Sub-regions.				Range beyond the Region.
	Europe.	Mediterranean	Siberia.	Japan.	
COLUMBÆ.					
84. Columbidæ ...	—	—	—	—	Cosmopolite
GALLINÆ.					
86. Pteroclidæ ...	—	–	—	—	Ethiopian, Indian
87. Tetraonidæ ...	—	—	—	—	Nearctic, Ethiopian, Oriental
88. Phasianidæ...		—	—	—	Oriental, Ethiopian, Nearctic
89. Turnicidæ ..		—		—	Ethiopian, Oriental, Australian
ACCIPITRES.					
94. Vulturidæ ...	—	—	—	—	All regions but Australian
96. Falconidæ ...	—	—	—	—	Cosmopolite
97. Pandionidæ...	—	—	—	—	Cosmopolite
98. Strigidæ ...	—	—	—	—	Cosmopolite
GRALLÆ.					
99. Rallidæ ...	—	—	—	—	Cosmopolite
100. Scolopacidæ...	—	—	—	—	Cosmopolite
104. Glareolidæ ...	—	—	—	—	Ethiopian, Oriental, Australian
105. Charadriidæ...	—	—	—	—	Cosmopolite
106. Otididæ ...	—	—	—	—	Ethiopian, Oriental, Australian
107. Gruidæ ...	—	—	—	—	Eastern Hemisphere, and N. America
113. Ardeidæ ...	—	—	—	—	Cosmopolite
114. Plataleidæ ...	—	—	—	—	Almost Cosmopolite
115. Ciconiidæ ...	—	—	—	—	Nearly Cosmopolite
117. Phænicopteridæ	—				Neotropical, Ethiopian, Indian
ANSERES.					
118. Anatidæ ...	—	—	—	—	Cosmopolite
119. Laridæ... ...	—	—	—	—	Cosmopolite
120. Procellariidæ	—	—	—	—	Cosmopolite
121. Pelecanidæ ...	—	—	—	—	Cosmopolite
123. Colymbidæ ...	—		—	—	Arctic and N. Temperate
124. Podicipidæ ..	—	—	—	—	Cosmopolite
125. Alcidæ... ...	—		—	—	N. Temperate zone
REPTILIA.					
OPHIDIA.					
1. Typhlopidæ..		—		—	All regions but Nearctic
5. Calamariidæ...		—			All other regions
6. Oligodontidæ				—	Oriental and Neotropical
7. Colubridæ ...	—	—	....	—	Almost Cosmopolite
8. Homalopsidæ	—	—		—	Oriental, and all other regions
9. Psammophidæ		—			Ethiopian and Oriental
18. Erycidæ... ...		—			Oriental and Ethiopian
20. Elapidæ... ...				—	Australian and all other regions
24. Crotalidæ ...			—	—	Nearctic, Neotropical, Oriental
25. Viperidæ ...	—	—	—	—	Ethiopian, Oriental

Order and Family.	Sub-regions.				Range beyond the Region.
	Europe.	Mediterranean.	Siberia.	Japan	
LACERTILIA.					
26. *Trogonophidæ*		—			
28. Amphisbænidæ		—			Ethiopian, Neotropical
30. Varanidæ ...		—			Oriental, Ethiopian, Australian
33. Lacertidæ ...	—	—	—	—	All continents but American
34. Zonuridæ ...		—			America, Africa, N. India
41. Gymnopthalmidæ	—	—	—		Ethiopian, Australian, Neotropical
45. Scincidæ ...	—	—	—	—	Almost Cosmopolite
46. *Ophiomoridæ* ..		—			
47. Scpidæ		—			Ethiopian
49. Geckotidæ ...	—	—		—	Almost Cosmopolite
51. Agamidæ ...	—	—		—	All continents but America
52. Chamæleonidæ		—			Ethiopian, Oriental
CHELONIA.					
57. Testudinidæ ...	...	--		—	All continents but Australia
59. Trionychidæ ...				—	Ethiopian, Oriental, Nearctic
60. Cheloniidæ ...					Marine
AMPHIBIA.					
URODELA.					
3. Proteidæ ...	—				Nearctic
5. Menopomidæ...			—		Nearctic
6. Salamandridæ	—	—	—	—	Nearctic to Andes of Bogota
ANOURA.					
10. Bufonidæ ...	—	—	—	—	All continents but Australia
13. Bombinatoridæ	—	—			Neotropical, New Zealand
15. Alytidæ	—				All regions but Oriental
17. Hylidæ	—	—		—	All regions but Ethiopian
18. Polypedatidæ			—		All the regions
19. Ranidæ	—	—	—	—	Almost Cosmopolite
20. Discoglossidæ	—	--	—	—	All regions but Nearctic
FISHES (FRESHWATER).					
ACANTHOPTERYGII.					
1. Gasterosteidæ	—	—	—	—	Nearctic
3. Percidæ	—	—	—	—	All regions but Australian
12. Scienidæ ...	—	—	—	—	All regions but Australian
26. *Comephoridæ*...			—		
37. Atherinidæ ...	—	—			N. America and Australia
PHYSOSTOMI.					
59. Siluridæ... ...	—	—	—	—	All warm regions
65. Salmonidæ ...	—	—	—	—	Nearctic, New Zealand
70. Esocidæ	—	—			Nearctic
71. Umbridæ ...	—				Nearctic
73. Cyprinodontidæ		—			All regions but Australia
75. Cyprinidæ ...	—	—	—	—	All regions but Australian and Neotropical

Order and Family.	Sub-regions.				Range beyond the Region.
	Europe.	Mediter-ranean.	Siberia.	Japan.	
GANOIDEI.					
96. Accipenseridæ	—	—	—		Nearctic
97. Polydontidæ ...				—	Nearctic
INSECTS. LEPI-DOPTERA (PART).					
DURINI (BUTTER-FLIES).					
1. Danaidæ ...	—	—		—	All tropical regions
2. Satyridæ ...	—	—	—	—	Cosmopolite
8. Nymphalidæ ...	—	—	—	—	Cosmopolite
9. Libytheidæ ...	—	—			All continents but Australia
10. Nemeobeidæ ..	—				Absent from Nearctic region and Australia
13. Lycænidæ ...	—	—	—	—	Cosmopolite
14. Pieridæ	—	—	—	—	Cosmopolite
15. Papilionidæ ...	—	—	—	—	Cosmopolite
16. Hesperidæ ...	—	—	—	—	Cosmopolite
SPHIRIGIDEA.					
17. Zygænidæ ...	—	—	—	—	Cosmopolite
21. Stygiidæ ...	—	—	—	—	Neotropical
22. Ægeriidæ ...	—	—	—	—	Absent only from Australia
23. Sphingidæ ...	—	—	—	—	Cosmopolite

COLEOPTERA.—Of about 80 families into which the Coleoptera are divided, all the more important are cosmopolite, or nearly so. It would therefore unnecessarily occupy space to give tables of the whole for each region.

LAND SHELLS.—The more important families being cosmopolite, and the smaller ones being somewhat uncertain in their limits, the reader is referred to the account of the families and genera under each region, and to the chapter on Mollusca in the concluding part of this work, for such information as can be given of their distribution.

TABLE II.

LIST OF THE GENERA OF TERRESTIAL MAMMALIA AND BIRDS INHABITING THE PALÆARCTIC REGION.

EXPLANATION.

Names in *italics* show genera peculiar to the region.
Names inclosed thus (...) show genera which just enter the region, but are not considered
　　properly to belong to it.
Genera which undoubtedly belong to the region are numbered consecutively.

MAMMALIA.

Order, Family, and Genus.	No. of Species.	Range within the Region.	Range beyond the Region.
PRIMATES.			
SEMNOPITHECIDÆ.			
(Semnopithecus	1	Eastern Thibet)	Oriental genus
CYNOPITHECIDÆ.			
1. Macacus	4	Gibraltar, N. Africa, E. Thibet to Japan	Oriental
CHIROPTERA.			
PTEROPIDÆ.			
(Pteropus ...	2	Egypt, Japan)	Tropics of the E. Hemis.
(Xantharpyia ...	1	N. Africa, Palestine)	Oriental, Austro-Malayan
RHINOLOPHIDÆ.			
2. Rhinolphus ...	9	Temperate & Southern parts of Region	Warmer parts E. Hemisphere
(Asellia	1	Egypt)	Ethiopian, Java
(*Rhinopoma* ...	1	Egypt, Palestine)	[?] India
(Nycteris... ...	1	Egypt)	Nubia, Himalaya
VESPERTILIONIDÆ.			
3. Vesperugo ...	1	Siberia, Amoorland	[?]
4. *Otonycteris* ...	1	Egypt	[?]
5. Vespertilio ...	35	The whole region	Cosmopolite
(Kerivoula ...	1	N. China)	Oriental, S. Africa
6. Miniopteris ...	1	S. Europe, N. Africa, Japan	S. Afric. Malaya, Austral.
7. Plecotus	1	S. Europe	Himalayas
8. Barbastellus ...	2	Mid. and S. Europe, Palestine	Darjeeling, Timor
NOCTILIONIDÆ.			
9 Molossus	2	S. Europe, N. Africa	Ethiop., Neotrop., Australian
INSECTIVORA.			
ERINACEIDÆ.			
10. Erinaceus ...	4	The whole region ; excl. Japan	Oriental, Africa.

Order, Family, and Genus.	No. of Species.	Range within the Region.	Range beyond the Region.
TALPIDÆ.			
11. *Talpa*	5	The whole region	N. India
12. *Scaptochirus* ...	1	N. China	
13. *Anurosorex* ...	1	N. China	
14. *Scaptonyx* ...	1	N. China	
15. *Myogale*	2	S. E. Russia, Pyrenees	
16. *Nectogale*	1	Thibet	
17. Urotrichus ...	1	Japan	N. W. America
18. *Uropsilus*	1	E. Thibet	
SORICIDÆ.			
19. Sorex	10	The whole region	Absent from Australia & S. America
20. Crocidura	4	W. Europe to N. China	[?]
CARNIVORA.			
FELIDÆ.			
21. Felis	12	The whole region ; excl. extreme North	All regions but Austral.
22. Lyncus	9	S. Europe to Arctic sea	America N. of 66° N.Lat.
VIVERRIDÆ.			
(Viverra	1	N. China)	Oriental and Ethiopian
23. Genetta	1	S. Europe & N. Africa, Palestine	Ethiopian
(Herpestes ...	1	N. Africa, Spain [?], Palestine)	Oriental and Ethiopian
HYÆNIDÆ.			
24. Hyæna	1	N. Africa and S. W. Asia	Ethiopian, India
CANIDÆ.			
25. Canis	4	The whole region	All reg. but Austral. [?]
26. *Nyctereutes* ...	1	Japan, Amoorland, N. China	
MUSTELIDÆ.			
27. Martes	7	N. Europe and Asia, E. Thibet	Oriental, Nearctic
28. *Putorius*	3	W. Europe to N. E. Asia	
29. Mustela	10	The whole region	Nearctic, Ethiop., Himalayas, Peru
30. Vison	2	Europe and Siberia	N. America, N. India, China
31. Gulo	1	The Arctic regions	Arctic America
32. Lutra	2	The whole region	Oriental
33. *Lutronectes* ...	1	Japan	
34. Enhydris ..	1	N. Asia and Japan	California
35. *Meles*	2	Cen. Europe, Palestine, N.China, Japan	China to Hongkong
ÆLURIDÆ.			
36. Ælurus	1	S. E. Thibet	Nepal
37. *Æluropus* ...	1	E. Thibet	
URSIDÆ.			
38. Thalassarctos ...	1	Arctic regions	Arctic America
39. Ursus	4	The whole region	Oriental, Nearctic, Chil

Order, Family, and Genus.	No. of Species.	Range within the Region.	Range beyond the Region.
OTARIIDÆ.			
40. Callorhinus ...	1	Kamschatka and Behring's Straits	
41. Zalophus— ...	1	Japan	California
42. Eumetopias ...	1	Japan, Behring's Straits	California
TRICHECHIDÆ.			
43. Trichechus ...	1	Polar Seas	Arctic America
PHOCIDÆ.			
44. Callocephalus ...	3	North Sea, Caspian, Lake Baikal	Greenland
45. Pagomys	2	North Sea, Japan	N. Pacific
46. Pagophilus ...	2	Northern Seas	N. Pacific
47. Phoca	2	Northern Seas	N. Pacific
48. Halichærus ...	1	North Sea and Baltic	Greenland
49. *Pelagius*	2	Madeira to Black Sea	
50. Cystophora ...	2	N. Atlantic	N. Atlantic
SIRENIA.		. . . . . .	Tropics & Behring's Strts.
CETACEA.		. . . . . .	Oceanic
UNGULATA.			
EQUIDÆ.			
51. Equus	4	Cent. & and W. Asia & N. Africa	Ethiopian
SUIDÆ.			
52. Sus	2	The whole region	Oriental, Austro-Malayan
CAMELIDÆ.			
53. *Camelus*	2	Deserts of Cent. and W. Asia and N. Africa	
CERVIDÆ.			
54. Alces	1	North Europe and Asia	N. America
55. Tarandus ...	1	Arctic Europe and Asia	Arctic America
56. Cervus	8	The whole region	All regions but Austral.
57. *Dama*	1	Mediterranean district	
58. *Elaphodus* ...	1	N. W. China	
59. *Lophotragus* ...	1	N. China	
60. *Capreolus* ...	2	Temp. Europe and W. Asia and N. China	
61. *Moschus*	1	Amoor R., N. China, to Himalayas	
62. *Hydropotes* ...	1	N. China	
BOVIDÆ.			
63. ⎰ Bos	1	Europe, (not wild)	Oriental
64. ⎱ Bison	1	Poland and Caucasus	Nearctic
65. ⎰ *Poephagus* ...	1	Thibet	
66. *Addax*	1	N. Africa to Syria	
67. Oryx	1	N. Africa to Syria	Ethiopian deserts
68. ⎰ Gazella	12	N. Africa to Persia, and Beloochistan	S. Africa, India
69. ⎱ *Procapra* ...	2	W. Thibet and Mongolia	

R

Order, Family, and Genus.	No. of Species.	Range within the Region.	Range beyond the Region.
70. { Saiga	1	E. Europe and W. Asia	
71. { Pantholops ...	1	W. Thibet	
(Alcephalus ...	1	Syria)	Ethiopian genus.
72. Budorcas ...	2	E. Himalayas to E. Thibet	
73. Rupicapra ...	2	Pyrenees to Caucasus	
74. Nemorhedus ...	7	E. Himalayas to E. China and Japan	Oriental to Sumatra, Formosa
75. Capra	20	Spain to Thibet and N. E. Africa	Nilgherries, RockyMtns.
HYRACOIDEA			
HYRACIDÆ.			
(Hyrax	1	Syria)	Ethiopian genus
RODENTIA.			
MURIDÆ.			
76. Mus	?15	The whole region	E. Hemisphere
77. Cricetus	9	The whole region	
78. Cricetulus ...	3	N. China	
79. Meriones ...	8	W. and Central Asia to N. China, N. Africa	Ethiopian, Indian.
80. Rhombomys ...	6	E. Europe, Cent. Asia, N. Africa	
81. Psammomys ...	3	Egypt and Palestine	
82. Sminthus ...	3	East Europe, Siberia	
83. Arvicola	?21	The whole region	Himalayas, Nearctic
84. Cuniculus ...	1	N.E. Europe, Siberia	Arctic America
85. Myodes	1	North of region	Nearctic
86. Myospalax ...	3	Altai Mountains and N. China	
SPALACIDÆ.			
87. Ellobius	1	S. Russia and S. W. Siberia	
88. Spalax	1	Hungary and Greece to W. Asia, Palestine	
DIPODIDÆ.			
89. Dipus	?15	S. E. Europe and N. Africa to N. China	Africa, India
MYOXIDÆ.			
90. Myoxus	12	Temperate parts of whole region	Ethiopian
CASTORIDÆ.			
91. Castor	1	Temperate zone, from France to Amoorland	N. America
SCIURIDÆ.			
92. Sciurus	8	The whole region	All regions but Austral.
93. Sciuropterus ...	4	Finland to Siberia and Japan	Oriental, Nearctic
94. Pteromys... ...	3	Japan and W. China	Oriental
95. Spermophilus ...	10	E. Europe to N. China and Kamschatka	Nearctic
96. Arctomys... ...	4	Alps to E. Thibet and Kamschatka	Nearctic
OCTODONTIDÆ.			
97. Ctenodactylus ...	1	N. Africa	
HYSTRICIDÆ.			
98. Hystrix	2	S. Europe, Palestine, N. China.	Ethiopian, Oriental

Order, Family, and Genus.	No of Species.	Range within the Region.	Range beyond the Region.
LAGOMYIDÆ.			
99. Lagomys ...	10	Volga to E. Thibet and Kamschatka	Nearctic
LEPORIDÆ.			
100. Lepus	12	The whole region	All regions but Austral.

BIRDS.

PASSERES.

Order, Family, and Genus.	No of Species.	Range within the Region.	Range beyond the Region.
TURDIDÆ.			
1. Turdus	18	The whole region (excluding Spitzbergen)	Almost cosmopolite
2. Oreocincla ...	1	N.E. Asia and Japan, straggler to Europe	Oriental and Australian
3. Monticola ...	3	S. Europe, N. Africa, Palestine, N. China	Oriental and S. African
(Bessornis ...	1	Palestine)	Tropical and S. Africa
SYLVIIDÆ.			
4. Cisticola — ...	1	S. W. Europe, N. Africa, Japan	Ethiop., Orient., Austral.
5. Acrocephalus...	10	W. Europe to Japan	Orient., Ethiop., Austral.
6. Dumeticola ...	4	Nepaul, Lake Baikal, E. Thibet, high	
7. Potamodus ...	3	W. and S. Europe, N. Africa, E. Thibet	
8. Lusciniola ...	1	S. Europe	
9. Locustella ...	7	W. Europe and N. Africa to Japan	India, winter migrants (?)
10. Bradyptetus ...	2	S. Europe and Palestine	E. and S. Africa
11. Calamodus ...	? 3	Europe, N. Africa, Palestine	
12. Phylloscopus...	6	The whole region (excluding western islands)	Oriental
13. Hypolais ...	9	Europe, N. Africa, Palestine, China	China, Moluccas, India, Africa
14. Abrornis ...	2	Cashmere, E. Thibet	Oriental region
15. Reguloides ...	2	Europe and China	N. India, Formosa
16. Regulus ...	4	The whole region (excluding Iceland, &c.)	N. and Central America
17. Aedon	2	S. Europe, W. Asia, N. Africa	E. and S. Africa
18. Pyrophthalma	2	E. Europe and Palestine	
19. Melizophilus ...	2	W. and S. Europe, Sardinia	
20. Sylvia	6	Madeira to W. India, N. Africa	N.E. Africa, Ceylon migrants (?)
21. Curruca ...	7	Madeira to India, N. Africa	E. Africa, India, migrants
22. Luscinia ...	2	W. Europe, N. Africa, Persia	
23. Cyanecula ...	3	Europe and N. Africa to Kamschatka	Abyssinia and India migrants
24. Calliope ...	2	N. Asia, Himalayas, China	Centl. India (? migrant)
25. Erithacus ...	3	Atlantic Islands to Japan	
26. Grandala ...	1	High Himalayas and E. Thibet	

R 2

Order, Family, and Genus.	No. of Species	Range within the Region.	Range beyond the Region.
27. { Ruticilla ...	10	Eu. to Japan, N. Afr., Himalayas	Abyssinia, India
28. { Larvivora ...	2	E. Thibet, Amoor, Japan	Oriental
29. Dromolæa ..	3	S. Europe, N. Africa, Palestine	Ethiopian
30. Saxicola— ...	10	The whole region	E. and S. Africa, India
31. Cercomela ...	2	Palestine (a desert genus)	N.E. Africa, N.W. India
32. Pratincola ...	3	W. Europe, N. Africa to India	Ethiopian to Oriental
33. *Accentor*	12	W. Europe to Japan; high Himalayas	Himalayas (?) in winter
TIMALIIDÆ.			
34. *Pterorhinus* ..	3	Thibet and N. W. China	
(Malacocercus ...	1	Palestine)	Oriental genus
(Crateropus —	2	N. Africa, Persia)	Ethiopian genus
(Trochalopteron	3	E. Thibet)	Oriental genus
(Ianthocincla —	3	E. Thibet)	Oriental genus
PANURIDÆ.			
(Paradoxornis	3	Himalayas and E. Thibet)	(?) Oriental genus
35. Conostoma ...	1	High Himalayas and E. Thibet	
36. Suthora	3	E. Thibet	Himalayas, China, Formosa
37. *Panurus*	1	W. Europe to W. Siberia	
38. *Heteromorpha*..	1	Nepaul and E. Thibet, from 10,000 feet altitude	
39. *Cholornis*... ...	1	E. Thibet	
CINCLIDÆ.			
40. Cinclus	5	The whole region (Atlantic Islands excluded)	American highlands
(Myiophonus ...	1	Turkestan, Thian-Shan Mountains, 6,000 feet	Oriental genus
TROGLODYTIDÆ.			
41. Troglodytes ...	3	Iceland and Britain to Japan	Neotropical and Nearctic, Himalayas
(Pnoepyga ...	2	E. Thibet)	Oriental genus
CERTHIIDÆ.			
42. Certhia	2	W. Europe to N. China	Himalayas, Nearctic
43. *Tichodroma* ...	1	S. Europe to N. China	Abyssinia, Nepaul, high
SITTIDÆ.			
44. Sitta	7	W. Europe to Himalayas and Japan	India, Nearctic
PARIDÆ.			
45. Parus	20	W. Europe to Kamschatka, N. Africa	Nearctic, Oriental, Ethiopian
46. Lophophanes ...	6	Europe and high Himalayas	Nearctic
47. *Acredula*	6	W. Europe to N. China and Kamschatka	
48. Ægithalus ...	1	S. E. Europe	Ethiopian
LIOTRICHIDÆ.			
(Proparus... ...	4	Moupin, in E. Thibet)	Oriental genus and fam.

Order, Family, and Genus.	No. of Species.	Range within the Region.	Range beyond the Region.
PYCNONOTIDÆ.			
49. Microscelis ...	1	Japan	Oriental genus
50. Pycnonotus ...	2	Palestine, N. China, Japan	Oriental and Ethiopian
ORIOLIDÆ.			
51. Oriolus	2	S. Europe, China	Ethiopian and Oriental
MUSCICAPIDÆ.			
52. Muscicapa ...	2	W. and Central Europe	Ethiopian.
53. Butalis	2	W. Europe to Japan and China	E. and S. Africa, Moluccas
54. Erythrosterna...	3	Central Europe to N. China and Japan	Oriental & Madagascar
(Xanthopygia ...	1	Japan)	Oriental genus
(Eumyias— ...	1	E. Thibet)	Oriental genus
(Cyanoptila ...	1	Japan and Amoor)	Oriental genus
(Siphia	1	Moupin, E. Thibet)	Oriental genus
55. Tchitrea	2	N. China and Japan	Ethiopian and Oriental
LANIIDÆ.			
56. Lanius	11	The whole region (excl. Atlantic Islands)	Nearctic, Ethiopian, Oriental
(Telephonus ...	1	N. Africa)	Ethiopian genus
CORVIDÆ.			
57. Garrulus	7	W. Europe, N. Africa, to Japan	Himalayas, Formosa
58. Perisoreus ...	1	N. Europe and Siberia	N. America
(Urocissa	2	Cashmere, Japan)	Oriental genus
59. Nucifraga ...	3	W. Europe to Japan, and Himalayas	Himalayan pine forests
60. Pica...	5	W. Europe to China and Japan	S. China and Formosa migrants [?]
61. Cyanopica ...	2	Spain, N. E. Asia and Japan	
62. Corvus	12	The whole region	Cosmopolite (excl. S. Am.)
63. Fregilus	3	W. Europe to N. China, Himalayas	Abyssinian mountains
NECTARINIIDÆ.			
(Arachnecthra	1	Palestine)	Oriental genus
DICÆIDÆ.			
(Zosterops ...	1	Amoor and Japan)	Ethiop., Orien., Austral.
AMPELIDÆ.			
64. Ampelis	2	Northern half of region	North America
HIRUNDINIDÆ.			
65. Hirundo	2	The whole region	Cosmopolite
66. Cotyle	2	The whole region (excl. Atlan. Is.)	Nearctic, Ethiop., Orien.
67. Chelidon	3	The whole region	Oriental
FRINGILLIDÆ.			
68. Fringilla		The whole region	Africa

Order, Family, and Genus.	No. of Species	Range within the Region.	Range beyond the Region.
69. *Acanthis* ...	3	Europe and N. Africa to Central Asia	
70. *Procarduelis* ...	1	High Himalayas and E. Thibet	
71. Chrysomitris...	2	W. Europe to Japan	N. and S. America
72. *Dryospiza* ...	4	Atlantic Islands to Palestine, N. Africa	
73. *Metoponia* ...	1	N. E. Europe to W. Himalayas	
74. Chlorospiza ...	5	W. Europe, N. Africa to Japan	China, E. Africa
75. Passer	8	The whole region	Ethiopian, Oriental
76. Montifringilla	4	Europe to Cashmere and Siberia	
77. *Fringillauda*..	1	N. W. Himalayas to E. Thibet, high	
78. Coccothraustes	3	W. Europe, High Himalayas to Japan	N. America
79. *Mycerobas* ...	2	Central Asia & High Himalayas	
80 Eophona... ...	2	E. Thibet, China, and Japan	China
81. *Pyrrhula* ...	9	Azores to Japan, High Himalayas	Alaska
(Crithagra ...	1	Palestine)	Ethiopian genus
82. Carpodacus ...	12	Cent. Eu. to Japan, High Himalayas	India & China, N. Amer.
83. *Erythrospiza*...	4	N. Africa to Afghanistan and Turkestan	
84. *Uragus*	2	Turkestan & E. Thibet to Japan	
85. Loxia	3	Europe, High Himalayas to Japan	N. America
86. Pinicola... ...	1	N. Europe, Siberia	N. America
87. *Propyrrhula* ...	1	High Himalayas	Darjeeling in winter
88. *Pyrrhospiza* ...	1	Snowy Himalayas	
89. Linota	6	The whole region	N. America
90. Leucosticte ...	4	Turkestan to Kamschatka	N. W. America
Emberizinæ			
91. (Euspiza ...	4	E. Europe to Japan	N. America
92.) *Emberiza*...	25	Europe to Japan	N. India, China
93.) Fringillaria...	2	S. Europe, N. Africa	African genus
94. (Plectrophanes	2	Northern half of region	N. America
STURNIDÆ.			
95. Pastor	1	East Europe, Central Asia	India
96. Sturnia	2	Amoor, Japan, N. China	Oriental
97. Sturnus	3	The whole region (excl. Atlantic Islands)	India, China
(Amydrus ...	1	Palestine)	N. E. African genus
98. *Podoces*	3	Cen. Asia, Turkestan, Yarkand	
ALAUDIDÆ.			
99. Otocorys ...	3	N. Europe to Japan, N. Africa, Arabia	India, N. America, Andes
100. Alauda	7	The whole region (excl. Iceland)	India, Africa
101. Galerita... ...	2	Central Europe to N. China, N. Africa	India, Central Africa
102. Calandrella ...	4	Central Europe to N. China, N Africa	ndia
103. *Melanocorypha*	5	S. Eu. N. Africa, N. & Cen. Asia	N. W. India
104. *Pallasia*... ...	1	Mongolia	
(Certhilauda ...	1	N. Africa)	S. African genus
(Alaemon ...	1	N. Africa, Arabia)	Ethiopian genus

Order, Family, and Genus.	No of Species.	Range within the Region	Range beyond the Region.
105. Ammomanes...	3	S. Europe, N. Africa, to Cashmere	Africa, India
MOTACILLIDÆ.			
106. Motacilla ...	6	The whole region	Oriental, Ethiopian
107. Budytes ...	4	Europe to China	Oriental, Moluccas
108. Calobates ...	2	Atlantic Is., W. Europe, to China	Malaisia, Madagascar
PITTIDÆ.			
(Pitta	1	Japan)	Oriental & Austral. genus
PICARIÆ.			
PICIDÆ.			
109. Picoides... ...	3	N. and Cen. Europe to Thibet & E. Asia	North America
110. Picus	16	The whole region (excl. Atlantic Islands)	India, China, N. and. S. America
111. Hypopicus ...	1	N. China	Himalayas
(Yungipicus ...	1	N. China)	Oriental genus
112. Dryocopus ...	1	N. & Cen. Europe to N. China	Neotropical
113. Gecinus	6	W. Europe to Thibet, Amoor & Japan	Oriental
YUNGIDÆ.			
114. Yunx	2	W. Europe to N. W. India, Thibet and Japan	N. E. Africa, S. Africa
CUCULIDÆ.			
115. Cuculus	2	The whole region (excl. Atlantic Islands)	Ethiop. Oriental Austral.
116. Coccystes ...	1	S. Europe and N. Africa	Ethiopian and Oriental
CORACIIDÆ.			
117. Coracias... ...	1	Cent. Europe to Cent. Asia	Ethiopian, Oriental
(Eurystomus ...	1	Amoor in summer)	Oriental & Austral. genus
MEROPIDÆ.			
118. Merops	2	S. Europe to Cashmere, N. Afric	Ethiopian and Oriental
ALCEDINIDÆ.			
(Halcyon ...	3	W. Asia, N. China, Japan)	Ethiop., Orien., Austral.
119. Alcedo	2	Europe, N. China	
120. Ceryle	2	S. E. Europe, Japan	Africa, India, America
UPUPIDÆ.			
121. Upupa	1	S. Europe, N. China	Ethiop. & Oriental genus
CAPRIMULGIDÆ.			
122. Caprimulgus...	5	Europe to Japan	Ethiopian and Oriental
CYPSELIDÆ.			
123. Cypselus ...	4	The whole region (excl. Iceland	Ethiopian, America
124. Chætura... ...	2	N. China, Dauria	Africa, India

Order, Family, and Genus.	No. of Species	Range within the Region.	Range beyond the Region.
COLUMBÆ.			
COLUMBIDÆ.			
125. Columba ...	6	The whole region	Africa, Asia, America
126. Turtur	4	W. Europe to Japan	Ethiopian and Oriental
(Alsæcomus ...	1	E. Thibet)	Oriental genus
GALLINÆ.			
PTEROCLIDÆ.			
127. Pterocles ...	2	S. Europe, N. Africa, to W. India	Ethiopian genus
128. *Syrrhaptes* ...	2	Central Asia, N. China	
TETRAONIDÆ.			
129. Francolinus ...	1	Borders of Mediterranean	Ethiopian, Oriental
130. *Perdix*	2	Europe to Mongolia	
131. Coturnix ...	1	Central and S. Europe to Japan	Ethiop., Orien., Austral.
132. *Lerwa*	1	Snowy Himalayas to E. Thibet	
133. *Caccabis*	5	Cen. Europe and N. Africa to N. W. Himalayas	Abyssinia, Arabia
134. *Tetraogallus* ...	4	Caucasus to E. Thibet and Altai Mountains	
135. Tetrao	4	Europe and N. Asia	N. America
136. Bonasa	1	Europe and N. Asia	N. America
137. Lagopus... ...	4	Iceland, W. Europe to Japan	N. America, Greenland
PHASIANIDÆ.			
138. *Crossoptilon* ...	4	Thibet, Mongolia, N. China	
139. *Lophophorus* ...	3	Cashmere to E. Thibet (highest woods)	
140. Tetraophasis ...	1	E. Thibet	E. Thibet (?)
141. Ceriornis ...	1	N. W. Himalayas (high)	Himalayas to W. China
142. Pucrasia— ...	3	N. W. Himalayas to N. W. China	Himalayas
143. *Phasianus* ...	10	Western Asia to Japan	W. Himalayas, Formosa
144. *Thaumalea* ...	3	E. Thibet to Amoor, N. China	West China
145. *Ithaginis* ...	2	Nepaul to E. Thibet (high)	
TURNICIDÆ			
146. Turnix	2	Spain and N. Africa, N. China	Ethiop., Orien., Austral.
ACCIPITRES.			
VULTURIDÆ.			
147. *Vultur*	1	Spain and N. Africa to N. China	
148. Gyps	1	S. Europe, Palestine, Cen. Asia	E. Africa, India
149. Otogyps ...	1	S. Europe, N. Africa	S. Africa, India
150. Neophron ...	1	Atlantic Isds. to Palestine	Africa, India
FALCONIDÆ.			
151. Circus	5	Europe to Japan	Almost Cosmopolite
152. Astur	1	Europe to N. China	Almost Cosmopolite
153. Accipiter ...	2	Europe to Japan	Almost Cosmopolite
154. Buteo	4	Europe to Japan	Cosmopolite (excl. Australia)

Order, Family, and Genus.	No. of Species.	Range within the Region.	Range beyond the Region.
155. Archibuteo ...	1	N. Europe to Japan	N. America
156. Gypaetus ...	1	S. Europe, N. Africa	Abyssinia, Himalayas
157. Aquila	5	Europe to Japan	Nearctic, Ethiop., Orien.
158. Nisaetus... ...	2	E. Europe, N. Africa, W. Asia	India, Australia
159. Circaetus ...	1	E. and S. Europe, N. Africa, W. Asia	Africa, India
160. Haliæetus ...	3	Iceland and S. Europe to Japan	Cosmopolite (excl. Neotropical region)
161. Milvus	4	Europe to Japan, N. Africa	The Old World &Austral.
162. Elanus	2	N. Africa, N. China to Amoor	Cosmopolite (excl. East U. S.)
163. Pernis	1	Europe to Japan	Ethiopian and Oriental
164. Falco	5	The whole region	Cosmopolite (excl. Pacific Islands)
165. Hierofalco ...	5	The whole region	N. America
166. Cerchneis ...	4	Atlantic Islands to Japan	Cosmop. (excl. Oceania)
PANDIONIDÆ.			
167. Pandion— ...	1	Europe to Japan	Cosmopolite
STRIGIDÆ.			
168. Surnia	1	N. Europe and Siberia	North America
169. Nyctea	1	Arctic regions	Arctic America
170. Athene	4	Central and S. Europe to Japan	Ethiop.,Orien., Austral.
(Ninox	1	N. China and Japan)	Oriental genus
171. Glaucidium ...	1	Europe to N. China	America
172. Bubo	2	Europe to N. China	Africa, India, America
173. Scops	3	S. Europe to Japan	African, Orien., Austral.
174. Syrnium... ...	5	Europe to Japan	African, Oriental, Amer.
175. Otus	2	Europe to Japan	Almost Cosmopolite
176. Nyctala	1	N. Europe to E. Siberia	N. America
177. Strix	1	Europe and N. Africa	All warm & temp. regions

Peculiar or very characteristic Genera of Wading and Swimming Birds.

GRALLÆ.			
RALLIDÆ.			
Ortygometra ...	8	Europe, N. E. Africa	
SCOLOPACIDÆ.			
Ibidorhyncha ..	1	Cashmere & Cen. Asia, N. China	Himalayan Valleys
Terekia	1	N. E. Europe and Siberia	India, Australia(migrant)
Helodromas ...	1	E. and N. Europe, N. India	
Machetes... ...	1	N. and Cen. Europe, Cen. Asia	India in winter
Eurinorhynchus	1	N.E. Asia	Bengal
GLAREOLIDÆ.			
Pluvianus ...	1	N. Africa, Spain	
CHARADRIIDÆ.			
Vanellus ...	8	Europe to the Punjaub	S. America

Order, Family, and Genus.	No. of Species	Range within the Region.	Range beyond the Region
OTIDIDÆ.			
Otis..	2	W. Europe to Mongolia, N. Africa	
ANSERES.			
ANATIDÆ.			
Aix...	1	N. China to Amoor	N. America
Bucephala ...	3	Iceland, N. Europe, and Asia	N. America
Histrionicus ...	1	Iceland, N. Siberia	N. America
Harelda	1	North of whole region	Arctic America
Somateria ...	3	North of whole region	N. America
Œdemia... ...	3	North of whole region	N. America
LARIDÆ.			
Rissa	1	North coasts of whole region	N. America
COLYMBIDÆ.			
Colymbus ...	3	North of whole region	N. America
ALCIDÆ.			
Alca	2	North coasts of whole region	N. America
Fratercula ...	3	North coasts of whole region	N. America
Uria	3	North coasts of whole region	N. America
Mergulus ...	1	Iceland and Arctic coasts	Arctic America

ETHIOPIAN REGION

Scale 1 inch = 1,000 miles

EXPLANATION

Terrestrial Contours

From Sea level to 1,000 feet .... White
1,000 feet to 2,500
2,500 „ 5,000
5,000 „ 10,000
10,000 „ 20,000
Above 20,000 feet

The Marine Contour of 1,000 feet
is shown by a dotted line

Pasture lands shown thus
Forest
Desert
The boundaries and reference numbers of the Sub-regions are shown in Red.

Stanford's Geographical Estab.t London.

CHAPTER XI.

THIS is one of the best defined of the great zoological regions, consisting of tropical and South Africa, to which must be added tropical Arabia, Madagascar, and a few other islands, all popularly known as African. Some naturalists would extend the region northwards to the Atlas Mountains and include the whole of the Sahara; but the animal life of the northern part of that great desert seems more akin to the Palæarctic fauna of North Africa. The Sahara is really a debatable land which has been peopled from both regions; and until we know more of the natural history of the great plateaus which rise like islands in the waste of sand, it will be safer to make the provisional boundary line at or near the tropic, thus giving the northern half to the Palæarctic, the southern to the Ethiopian region. The same line may be continued across Arabia.

With our present imperfect knowledge of the interior of Africa, only three great continental sub-regions can be well defined. The open pasture lands of interior tropical Africa are wonderfully uniform in their productions; a great number of species ranging from Senegal to Abyssinia and thence to the Zambesi, while almost all the commoner African genera extend over the whole of this area. Almost all this extensive tract of country is a moderately elevated plateau, with a hot and dry climate, and characterised by a grassy vegetation interspersed with patches of forest. This forms our first or East African sub-region. The whole of the west coast from the south side of the Gambia River to about 10° or 12° south latitude, is a very different kind of country; being almost wholly dense forests where not cleared by man, and having the hot moist uniform climate, and perennial luxuriance of vegetation, which characterise the great equatorial belt of forest all round the globe. This forest country extends to an unknown distance inland, but it was found, with its features well marked, by Dr. Schweinfurth directly he crossed the south-western watershed of the Nile; and far to the south we find it again unmistakably indicated, in the excessively moist forest country about the head waters of the Congo, where the heroic Livingstone met his death. In this forest district many of the more remarkable African types are alone found, and its productions occasionally present us with curious similarities to those of the far removed South American or Malayan forests. This is our second or West African sub-region.

Extra-tropical South Africa possesses features of its own, quite distinct from those of both the preceding regions (although it has also much in common with the first). Its vegetation is known to be one of the richest, most peculiar, and most remarkable on the globe; and in its zoology it has a speciality, similar in kind but less in degree, which renders it both natural and convenient to separate it as our third, or South African sub-region. Its limits are not very clearly ascertained, but it is probably bounded by the Kalahari desert on the north-west, and by the Limpopo Valley, or the mountain range beyond, on the north-east, although some of its peculiar forms extend to Mozambique. There remains the great Island of Madagascar, one of the most isolated and most interesting on the globe, as regards its animal productions; and to this must be added, the smaller islands of Bourbon, Mauritius and Rodriguez, the Seychelles and the Comoro Islands, forming together the Mascarene Islands,—the whole constituting our fourth sub-region.

Zoological Characteristics of the Ethiopian Region.—We have now to consider briefly, what are the peculiarities and characteristics of the Ethiopian Region as a whole,—those which give it its distinctive features and broadly separate it from the other primary zoological regions.

Mammalia.—This region has 9 peculiar families of mammalia. Chiromyidæ (containing the aye-aye); Potamogalidæ and Chrysochloridæ (Insectivora); Cryptoproctidæ and Protelidæ (Carnivora); Hippopotamidæ and Camelopardalidæ (Ungulata); and Orycteropodidæ (Edentata). Besides these it possesses 7 peculiar genera of apes, *Troglodytes, Colobus, Myiopithecus Cercopithecus, Cercocebus, Theropithecus,* and *Cynocephalus;* 2 subfamilies of lemurs containing 6 genera, confined to Madagascar, with 3 genera of two other sub-families confined to the continent; of Insectivora a family, Centetidæ, with 5 genera, peculiar to Madagascar, and the genera *Petrodromus* and *Rhynchocyon* belonging to the Macroscelididæ, or elephant-shrews, restricted to the continent; numerous peculiar genera or sub-genera of civets; *Lycaon* and *Megalotis,* remarkable genera of Canidæ; *Ictonyx,* the zorilla, a genus allied to the weasels; 13 peculiar genera of Muridæ; *Pectinator,* a genus of the South American family Octodontidæ; and 2 genera of the South American Echimyidæ or spiny rats. Of abundant and characteristic groups it possesses *Macroscelides, Felis, Hyæna, Hyrax, Rhinoceros,* and *Elephas,* as well as several species of zebra and a great variety of antelopes.

The great speciality indicated by these numerous peculiar families and genera, is still farther increased by the absence of certain groups dominant in the Old-World continent, an absence which we can only account for by the persistence, through long epochs, of barriers isolating the greater part of Africa from the rest of the world. These groups are, Ursidæ, the bears; Talpidæ the moles; Camelidæ, the camels; Cervidæ, the deer; Caprinæ, the goats and sheep; and the genera *Bos* (wild ox); and *Sus* (wild boar). Combining these striking deficiencies, with the no less striking peculiarities above enumerated, it seems hardly possible to have a region more sharply divided from the rest of the globe than this is, by its whole assemblage of mammalia.

Birds.—In birds the Ethiopian region is by no means so strikingly peculiar, many of these having been able to pass the ancient barriers which so long limited the range of mammalia.

It is, however, sufficiently rich, possessing 54 families of land
birds, besides a few genera whose position is not well ascertained,
and which may constitute distinct families. Of these 6 are
peculiar, Musophagidæ (the plantain eaters); Coliidæ (the colies);
Leptosomidæ, allied to the cuckoos; Irrisoridæ, allied to the
hoopoes; and Serpentaridæ, allied to the hawks. Only one
Passerine family is peculiar—Paictidæ, while most of the other
tropical regions possess several; but *Euryceros* and *Buphaga,*
here classed with the Sturnidæ, ought, perhaps, to form two
more. It has, however, many peculiar genera, especially among the
fruit-thrushes, Pycnonotidæ; flycatchers, Muscicapidæ; shrikes,
Lanidæ; crows, Corvidæ; starlings, Sturnidæ; and weaver-birds,
Ploceidæ; the latter family being very characteristic of the region.
It is also rich in barbets, Megalæmidæ (7 peculiar genera);
cuckoos, Cuculidæ; rollers, Coraciidæ; bee-eaters, Meropidæ;
hornbills, Bucerotidæ; and goat-suckers, Caprimulgidæ. It is
poor in parrots and rather so in pigeons; but it abounds in
Pterocles and *Francolinus,* genera of Gallinæ, and possesses 4
genera of the peculiar group of the guinea-fowls, forming part of
the pheasant family. It abounds in vultures, eagles, and other
birds of prey, among which is the anomalous genus *Serpentarius,*
the secretary-bird, constituting a distinct family. Many of the
most remarkable forms are confined to Madagascar and the
adjacent islands, and will be noticed in our account of that sub-
region.

Reptiles.—Of the reptiles there are 4 peculiar Ethiopian
families;—3 of snakes, Rachiodontidæ, Dendraspidæ, and Atrac-
taspidæ and 1 of lizards, Chamæsauridæ.

Psammophidæ (desert snakes) are abundant, as are Lycodontidæ
(fanged ground-snakes), and Viperidæ (vipers). The following
genera of snakes are peculiar or highly characteristic:—*Lepto-
rhynchus, Rhamnophis, Herpetethiops* and *Grayia* (Colubridæ);
Hopsidrophis and *Bucephalus* (Dendrophidæ); *Langalia* (Dryo-
phidæ); *Pythonodipsas* (Dipsadidæ); *Boedon, Lycophidion, Holu-
ropholis, Simocephalus* and *Lamprophis* (Lycodontidæ); *Hortulia*
and *Sanzinia* (Pythonidæ); *Cyrptophis, Elapsoidea* and *Pœcilophis* (Elapidæ); and *Atheris* (Viperidæ). The following genera

of lizards are the most characteristic:—*Monotrophis* (Lepidos-ternidæ) ; *Cordylus, Pseudocordylus, Platysaurus, Cordylosaurus, Pleurostichus, Saurophis* and *Zonurus* (Zonuridæ) ; *Sphænops, Scelotes, Sphænocephalus* and *Sepsina* (Sepidæ) ; *Pachydactylus* (Geckotidæ); *Agama* (Agamidæ); and *Chameleon* (Chameleonidæ). Of tortoises, *Cynyxis, Pyxis* and *Chersina* (Testudinidæ), and *Cycloderma* (Trionychidæ) are the most characteristic.

Amphibia.—Of the 9 families of amphibia there is only 1 peculiar, the Dactylethridæ, a group of toads ; but the Alytidæ, a family of frogs, are abundant.

Fresh-water Fish.—Of the 14 families of fresh-water fishes 3 are peculiar: Mormyridæ and Gymnarchidæ, small groups not far removed from the pikes ; and Polypteridæ, a small group of ganoid fishes allied to the gar-pikes (Lepidosteidæ) of North America.

Summary of Ethiopian Vertebrates.—Combining the results here indicated and set forth in greater detail in the tables of distribution, we find that the Ethiopian region possesses ex-amples of 44 families of mammalia, 72 of birds, 35 of reptiles, 9 of amphibia, and 15 of fresh-water fishes. It has 23 (or perhaps 25) families of Vertebrata altogether peculiar to it out of a total of 175 families, or almost exactly one-eighth of the whole. Out of 142 genera of mammalia found within the region, 90 are peculiar to it ; a proportion not much short of two-thirds. Of land birds there are 294 genera, of which 179 are peculiar ; giving a proportion of a little less than three-fifths.

Compared with the Oriental region this shows a con-siderably larger amount of speciality under all the heads; but the superiority is mainly due to the wonderful and iso-lated fauna of Madagascar, to which the Oriental region has nothing comparable. Without this the regions would be nearly equal.

Insects: Lepidoptera.—11 out of the 16 families of butter-flies have representatives in Africa, but none are peculiar. Acræidæ is one of the most characteristic families, and there are many interesting forms of Nymphalidæ, Lycænidæ, and Papilionidæ. The peculiar or characteristic forms are *Amauris* (Danaidæ); *Gnophodes, Leptoneura, Bicyclus, Heteropsis* and *Cœnyra* (Satyridæ); *Acrœa* (Acræidæ); *Lachnoptera, Precis, Salamis, Crenis, Godartia, Amphidema, Pseudacrœa, Catuna, Euryphene, Romalœosoma, Hamanumida, Aterica, Harma, Meneris, Charaxes,* and *Philognoma* (Nymphalidæ); *Pentila, Liptena, Durbania, Zeritis, Capys, Phytala, Epitola, Hewitsonia* and *Deloneura* (Lycænidæ); *Pseudopontia, Idmais, Teracolus, Callosune* (Pieridæ); *Abantis, Ceratrichia* and *Caprona* (Hesperidæ). The total number of species known is about 750; which is very poor for an extensive tropical region, but this is not to be wondered at when the nature of much of the country is considered. It is also, no doubt, partly due to our comparative ignorance of the great equatorial forest district, which is the only part likely to be very productive in this order of insects.

Colcoptera.—In our first representative family, Cicindelidæ or tiger-beetles, the Ethiopian region is rather rich, having 13 genera, 11 of which are peculiar to it; and among these are such remarkable forms as *Manticora, Myrmecoptera* and *Dromica;* with *Megacephala,* a genus only found elsewhere in Australia and South America.

In Carabidæ or carnivorous ground beetles, there are about 75 peculiar genera. Among the most characteristic are *Anthia, Polyrhina, Graphipterus* and *Piezia,* which are almost all peculiar; while *Orthogonius, Hexagonia, Macrochilus, Thyreopterus, Eudema,* and *Abacetus* are common to this and the Oriental region; and *Hypolithus* to the Neotropical.

Out of 27 genera of Buprestidæ, or metallic beetles, only 6 are peculiar to the region, one of the most remarkable being *Polybothrus,* confined to Madagascar. *Sternocera* and *Chrysochroa* are characteristic of this region and the Oriental; it has *Julodis* in common with the Mediterranean sub-region, and *Belionota* with the Malayan.

The region is not rich in Lucanidæ, or stag-beetles, possessing only 10 genera, 7 of which are peculiar, but most of them consist of single species. The other three genera, *Cladognathus*, *Nigidius*, and *Figulus*, are the most characteristic, though all have a tolerably wide range in the Old World.

In the elegant Cetoniidæ, or rose-chafers, this region stands preeminent, possessing 76 genera, 64 of which are peculiar to it. The others are chiefly Oriental, except *Oxythræa* which is European, and *Stethodesma* which is Neotropical. Preeminent in size and beauty is *Goliathus*, comprising perhaps the most bulky of all highly-coloured beetles. Other large and characteristic genera are *Ceratorhina*, *Ischnostoma*, *Anochilia*, *Diplognatha*, *Agenius*, and many others of less extent.

In the enormous tribe of Longicorns, or long-horned beetles, the Ethiopian is not so rich as the other three tropical regions; but this may be, in great part, owing to its more productive districts having never been explored by any competent entomologists. It nevertheless possesses 262 genera, 216 of which are peculiar, the others being mostly groups of very wide range. Out of such a large number it is difficult to select a few as most characteristic, but some of the peculiarities of distribution as regards other regions may be named. Among Prionidæ, *Tithoes* is a characteristic Ethiopian genus. A few species of the American genera *Parandra* and *Mallodon* occur here, while the North Temperate genus *Prionus* is only found in Madagascar. Among Cerambycidæ, *Promeces* is the most characteristic. The American genera *Oeme* and *Cyrtomerus* occur; while *Homalachnus* and *Philagathes* are Malayan, and *Leptocera* occurs only in Madagascar, Ceylon, Austro-Malaya, and Australia. The Lamiidæ are very fine; *Sternotomis*, *Tragocephala*, *Ceroplesis*, *Phryneta*, *Volumnia*, and *Nitocris*, being very abundant and characteristic. Most of the non-peculiar genera of this family are Oriental, but *Spalacopsis* and *Acanthoderes* are American, while *Tetraglenes* and *Schœnionta* have been found only in East and South Africa and in Malaya.

Terrestrial Mollusca—In the extensive family of the Helicidæ or snails, 13 genera are represented, only one of which, *Columna*, is peculiar. This region is however the metropolis of *Achatina*, some of the species being the largest land-shells

known. *Buliminus, Stenogyra,* and *Pupa* are characteristic
genera. *Bulimus* is absent, though one species inhabits St.
Helena. The operculated shells are not very well represented,
the great family of Cyclostomidæ having here only nine genera,
with but one peculiar, *Lithidion,* found in Madagascar, Socotra,
and Arabia. None of the genera appear to be well represented
throughout the region, and they are almost or quite absent from
West Africa.

According to Woodward's *Manual* (1868) West Africa has
about 200 species of land-shells, South Africa about 100,
Madagascar nearly 100, Mauritius about 50. All the islands
have their peculiar species; and are, in proportion to their
extent, much richer than the continent; as is usually the case.

THE ETHIOPIAN SUB-REGIONS.

It has been already explained that these are to some extent
provisional; yet it is believed that they represent generally the
primary natural divisions of the region, however they may be
subdivided when our knowledge of their productions becomes
more accurate.

I. *The East African Sub-region, or Central and East Africa.*

This division includes all the open country of tropical Africa
south of the Sahara, as well as an undefined southern margin of
that great desert. With the exception of a narrow strip along
the east coast and the valleys of the Niger and Nile, it is a vast
elevated plateau from 1,000 to 4,000 feet high, hilly rather than
mountainous, except the lofty table land of Abyssinia, with
mountains rising to 16,000 feet and extending south to the
equator, where it terminates in the peaks of Kenia and Kili-
mandjaro, 18,000 and 20,000 feet high. The northern portion
of this sub-region is a belt about 300 miles wide between the
Sahara on the north and the great equatorial forest on the south,
extending from Cape Verd, the extreme western point of Africa,
across the northern bend of the Niger and Lake Tchad to the
mountains of Abyssinia. The greater part of this tract has a

moderate elevation.   The eastern portion reaches from about the second cataract of the Nile, or perhaps from about the parallel of 20° N. Latitude, down to about 20° S. Latitude, and from the east coast to where the great forest region commences, or to Lake Tanganyika and about the meridian of 28° to 30° E. Longitude. The greater part of this tract is a lofty plateau.

The surface of all this sub-region is generally open, covered with a vegetation of high grasses or thorny shrubs, with scattered trees and isolated patches of forest in favourable situations. The only parts where extensive continuous forests occur, are on the eastern and western slopes of the great Abyssinian plateau, and on the Mozambique coast from Zanzibar to Sofala.   The whole of this great district has one general zoological character. Many species range from Senegal to Abyssinia, others from Abyssinia to the Zambesi, and a few, as *Mungos fasciatus* and *Phacochœrus œthiopicus*, range over the entire sub-region.   *Fennecus, Ictonyx*, and several genera of antelopes, characterise every part of it, as do many genera of birds.   *Coracias nævia, Corythornis cyanostigma, Tockus nasutus, T. erythrorhynchus, Parus leucopterus, Buphaga africana, Vidua paradisea*, are examples of *species*, which are found in the Gambia, Abyssinia and South East Africa, but not in the West African sub-region ; and considering how very little is known of the natural history of the country immediately south of the Sahara, it may well be supposed that these are only a small portion of the species really common to the whole area in question, and which prove its fundamental unity.

Although this sub-region is so extensive and so generally uniform in physical features, it is by far the least peculiar part of Africa.   It possesses, of course, all those wide-spread Ethiopian types which inhabit every part of the region, but it has hardly any special features of its own.   The few genera which are peculiar to it have generally a limited range, and for the most part belong, either to the isolated mountain-plateau of Abyssinia which is almost as much Palæarctic as Ethiopian, or to the woody districts of Mozambique where the fauna has more of a West or South African character.

s 2

Mammalia.—The only forms of Mammalia peculiar to this sub-region are *Theropithecus*, one of the Cynopithecidæ confined to Abyssinia; *Petrodromus* and *Rhynchocyon*, belonging to the insectivorous Macroscelididæ, have only been found in Mozambique; the Antelopine genus *Neotragus*, from Abyssinia southward; *Saccostomus* and *Pelomys* genera of Muridæ inhabiting Mozambique; *Heterocephalus* from Abyssinia, and *Heliophobius* from Mozambique, belonging to the Spalacidæ; and *Pectinator* from Abyssinia, belonging to the Octodontidæ. *Cynocephalus*, *Rhinoceros*, *Camelopardalis*, and antelopes of the genera *Oryx*, *Cervicapra*, *Kobus*, *Nanotragus*, *Cephalophus*, *Hippotragus*, *Alcephalus*, and *Catoblepas*, are characteristic; as well as *Felis*, *Hyœna*, and numerous civets and ichneumons.

Birds.—Peculiar forms of birds are hardly to be found here; we only meet with two—*Hypocolius*, a genus of shrikes in Abyssinia; and *Balœniceps*, the great boat-billed heron of the Upper Nile. Yet throughout the country birds are abundant, and most of the typical Ethiopian forms are well represented.

Reptiles.—Of reptiles, the only peculiar forms recorded are *Xenocalamus*, a genus of snakes, belonging to the Calamariidæ; and *Pythonodipsas*, one of the Dipsadidæ, both from the Zambesi; and among lizards, *Pisturus*, one of the Geckotidæ, from Abyssinia.

Amphibia and Fishes.—There are no peculiar forms of amphibia or of fresh-water fishes.

Insects.—Insects are almost equally unproductive of peculiar forms. Among butterflies we have *Abantis*, one of the Hesperidæ, from Mozambique; and in Coleoptera, 2 genera of Cicindelidæ, 8 of Carabidæ, 1 or 2 of Cetoniidæ, and about half-a-dozen of Longicorns: a mere nothing, as we shall see, compared with the hosts of peculiar genera that characterise each of the other sub-regions. Neither do land-shells appear to present any peculiar forms.

The fact that so very few special types characterise the extensive area now under consideration is very noteworthy. It justifies us in uniting this large and widespread tract of country as forming essentially but one sub-division of the great Ethiopian

PLATE IV.

CHARACTERISTIC ANIMALS OF EAST AFRICA.

region, and it suggests some curious speculations as to-the former history of that region, a subject which must be deferred to the latter part of this chapter. In none of the other great tropical regions does it occur, that the largest portion of their area, although swarming with life, yet possesses hardly any distinctive features except the absence of numerous types characteristic of the other sub-regions.

Plate IV.—Illustrating the Zoology of East Africa.—Although this sub-region has so little speciality, it is that which abounds most in large animals, and is, perhaps, the best representative of Africa as regards zoology. Some of the most distinctive of African animals range over the whole of it, and as, from recent explorations, many parts of this wide area have been made known to the reading public, we devote one of our plates to illustrate the especially African forms of life that here abound. The antelopes represented are the koodoo (*Tragelaphus strepsiceros*) one of the handsomest of the family, which ranges over all the highlands of Africa from Abyssinia to the southern districts. To the left is the aardvark, or earth pig, of North Eastern Africa (*Orycteropus æthiopicus*) which, to the north of the equator in East Africa, represents the allied species of the Cape of Good Hope. These Edentata are probably remnants of the ancient fauna of Africa, when it was completely isolated from the northern continents and few of the higher types had been introduced. The large bird in the foreground is the secretary-bird, or serpent-killer (*Serpentarius reptilivorus*), which has affinities both for the birds-of-prey and the waders. It is common over almost all the open country of Africa, destroying and feeding on the most venomous serpents. The bird on the wing is the red-billed promerops (*Irrisor erythrorhynchus*), a handsome bird with glossy plumage and coral-red bill. It is allied to the hoopoes, and feeds on insects which it hunts for among the branches of trees. This species also ranges over a large part of east and central Africa to near the Cape of Good Hope. Other species are found in the west; and the genus, which forms a distinct family, *Irrisoridæ*, is one of the best marked Ethiopian types of birds. In the distance is a rhinoceros, now one of the characteristic features of African zoology, though there is reason to believe that it is a comparatively recent intruder into the country.

II. The West-African Sub-region.

This may be defined as the equatorial-forest sub-region, since it comprises all that portion of Africa, frcm the west coast inland, over which the great equatorial forests prevail more or less uninterruptedly. These commence to the south of the Gambia River, and extend eastwards in a line roughly parallel to the southern margin of the great desert, as far as the sources of the upper Nile and the mountains forming the western boundary of the basin of the great lakes ; and southward to that high but marshy forest-country in which Livingstone was travelling at the time of his death. Its southern limits are undetermined, but are probably somewhere about the parallel of 11° S. Latitude.[1]

This extensive and luxuriant district has only been explored zoologically in the neighbourhood of the West coast. Much, no doubt, remains to be done in the interior, yet its main features are sufficiently well known, and most of its characteristic types of animal life have, no doubt, been discovered,

Mammalia.—Several very important groups of mammals are peculiar to this sub-region. Most prominent are the great anthropoid apes—the gorilla and the chimpanzee—forming the genus *Troglodytes ;* and monkeys of the genera *Myiopithecus* and *Cercocebus.* Two remarkable forms of lemurs, *Perodicticus* and *Arctocebus*, are also peculiar to West Africa. Among the Insectivora is *Potamogale*, a semi-aquatic animal, forming a distinct family ; and three peculiar genera of civets (Viverridæ) have been described. *Hyomoschus*, a small, deer-like animal, belongs to the Tragulidæ, or chevrotains, a family otherwise

---

[1] Dr. Schweinfurth has accurately determined the limits of the sub-region at the point where he crossed the watershed between the Nile tributaries and those of the Shari, in 4½° N. Lat. and 28½° E. Long. He describes a sudden change in the character of the vegetation, which to the southward of this point assumes a West-African character. Here also the chimpanzee and grey parrot first appear, and certain species of plants only known elsewhere in Western Africa.

confined to the Oriental region; and in the squirrel family is a curious genus, *Anomalurus*, which resembles the flying squirrels of other parts of the world, without being directly allied to them.

Birds.—In this class we find a larger proportionate number of peculiar forms. *Hypergerus* and *Alethe*, belonging to the Timaliidæ, or babblers, are perhaps allied to Malayan groups; *Parinia*, a peculiar form of tit, is found only in Prince's Island; *Ixonotus* is an abundant and characteristic form of Pycnonotidæ; *Fraseria*, *Hypodes*, *Cuphopterus*, and *Chaunonotus*, are peculiar genera of shrikes; *Picathartes* is one of the many strange forms of the crow family; *Cinnyricinclus* is a peculiar genus of sunbirds; *Pholidornis* is supposed to belong to the Oriental Dicæidæ, or flower-peckers; *Waldenia* is a recently-described new form of swallow; *Ligurnus*, a finch, *Spermospiga*, a weaver bird, and *Onychognathus* a starling, are also peculiar West African genera. Coming to the Picariæ we have *Verreauxia*, a peculiar woodpecker; three peculiar genera of barbets (Megalæmidæ); the typical plantain-eaters (Musophaga); *Myioceyx*, a peculiar genus of kingfishers; while *Berenicornis* is a genus of crested hornbills, only found elsewhere in Malaya. The grey parrots, of the genus *Psittacus*, are confined to this sub-region, as are two peculiar genera of partridges, and three of guineafowl. We have also here a species of *Pitta*, one of the Oriental family of ground-thrushes; and the Oriental paroquets, *Palæornis*, are found here as well as in Abyssinia and the Mascarene Islands.

We thus find, both in the Mammalia and birds of West Africa, a special Oriental or even Malayan element not present in the other parts of tropical Africa, although appearing again in Madagascar. In the Mammalia it is represented by the anthropoid apes; by *Colobus* allied to *Semnopithecus*, and by *Cercocebus* allied to *Macacus*; and especially by a form of the Malayan family of chevrotains (Tragulidæ). The Malayan genus of otters, *Aonyx*, is also said to occur in West and South Africa. In birds we have special Oriental and Malayan affinities in *Alethe*, *Pholidornis*, *Berenicornis*, *Pitta*, and *Palæornis*; while the Oriental genus *Treron* has a wide range in Africa. We shall endeavour to ascertain the meaning of this special relation at a subsequent stage of our inquiries.

Plate V.—River Scene in West Africa, with Characteristic Animals.—Our artist has here well represented the luxuriance and beauty of a tropical forest; and the whole scene is such as might be witnessed on the banks of one of the rivers of equatorial West Africa. On the right we see a red river-hog (*Potamochœrus penicillatus*), one of the handsomest of the swine family, and highly characteristic of the West African sub-region. In a tree overhead is the potto (*Perodicticus potto*), one of the curious forms of lemur confined to West Africa. On the left is the remarkable *Potamogale velox*, first discovered by Du Chaillu,—an Insectivorous animal, with the form and habits of an otter. On the other side of the river are seen a pair of gorillas (*Troglodytes gorilla*), the largest of the anthropoid apes.

The bird on the wing is the Whydah finch (*Vidua paradisea*), remarkable for the enormous plumes with which the tail of the male bird is decorated during the breeding season. The crested bird overhead is one of the beautiful green touracos (*Turacus macrorhynchus*), belonging to the Musophagidæ, or plantain-eaters, a family wholly African, and most abundant in the western sub-region.

Reptiles.—In this class we find a large number of peculiar forms ; 13 genera of snakes, 3 of lizards, and 2 of tortoises being confined to the sub-region. The snakes are *Pariaspis, Elapops,* and *Prosymna* (Calamariidæ), *Rhamnophis, Herpetethiops,* and *Grayia* (Colubridæ), *Neusterophis* and *Limnophis* (Homalopsidæ), *Simocephalus* and *Holurophis* (Lycodontidæ) ; *Pelophilus* (Pythonidæ) ; *Elapsoidea* (Elapidæ) ; and *Atheris* (Viperidæ). The lizards are *Dalophia* (Lepidosternidæ) ; *Otosaurus* (Scincidæ); *Psilodactylus* (Geckotidæ). The tortoises, *Cinyxis* (Testudinidæ) and *Tetrathyra* (Trionichidæ).

Amphibia.—Of Amphibia, there are 2 peculiar genera of tree-frogs, *Hylambatis* and *Hemimantis,* belonging to the Polypedatidæ.

PLATE V.

SCENE IN WEST AFRICA, WITH CHARACTERISTIC ANIMALS.

Here, too, we find some interesting relations with the Oriental region on the one side, and the Neotropical on the other. The snakes of the family Homalopsidæ have a wide range, in America, Europe, and all over the Oriental region, but are confined to West Africa in the Ethiopian region. *Dryiophis* (Dryiophidæ) and *Dipsadoboa* (Dipsadidæ) on the other hand, are genera of tropical America which occur also in West Africa. The family of lizards, Acontiadæ, are found in West and South Africa, Ceylon, and the Moluccas. The family of toads, Engystomidæ, in West and South Africa and the whole Oriental region; while the Phryniscidæ inhabit tropical Africa and Java.

Insects.—We have here a large number of peculiar genera. There are 10 of butterflies, *Lachnoptera, Amphidema,* and *Catuna* belonging to the Nymphalidæ, while four others are Lycænidæ. The genus *Euxanthe* is common to West Africa and Madagascar.

Of Coleoptera there are 53 peculiar genera; 20 are Carabidæ, 2 Lucanidæ, 12 Cetoniidæ, 3 Prionidæ, 16 Cerambycidæ, and 34 Lamiidæ. Besides these there are 4 or 5 genera confined to West Africa and Madagascar.

Land Shells.—West Africa is very rich in land shells, but it does not appear to possess any well-marked genera, although several of the smaller groups or sub-genera are confined to it. Helicidæ of the genera *Nanina Buliminus* and *Achatina* are abundant and characteristic.

Islands of the West African Sub-region.—The islands in the Gulf of Guinea are, Fernando Po, very near the main land, with Prince's Island and St. Thomas, considerably further away to the south-west. Fernando Po was once thought to be a remarkable instance of an island possessing a very peculiar fauna, although close to the main land and not divided from it by a deep sea. This, however, was due to our having obtained considerable collections from Fernando Po, while the opposite coast was almost unknown. One after another the species supposed to be peculiar have been found on the continent, till it becomes probable, that, as in the case of other islands similarly situated, it contains no peculiar species whatever. The presence of numerous mammalia, among which are baboons, lemurs, *Hyrax,* and

Anomalurus, shows that this island has probably once been united to the continent.

Prince's Island, situated about 100 miles from the coast, has no mammals, but between 30 and 40 species of birds. Of these 7 are peculiar species, viz., *Zosterops ficedulina, Cuphopterus dohrni* (a peculiar genus of Sylviidæ), *Symplectes princeps, Crithagra rufilata, Columba chlorophœa, Peristera principalis,* and *Strix thomensis.*

In the Island of St. Thomas, situated on the equator about 150 miles from the coast, there are 6 peculiar species out of 30 known birds, viz., *Scops leucopsis, Zosterops lugubris, Turdus olivaceo-fuscus, Oriolus crassirostris, Symplectes sancti-thomœ* and *Aplopelia simplex;* also *Strix thomensis* in common with Prince's Island. The remainder are all found on the adjacent coasts. It is remarkable that in Prince's Island there are no birds of prey, any that appear being driven off by the parrots (*Psittacus erithacus*) that abound there; whereas in St. Thomas and Fernando Po they are plentiful.

III. South-African Sub-region.

This is the most peculiar and interesting part of Africa, but owing to the absence of existing barriers its limits cannot be well defined. The typical portion of it hardly contains more than the narrow strip of territory limited by the mountain range which forms the boundary of the Cape Colony and Natal, while in a wider sense it may be extended to include Mozambique. It may perhaps be best characterised as bounded by the Kalahari desert and the Limpopo river. It is in the more limited district of the extreme south, that the wonderful Cape flora alone exists. Here are more genera and species, and more peculiar types of plants congregated together, than in any other part of the globe of equal extent. There are indications of a somewhat similar richness and specialization in the zoology of this country; but animals are so much less closely dependent on soil and climate, that much of the original peculiarity has been obliterated, by long continued interchange of species with so vast an area as that of Africa south of the equator. The extreme peculiarity and isolation of the flora must not, however, be lost sight of, if we would correctly interpret the phenomena afforded by the distribution of animal life on the African continent.

Mammalia.—A much larger number of peculiar forms of mammals are found here than in any of the other sub-regions, although it is far less in extent than either of the three divisions of the continent. Among Insectivora we have the Chrysochloridæ, or golden moles, consisting of two genera confined to South Africa; while the Macroscelididæ, or elephant shrews, are also characteristically South African, although ranging as far as Mozambique and the Zambezi, with one outlying species in North Africa. The Viverridæ are represented by three peculiar genera, *Ariela, Cynictis,* and *Suricata.* The Carnivora present some remarkable forms: *Proteles,* forming a distinct family allied to the hyænas and weasels; and two curious forms of Canidæ— *Megalotis* (the long-eared fox) and *Lycaon* (the hyæna-dog), the latter found also in parts of East Africa. *Hydrogale* is a peculiar form of Mustelidæ; *Pelea* one of the antelopes; *Dendromys, Malacothrix,* and *Mystromys* are peculiar genera of the mouse family (Muridæ); *Bathyerges* one of the mole-rats (Spalacidæ); *Pedetes,* the Cape-hare, a remarkable form of jerboa; and *Petromys,* one of the spiny-rats (Echimyidæ). The remarkable *Orycteropus,* or earth-pig, has one species in South and one in North East Africa. We have thus eighteen genera of mammalia almost or quite peculiar to South Africa.

Birds.—These do not present so many peculiar forms, yet some are very remarkable. *Chætops* is an isolated genus of thrushes (Turdidæ). *Lioptilus,* one of the fruit-thrushes (Pycnonotidæ). *Pogonocichla,* one of the fly-catchers; *Urolestes,* a shrike; *Promerops,* a sun-bird; *Philetærus* and *Chera,* weaverbirds; and three peculiar genera of larks—*Spizocorys, Heterocorys,* and *Tephrocorys,* complete the list of peculiar types of Passeres. A wood-pecker, *Geocolaptes,* is nearly allied to a South American genus. The Cape-dove, *Œna,* is confined to South and East Africa and Madagascar; and *Thalassornis* is a peculiar form of duck. Several genera are also confined to West and South Africa;—

as *Phyllastrephus* (Pycnonotidæ), *Smithornis* (Muscicapidæ),
Corvinella (Laniidæ) ; *Barbatula* and *Xylobucco* (Megalæmidæ) ;
Ceuthmochares, also in Madagascar, (Cuculidæ) ; *Typanistria*
(Columbidæ). Other remarkable forms, though widely spread
over Africa, appear to have their metropolis here, as *Colius* and
Indicator. Others seem to be confined to South Africa and
Abyssinia, as the curious *Buphaga* (Sturnidæ) ; and *Apalo-
derma* (Trogonidæ). *Machærhamphus* (Falconidæ) is found only
in South-West Africa, Madagascar, and the Malay Peninsula.

Reptiles.—There are 4 peculiar genera of snakes,—*Typhline*,
belonging to the blind burrowing snakes, Typhlopidæ; *Lamprophis* (Lycodontidæ) ; *Cyrtophis* and *Pœcilophis* (Elapidæ), a
family which is chiefly Oriental and Australian. Of Lizards
there are 10 peculiar genera ; *Monotrophis* (Lepidosternidæ), but
with an allied form in Angola; *Cordylus, Pseudocordylus, Platysaurus, Cordylosaurus, Pleurostichus,* and *Saurophis*, all peculiar
genera of Zonuridæ ; *Chamæsaura*, forming the peculiar family
Chamæsauridæ ; *Colopus* and *Rhopitropus* (Geckotidæ).

Amphibia.—Of Amphibia there are 4 peculiar genera :
Schismaderma (Bufonidæ) ; *Brachymerus* (Engystomidæ) ; *Phry-
nobatrachus* and *Stenorhynchus* (Ranidæ). These last are allied
to Oriental genera, and the only other Engystomidæ are Oriental
and Neotropical.

Fresh-water Fish.—Of fresh-water fishes there is 1 genus—*Ab-
rostomus*—belonging to the carp family, peculiar to South Africa.

Insects.—South Africa is excessively rich in insects, and the
number of peculiar types surpasses that of any other part of the
region. We can only here summarize the results.

Lepidoptera.—Of butterflies there are 7 peculiar genera ; 2
belonging to the Satyridæ, 1 to Acræidæ, 3 to Lycænidæ,
and 1 to Hesperidæ. *Zeritis* (Lycænidæ) is also characteristic
of this sub-region, although 1 species occurs in West Africa.

Coleoptera.—These are very remarkable. In the family of
Cicindelidæ, or tiger-beetles, we have the extraordinary *Manticora*
and *Platychile*, forming a sub-family, whose nearest allies are in
North America ; as well as *Ophryodera* and *Dromica*, the latter
an extensive genus, which ranges as far north as Mozambique

and Lake Ngami. Another genus of this family, *Jansenia*, is common to South Africa and South India.

In the large family of Carabidæ, or ground-beetles, there are 17 peculiar South African genera, the most important being *Crepidogaster, Hytrichopus, Arsinoe*, and *Piezia*. Three others— *Eunostus, Glyphodactyla*, and *Megalonychus*—are common to South Africa and Madagascar only. There is also a genus in common with Java, and one with Australia.

Of Lucanidæ, or stag-beetles, there are 3 peculiar genera; of Cetoniidæ, or rose-chafers, 14; and of Buprestidæ, 2.

In the great family of Longicorns there are no less than 67 peculiar genera—an immense number when we consider that the generally open character of the country, is such as is not usually well suited to this group of insects. They consist of 5 peculiar genera of Prionidæ, 25 of Cerambycidæ, and 37 of Lamiidæ.

Summary of South-African Zoology.—Summarizing these results, we find that South Africa possesses 18 peculiar genera of Mammalia, 12 of Birds, 18 of Reptiles, 1 of Fishes, 7 of Butterflies, and 107 of the six typical families of Coleoptera. Besides this large amount of speciality it contains many other groups, which extend either to West Africa, to Abyssinia, or to Madagascar only, a number of which are no doubt to be referred as originating here. We also find many cases of direct affinity with the Oriental region, and especially with the Malay districts, and others with Australia; and there are also less marked indications of a relation to America.

Atlantic Islands of the Ethiopian Region. St. Helena.—The position of St. Helena, about 1,000 miles west of Africa and 16° south of the equator, renders it difficult to place it in either of the sub-regions; and its scanty fauna has a general rather than any special resemblance to that of Africa. The entire destruction of its luxuriant native forests by the introduction of goats which killed all the young trees (a destruction which was nearly completed two centuries ago) must have led to the extermination of most of the indigenous birds and insects. At present there is no land bird that is believed to be really indigenous, and but one

wader, a small plover (*Ægialitis sanctæ-helenæ*) which is peculiar
to the island, but closely allied to African species. Numerous
imported birds, such as canaries, Java sparrows, some African
finches, guinea-fowls, and partridges, are now wild. There are
no native butterflies, but a few introduced species of almost
world-wide range. The only important remnant of the original
fauna consists of beetles and land shells. The beetles are the
more numerous and have been critically examined and described
by Mr. T. V. Wollaston, whose researches in the other Atlantic
islands are so well known.

Coleoptera of St. Helena.—Omitting those beetles which get
introduced everywhere through man's agency, there are 59 species
of Coleoptera known from St. Helena; and even of these there
are a few widely distributed species that may have been introduced by man. It will be well, therefore, to confine ourselves
almost wholly to the species peculiar to the island, and, therefore,
almost certainly forming part of the endemic or original fauna.
Of these we find that 10 belong to genera which have a very
wide range, and thus afford no indication of geographical affinity;
2 belong to genera which are characteristic of the Palæarctic
fauna (*Bembidium, Longitarsus*); 3 to African genera (*Adoretus,
Sciobius, Aspidomorpha*); and two species of *Calosoma* are most
allied to African species. There are also 4 African species,
which may be indigenous in St. Helena. The peculiar genera,
7 in number, are, however, the most interesting. We have first
Haplothorax, a large beetle allied to *Carabus* and *Calosoma*, though
of a peculiar type. This may be held to indicate a remote
Palæarctic affinity. *Melissius*, one of the Dynastidæ, is allied to
South African forms. *Microxylobius*, one of the Cossonides (a
sub-family of Curculionidæ) is the most important genus, comprising as it does 13 species. It is, according to Mr. Wollaston,
an altogether peculiar type, most allied to *Pentarthrum*, a genus
found in St. Helena, Ascension, and the south of England, and
itself very isolated. *Nesiotes*, another genus of Curculionidæ,
belongs to a small group, the allied genera forming which inhabit
Europe, Madeira, and Australia. A third peculiar and isolated
genus is *Trachyphlæosoma*. The Anthribidæ are represented by

2 genera, *Notioxenus* and *Homœodera*, which are altogether peculiar and isolated, and contain 9 species. Thus no less than 27 species, or more than half of the undoubtedly indigenous beetles, belong to 5 peculiar and very remarkable genera of Rhyncophora.

It appears from this enumeration, that the peculiar species as a whole, exhibit most affinity to the Ethiopian fauna; next to the South European fauna; and lastly to that of the islands of the North Atlantic; while there is such a large amount of peculiarity in the most characteristic forms, that no special geographical affinity can be pointed out.

Land Shells.—These consist of about a dozen living species, and about as many extinct found in the surface soil, and probably exterminated by the destruction of the forests. The genera are *Succinea*, *Zonites*, *Helix*, *Bulimus*, *Pupa*, and *Achatina*. The *Bulimi* (all now extinct but one) comprise one large, and several small species, of a peculiar type, most resembling forms now inhabiting South America and the islands of the Pacific. *Zonites* is chiefly South European, but the other genera are of wide range, and none are peculiar to the island.

The marine shells are mostly Mediterranean, or West Indian species, with some found in the Indian Ocean; only 4 or 5 species being peculiar to the island.

Tristan d'Acunha.—This small island is situated nearly midway between the Cape of Good Hope and the mouth of the La Plata, but it is rather nearer Africa than America, and a little nearer still to St. Helena. An island so truly oceanic and of whose productions so little is known, cannot be placed in any region, and is only noticed here because it comes naturally after St. Helena. It is known to possess three peculiar land birds. One is a thrush (*Nesocichla eremita*) whose exact affinities are not determined; the other a small water-hen (*Gallinula nesiotis*) allied to our native species, but with shorter and softer wings, which the bird does not use for flight. A finch of the genus *Crithagra* shows African affinities; while another recently described as *Nesospiza acunhæ* (Journ. für Orn. 1873, p. 154) forms a new genus said to resemble more nearly some American forms.

The only known land-shells are 2 peculiar species of *Balea*, a genus only found elsewhere in Europe and Brazil.

IV. Madagascar and the Mascarene Islands, or the Malagasy Sub-region.

This insular sub-region is one of the most remarkable zoological districts on the globe, bearing a similar relation to Africa as the Antilles to tropical America, or New Zealand to Australia, but possessing a much richer fauna than either of these, and in some respects a more remarkable one even than New Zealand. It comprises, besides Madagascar, the islands of Mauritius, Bourbon, and Rodriguez, the Seychelles and Comoro islands. Madagascar itself is an island of the first class, being a thousand miles long and about 250 miles in average width. It lies parallel to the coast of Africa, near the southern tropic, and is separated by 230 miles of sea from the nearest part of the continent, although a bank of soundings projecting from its western coast reduces this distance to about 160 miles. Madagascar is a mountainous island, and the greater part of the interior consists of open elevated plateaus; but between these and the coast there intervene broad belts of luxuriant tropical forests. It is this forest-district which has yielded most of those remarkable types of animal life which we shall have to enumerate; and it is probable that many more remain to be discovered. As all the main features of this sub-region are developed in Madagascar, we shall first endeavour to give a complete outline of the fauna of that country, and afterwards show how far the surrounding islands partake of its peculiarities.

Mammalia.—The fauna of Madagascar is tolerably rich in genera and species of mammalia, although these belong to a very limited number of families and orders. It is especially characterized by its abundance of Lemuridæ and Insectivora; it also possesses a few peculiar Carnivora of small size; but most of the other groups in which Africa is especially rich—apes and monkeys, lions, leopards and hyænas, zebras, giraffes, antelopes, elephants and rhinoceroses, and even porcupines and squirrels, are wholly wanting. No less than 40 distinct families of land mammals are represented on the continent of Africa, only 11 of which occur in Madagascar, which also possesses 3 families peculiar to itself. The following is a list of all the genera of Mammalia as yet known to inhabit the island :—

PRIMATES.
LEMURIDÆ.

Indrisinæ.

	Species.
Indris	6

Lemurinæ.

Lemur	15
Hapalemur	2
Microcebus	4
Chirogaleus	5
Lepilemur	2

CHIROMYIDÆ.

Chiromys	1

BATS—(Chiroptera).
PTEROPIDÆ.

Pteropus	2

RHINOLOPHIDÆ.

Rhinolophus	1

VESPERTILIONIDÆ.

Vespertilio	1
Taphozous	1

NOCTILIONIDÆ.

Nyctinomus	1

INSECTIVORA.
CENTETIDÆ.

	Species.
Centetes	2
Hemicentetes	2
Ericulus	2
Oryzorictes	1
Echinops	3

SORICIDÆ.

Sorex	1

CARNIVORA.
CRYPTOPROCTIDÆ.

Cryptoprocta	1

VIVERRIDÆ.

Fossa	2
Galidia	3
Galidictis	2
Eupleres	1

UNGULATA.
SUIDÆ.

Potamochœrus	1

RODENTIA.
MURIDÆ.

Nesomys	1
Hypogeomys	1
Brachytarsomys	1

We have here a total of 12 families, 27 genera, and 65 species of Mammals ; 3 of the families and 20 of the genera (indicated by italics) being peculiar. All the species are peculiar, except perhaps one or two of the wandering bats. Remains of a *Hippopotamus* have been found in a sub-fossil condition, showing that this animal probably inhabited the island at a not very remote epoch.

The assemblage of animals above noted is remarkable, and seems to indicate a very ancient connection with the southern portion of Africa, before the apes, ungulates, and felines had entered it. The lemurs, which are here so largely developed, are repre-

T

sented by a single group in Africa, with two peculiar forms on
the West coast. They also re-appear under peculiar and isolated
forms in Southern India and Malaya, and are evidently but the
remains of a once wide-spread group, since in Eocene times they
inhabited North America and Europe, and very probably the
whole northern hemisphere. The Insectivora are another group
of high antiquity, widely scattered over the globe under a
number of peculiar forms; but in no equally limited area represented by so many peculiar types as in Madagascar. South and
West Africa are also rich in this order.

The Carnivora of Madagascar are mostly peculiar forms of
Viverridæ, or civets, a family now almost confined to the
Ethiopian and Oriental regions, but which was abundant in
Europe during the Miocene period.

The *Potamochœrus* is a peculiar *species* only, which may be
perhaps explained by the unusual swimming powers of swine,
and the semi-aquatic habits of this genus, leading to an immigration at a later period than in the case of the other Mammalia.
The same remark will apply to the small *Hippopotamus*, which
was coeval with the great Struthious bird Æpiornis.

Rodents are only represented by three peculiar forms of
Muridæ, but it is probable that others remain to be discovered.

Birds.—Madagascar is exceedingly rich in birds, and especially in remarkable forms of Passeres. No less than 88 genera
and 111 species of land-birds have been discovered, and every
year some additions are being made to the list. The African
families of Passeres are almost all represented, only two being
absent—Paridæ and Fringillidæ, both very poorly represented in
Africa itself. Among the Picariæ, however, the case is very
different, no less than 7 families being absent, viz.—Picidæ,
or woodpeckers; Indicatoridæ, or honey-guides; Megalæmidæ,
or barbets; Musophagidæ, or plantain-eaters; Coliidæ, or colies;
Bucerotidæ, or hornbills; and Irrisoridæ, or mockers. Three of
these are peculiar to Africa, and all are well represented there,
so that their absence from Madagascar is a very remarkable fact.
The number of peculiar genera in Madagascar constitutes one of
the main features of its ornithology, and many of these are so

isolated that it is very difficult to classify them, and they remain to this day a puzzle to ornithologists. In order to exhibit clearly the striking characteristics of the bird-fauna of this island, we shall first give a list of all the peculiar genera; another, of the genera of which the species only are peculiar; and, lastly, a list of the species which Madagascar possesses in common with the African continent.

GENERA OF BIRDS PECULIAR TO MADAGASCAR, OR FOUND ELSEWHERE
ONLY IN THE MASCARENE ISLANDS.

SYLVIIDÆ.	Species.	STURNIDÆ.	Species.
1. Bernieria	2	19. Euryceros (?)	1
2. Ellisia	1	20. Hartlaubia	1
3. Mystacornis	1	21. Falculia	1
4. Eroessa	1		
5. Gervasia	1	PAICTIDÆ.	
		22. Philepitta	1
TIMALIIDÆ.			
6. Oxylabes	2	CUCULIDÆ.	
		23. Coua	9
CINCLIDÆ (?).		24. Cochlothraustes	1
7. Mesites	1		
		LEPTOSOMIDÆ.	
SITTIDÆ.		25. Leptosomus	1
8. *Hypherpes*	1		
		CORACIIDÆ.	
PYCNONOTIDÆ (?)		26. Atelornis	2
9. Tylas	1	27. Brachypteracias	1
		28. Geobiastes	1
ORIOLIDÆ.			
10. Artamia	3	PSITTACIDÆ.	
11. Cyanolanius	1	29. Coracopsis	2
MUSCICAPIDÆ.		COLUMBIDÆ.	
12. Newtonia	1	30. *Alectrænas*	1
13. Pseudobias	1		
		TETRAONIDÆ.	
LANIIDÆ.		31. *Margaroperdix*	1
14. Calicalicus (?)	1		
15. Vanga	4	FALCONIDÆ.	
		32. Nisoides	1
NECTARINIIDÆ.		33. Eutriorchis	1
16. Neodrepanis	1		
		Total species of peculiar genera	50
HIRUNDINIDÆ.			
17. Phedina	1	ÆPYORNITHIDÆ (extinct).	
		34. Æpyornis	1
PLOCEIDÆ.			
18. Nelicurvius	1		

ETHIOPIAN OR ORIENTAL GENERA WHICH ARE REPRESENTED IN
MADAGASCAR BY PECULIAR SPECIES.

TURDIDÆ.	Species.
1. Bessonornis	1

SYLVIIDÆ.	
2. Acrocephalus	1
3. *Copsychus* (Or.) ...	1
4. Pratincola	1

PYCNONOTIDÆ.	
5. *Hypsipetes* (Or.) ...	1
6. Andropadus	1

CAMPEPHAGIDÆ.	
7. Campephaga	1

DICRURIDÆ.	
8. Dicrurus	1

MUSCICAPIDÆ.	
9. Tchitrea	1

LANIIDÆ.	
10. Laniarius	1

NECTARINIIDÆ.	
11. Nectarinia	1

PLOCEIIDÆ.	
12. Foudia	2
13. Hypargos	1
14. Spermestes	1

ALAUDIDÆ.	
15. Mirafra	1

MOTACILLIDÆ.	
16. Motacilla	1

CUCULIDÆ.	
17. Ceuthmochares ...	1
18. Centropus	1
19. Cuculus	1

CORACIIDÆ.	
20. Eurystomus	1

ALCEDINIDÆ.	Species.
21. Corythornis	1
22. Ispidina	1

UPUPIDÆ.	
23. Upupa (?)	1

CAPRIMULGIDÆ.	
24. Caprimulgus ...	1

CYPSELIDÆ.	
25. Cypselus	2
26. Chætura	1

PSITTACIDÆ.	
27. Poliopsitta	1

COLUMBIDÆ.	
28. Treron	1
29. Columba	1
30. Turtur	1

PTEROCLIDÆ.	
31. Pterocles	1

TETRAONIDÆ.	
32. Francolinus	1

PHASIANIDÆ.	
33. Numida	1

TURNICIDÆ.	
34. Turnix	1

FALCONIDÆ.	
35. Polyboroides ...	1
36. Circus	1
37. Astur	3
38. Accipiter	1
39. Buteo	1
40. Haliæetus	1
41. Pernis	1
42. Baza	1
43. Cerchneis	1

STRIGIDÆ.	Species.	PLATALEIDÆ.	Species.
44. Athene	1	49. Ibis	1
45. Scops	1		

RALLIDÆ.		PODICIPIDÆ.	
46. Rallus	3	50. Podiceps	1
47. Porzana	1		

SCOLOPACIDÆ.		Total peculiar species of Eth. or Or. genera	56
48. Gallinago	1		

SPECIES OF BIRDS COMMON TO MADAGASCAR AND AFRICA OR ASIA.

1. Cisticola cursitans.
2. Corvus scapulatus.
3. Crithagra canicollis.
4. Merops superciliosus.
5. Collocalia fuciphaga.
6. Œna capensis.
7. Aplopelia tympanistria.
8. Falco minor.
9. Falco concolor.
10. Milvus ægyptius.
11. Milvus migrans.
12. Strix flammea.

These three tables show us an amount of speciality hardly to be found in the birds of any other part of the globe. Out of 111 land-birds in Madagascar, only 12 are identical with species inhabiting the adjacent continents, and most of these belong to powerful-winged, or wide-ranging forms, which probably now often pass from one country to the other. The peculiar species —49 land-birds and 7 waders, or aquatics—are mostly well-marked forms of African genera. There are, however, several genera (marked by italics) which have Oriental or Palæarctic affinities, but not African, viz.—*Copsychus, Hypsipetes, Hypherpes, Alectrœnas,* and *Margaroperdix.* These indicate a closer approximation to the Malay countries than now exists.

The table of 33 peculiar genera is of great interest. Most of these are well-marked forms, belonging to families which are fully developed in Africa; though it is singular that not one of the exclusively African families is represented in any way in Madagascar. Others, however, are of remote or altogether doubtful affinities. *Sittidæ* is Oriental and Palæarctic, but not Ethiopian. *Oxylabes* and *Mystacornis* are of doubtful affinities. *Artamia* and *Cyanolanius* still more so, and it is quite undecided what family they belong to. *Calicalicus* is almost equally obscure. *Neodrepanis,* one of the most recent discoveries, seems to connect the Nectariniidæ with the Pacific

Depanididæ. *Euryceros* is a complete puzzle, having been placed with the hornbills, the starlings, or as a distinct family. *Falculia* is an exceedingly aberrant form of starling, long thought to be allied to *Irrisor*. *Philepitta*, forming a distinct family, (Paictidæ), is most remarkable and isolated, perhaps with remote South American affinities. *Leptosoma* is another extraordinary form, connecting the cuckoos with the rollers. *Atelornis*, *Brachypteracias*, and *Geobiastes*, are terrestrial rollers, with the form and colouring of *Pitta*. So many perfectly isolated and remarkable groups are certainly nowhere else to be found; and they fitly associate with the wonderful aye-aye (*Chiromys*), the insectivorous Centetidæ, and carnivorous *Cryptoprocta* among the Mammalia. They speak to us plainly of enormous antiquity, of long-continued isolation; and not less plainly of a lost continent or continental island, in which so many, and various, and peculiarly organized creatures, could have been gradually developed in a connected fauna, of which we have here but the fragmentary remains.

Plate VI.—Illustrating the characteristic features of the Zoology of Madagascar.—The lemurs, which form the most prominent feature in the zoology of Madagascar, being comparatively well-known from the numerous specimens in our zoological gardens; and good figures of the Insectivorous genera not being available, we have represented the nocturnal and extraordinary aye-aye (*Chiromys madagascariensis*) to illustrate its peculiar and probably very ancient mammalian fauna; while the river-hogs in the distance (*Potamochærus edwardsii*) allied to African species, indicate a later immigration from the mainland than in the case of most of the other Mammalia. The peculiar birds being far less generally known, we have figured three of them. The largest is the *Euryceros prevosti*, here classed with the starlings, although its remarkable bill and other peculiarities render it probable that it should form a distinct family. Its colours are velvety black and rich brown with the bill of a pearly grey. The bird beneath (*Vanga curvirostris*) is one of the peculiar Madagascar shrikes whose plumage, variegated with green-black and pure white is very conspicuous; while that in

PLATE VI.

SCENE IN MADAGASCAR, WITH CHARACTERISTIC ANIMALS.

the right hand corner is the *Leptosoma discolor*, a bird which appears to be intermediate between such very distinct families as the cuckoos and the rollers, and is therefore considered to form a family by itself. It is a coppery-green above and nearly white beneath, with a black bill and red feet. The fan-shaped plant on the left is the traveller's tree (*Urania speciosa*), one of the peculiar forms of vegetation in this marvellous island.

Reptiles.—These present some very curious features, comparatively few of the African groups being represented, while there are a considerable number of Eastern and even of American forms. Beginning with the snakes, we find, in the enormous family of Colubridæ, none of the African types; but instead of them three genera—*Herpetodryas*, *Philodryas*, and *Heterodon*—only found elsewhere in South and North America. The Psammophidæ, which are both African and Indian, are represented by a peculiar genus, *Mimophis*. The Dendrophidæ are represented by *Ahætulla*, a genus which is both African and American. The Dryiophidæ, which inhabit all the tropics but are most developed in the Oriental region, are represented by a peculiar genus, *Langaha*. The tropical Pythonidæ are represented by another peculiar genus, *Sanzinia*. The Lycodontidæ and Viperidæ, so well developed in Africa, are entirely absent.

The lizards are no less remarkable. The Zonuridæ, abundantly developed in Africa, are represented by one peculiar genus, *Cicigna*. The wide-spread Scincidæ by another peculiar genus, *Pygomeles*. The African Sepsidæ, are represented by three genera, two of which are African, and one, *Amphiglossus*, peculiar. The Acontiadæ are represented by a species of the African genus *Acontias*. Of Scincidæ there is the wide-spread *Euprepes*. The Sepidæ are represented by the African genera *Seps* and *Scelotes*. The Geckotidæ are not represented by any purely African genera, but by *Phyllodactylus*, which is American and Australian; *Hemidactylus*, which is spread over all the tropics; by two peculiar genera; and by *Uroplatis*, *Geckolepis*, and *Phelsuma*, confined to Madagascar, Bourbon, and the Andaman Islands. The Agamidæ, which are mostly Oriental and are represented in

Africa by the single genus *Agama*, have here three peculiar genera, *Tracheloptychus, Chalarodon,* and *Hoplurus.* Lastly, the American Iguanidæ are said to be represented by a species of the South American genus *Oplurus.* The classification of Reptiles is in such an unsettled state that some of these determinations of affinities are probably erroneous; but it is not likely that any corrections which may be required will materially affect the general bearing of the evidence, as indicating a remarkable amount of Oriental and American relationship.

The other groups are of less interest. Tortoises are represented by two African or wide-spread genera of Testudinidæ, *Testudo* and *Chersina,* and by one peculiar genus, *Pyxis;* and there are also two African genera of Chelydidæ.

The Amphibia are not very well known. They appear to be confined to species of the wide-spread Ethiopian and Oriental genera—*Hylarana, Polypedates,* and *Rappia* (Polypedatidæ); and *Pyxicephalus* (Ranidæ).

Fresh-water Fishes.—These appear to be at present almost unknown. When carefully collected they will no doubt furnish some important facts.

The Mascarene Islands.

The various islands which surround Madagascar—Bourbon, Mauritius, Rodriguez, the Seychelles, and the Comoro Islands —all partake in a considerable degree of its peculiar fauna, while having some special features of their own.

Indigenous Mammalia (except bats) are probably absent from all these islands (except the Comoros), although *Lemur* and *Centetes* are given as natives of Bourbon and Mauritius. They have, however, perhaps been introduced from Madagascar. *Lemur mayottensis,* a peculiar species, is found in the Comoro Islands, where a Madagascar species of *Viverra* also occurs.

Bourbon and Mauritius may be taken together, as they much resemble each other. They each possess species of a peculiar genus of Campephagidæ, or caterpillar shrikes, *Oxynotus;* while the remarkable *Fregilupus,* belonging to the starling family, inhabits Bourbon, if it is not now extinct. They also have peculiar species of *Pratincola, Hypsipetes, Phedina, Tchitrea, Zosterops, Foudia, Collocalia,* and *Coracopsis;* while Mauritius has a very peculiar form of dove of the sub-genus *Trocaza;* an *Alectrœnas,* extinct within the last thirty years; and a species of the Oriental genus of parroquets, *Palæornis.* The small and remote island of Rodriguez has another *Palæornis,* as well as a peculiar *Foudia,* and a *Drymœca* of apparently Indian affinity.

Coming to the Seychelle Islands, far to the north, we find the only mammal an Indian species of bat (*Pteropus edwardsii*). Of the twelve land-birds all but one are peculiar species, but all belong to genera found also in Madagascar, except one—a peculiar species of *Palæornis.* This is an Oriental genus, but found also in several Mascarene Islands and on the African continent. A species of black parrot (*Coracopsis barklayi*) and a weaver bird of peculiar type (*Foudia seychellarum*) show, however, a decided connection with Madagascar. There are also two peculiar pigeons—a short-winged *Turtur* and an *Alectrœnas.*

Most of the birds of the Comoro Islands are Madagascar species, only two being African. Five are peculiar, belonging to the genera *Nectarinia, Zosterops, Dicrurus, Foudia,* and *Alectrœnas.*

Reptiles are scarce. There appear to be no snakes in Mauritius and Bourbon, though some African species are said to be found in the Seychelle Islands. Lizards are fairly represented. Mauritius has *Cryptoblepharus,* an Australian genus of Gymnopthalmidæ; *Hemidactylus* (a wide-spread genus); *Percpus* (Oriental and Australian)—both belonging to the Geckotidæ. Bourbon has *Heteropus,* a Moluccan and Australian genus of Scincidæ; *Phelsuma* (Geckotidæ), and *Chameleo,* both found also in Madagascar; as well as *Pyxis,* one of the tortoises. The Seychelles have *Theconyx,* a peculiar genus of Geckotidæ, and *Chameleo.* Gigantic land-tortoises, which formerly inhabited most of the Mascarene Islands, now only survive in Aldabra, a small island north of the Seychelles. These will be noticed again further on. Amphibia seem only to be recorded from the Seychelles, where two genera of tree-frogs of the family Polypedatidæ are found; one (*Megalixalus*) peculiar, the other (*Rappia*) found also in Madagascar and Africa.

The few insect groups peculiar to these islands will be noted when we deal with the entomology of Madagascar.

Extinct fauna of the Mascarene Islands and Madagascar.—Before quitting the vertebrate groups, we must notice the remarkable birds which have become extinct in these islands little more than a century ago. The most celebrated is the dodo of the Mauritius (*Didus ineptus*), but an allied genus, *Pezophaps*, inhabited Rodriguez, and of both of these almost perfect skeletons have been recovered. Other species probably existed in Bourbon. Remains of two genera of flightless rails have also been found, *Aphanapteryx* and *Erythromachus*; and even a heron (*Ardea megacephala*) which was short-winged and seldom flew; while in Madagascar there lived a gigantic Struthious bird, the *Æpyornis*. Some further details as to these extinct forms will be found under the respective families, Dididæ, Rallidæ, and Æpyornithidæ, in the fourth part of this work; and their bearing on the past history of the region will be adverted to in the latter part of this chapter. Dr. Günther has recently distinguished five species of fossil tortoises from Mauritius and Rodriguez,—all of them quite different from the living species of Aldabra.

Insects.—The butterflies of Madagascar are not so remarkable as some other orders of insects. There seems to be only one peculiar genus, *Heteropsis* (Satyridæ). The other genera are African, *Leptoneura* being confined to Madagascar and South Africa. There are some fine *Papilios* of uncommon forms. The most interesting lepidopterous insect, however, is the fine diurnal moth (*Urania*), as all the other species of the genus inhabit tropical America and the West Indian Islands.

The Coleoptera have been better collected, and exhibit some very remarkable affinities. There is but one peculiar genus of Cicindelidæ, *Pogonostoma*, which is allied to the South American genus, *Ctenostoma*. Another genus, *Peridexia*, is common to Madagascar and South America. None of the important African genera are represented, except *Eurymorpha*; while *Meglaomma* is common to Madagascar and the Oriental region.

In the Carabidæ we have somewhat similar phenomena on a

wider scale. Such large and important African genera as *Polyhirma* and *Anthia*, are absent; but there are four genera in common with South Africa, and two with West Africa; while three others are as much Oriental as African. One genus, *Distrigus*, is wholly Oriental; and another, *Homalosoma*, Australian. *Colpodes*, well developed in Bourbon and Mauritius, is Oriental and South American. Of the peculiar genera, *Sphærostylis* has South American affinities; *Microchila*, Oriental; the others being related to widely distributed genera.

The Lucanidæ are few in number, and all have African affinities. Madagascar is very rich in Cetoniidæ, and possesses 20 peculiar genera. *Bothrorhina*, and three other genera belonging to the *Ichnostoma* group, have wholly African relations. *Doryscelis* and *Chromoptila* are no less clearly allied to Oriental genera. A series of eight peculiar genera belong to the Schizorhinidæ, a family the bulk of which are Australian, while there are only a few African forms. The remaining genera appear to have African affinities, but few of the peculiarly African genera are represented. *Glyciphana* is characteristic of the Oriental region.

The Buprestidæ of Madagascar consist mainly of one large and peculiar genus, *Polybothris*, allied to the almost cosmopolite *Psiloptera*. Most of the other genera are both Ethiopian and Oriental; but *Polycesta* is mainly South American, and the remarkable and isolated genus *Sponsor* is confined to the Mauritius with a species in Celebes and New Guinea.

The Longicorns are numerous and interesting, there being no less than 24 peculiar genera. Two of the genera of Prionidæ are very isolated, while a third, *Closterus*, belongs to a group which is Malayan and American.

Of the Cerambycidæ, *Philematium* ranges to Africa and the West Indies; *Leptocera* is only found eastward in Ceylon and the New Hebrides; while *Euporus* is African. Of the peculiar genera, 2 are of African type; 3 belong to the *Leptura* group, which are mostly Palæarctic and Oriental, with a few in South Africa; while *Philocalocera* is allied to a South American genus.

Among the Lamiidæ there are several wide-ranging and 7

African genera; but *Coptops* is Oriental, and the Oriental
Praonetha occurs in the Comoro Islands. Among the peculiar
genera several have African affinities, but *Tropidema* belongs to
a group which is Oriental and Australian; *Oopsis* is found also
in the Pacific Islands; *Mythergates, Sulemus,* and *Coedomœa,* are
allied to Malayan and American genera.

General Remarks on the Insect-fauna of Madagascar.—Taking
the insects as a whole, we find the remarkable result that their
affinities are largely Oriental, Australian, and South American:
while the African element is represented chiefly by special
South African or West African forms, rather than by such as
are widely spread over the Ethiopian region.[1] In some
families—as Cetoniidæ and Lamiidæ—the African element
appears to preponderate; in others—as Cicindelidæ—the South
American affinity seems strongest; in Carabidæ, perhaps the
Oriental; while in Buprestidæ and Cerambycidæ the African
and foreign elements seem nearly balanced. We must not impute too much importance to these foreign alliances among
insects, because we find examples of them in every country on
the globe. The reason they are so much more pronounced in
Madagascar may be, that during long periods of time this island
has served as a refuge for groups that have been dying out on
the great continents; and that, owing to the numerous deficiencies of a somewhat similar kind in the series of vertebrata
in Australia and South America, the same groups have often
been able to maintain themselves in all these countries as well
as in Madagascar. It must be remembered too, that these peculiarities in the Malagasy and Mascarene insect-fauna are but ex-
aggerations of a like phenomenon on the mainland. Africa also
has numerous affinities with South America, with the Malay
countries, and with Australia; but they do not bear anything like
so large a proportion to the whole fauna, and do not, therefore,
attract so much attention. The special conditions of existence
and the long-continued isolation of Madagascar, will account for
much of this difference; and it will evidently not be necessary

[1] There are also some special resemblances between the plants of Madagascar and South Africa, according to Dr. Kirk.

to introduce, as some writers are disposed to do, a special land connection or near approach between Madagascar and all these countries, independently of Africa; except perhaps in the case of the Malay Islands, as will be discussed further on.

Land-shells.—Madagascar and the adjacent islands are all rich in land-shells. The genera of Helicidæ are *Vitrina, Helix, Achatina, Columna* (peculiar to Madagascar and West Africa), *Buliminus, Cionella* (chiefly Oriental and South American, but not African), *Pupa, Streptaxis,* and *Succinea.* Among the Operculata we have *Truncatella* (widely scattered, but not African); *Cyclotus* (South American, Oriental, and South African); *Cyclophorus* (mostly Oriental, with a few South African); *Leptopoma* (Oriental); *Megalomastoma* (Malayan and South American); *Lithidion* (peculiar to Madagascar, Socotra, and South-West Arabia); *Otopoma* (with the same range, but extending to West India and New Ireland); *Cyclostomus* (widely spread but not African); and *Omphalotropis* (wholly Oriental and Australian). We thus find the same general features reproduced in the landshells as in the insects, and the same remarks will to a great extent apply to both. The classification of the former is, however, by no means so satisfactory, and we have no extensive and accurate general catalogues of shells, like those of Lepidoptera and Coleoptera, which have furnished us with such valuable materials for the comparison of the several faunas.

On the probable Past History of the Ethiopian Region.

Perhaps none of the great zoological regions of the earth present us with problems of greater difficulty or higher interest than the Ethiopian. We find in it the evidence of several distinct and successive faunas, now intermingled; and it is very difficult, with our present imperfect knowledge, to form an adequate conception of how and when the several changes occurred. There are, however, a few points which seem sufficiently clear, and these afford us a secure foundation in our endeavour to comprehend the rest.

Let us then consider what are the main facts we have to account for.—1. In Continental Africa, more especially in the south and west, we find, along with much that is peculiar, a number of genera showing a decided Oriental, and others with an equally strong South American affinity; this latter more particularly showing itself among reptiles and insects. 2. All over Africa, but more especially in the east, we have abundance of large ungulates and felines—antelopes, giraffes, buffaloes, elephants, and rhinoceroses, with lions, leopards, and hyænas, all of types now or recently found in India and Western Asia. 3. But we also have to note the absence of a number of groups which abound in the above-named countries, such as deer, bears, moles, and true pigs; while camels and goats—characteristic of the desert regions just to the north of the Ethiopian—are equally wanting. 4. There is a wonderful unity of type and want of speciality in the vast area of our first sub-region extending from Senegal across to the east coast, and southward to the Zambezi; while West Africa and South Africa each abound in peculiar types. 5. We have the extraordinary fauna of Madagascar to account for, with its evident main derivation from Africa, yet wanting all the larger and higher African forms; its resemblances to Malaya and to South America; and its wonderful assemblage of altogether peculiar types.

Here we find a secure starting-point, for we are sure that Madagascar must have been separated from Africa before the assemblage of large animals enumerated above, had entered it. Now, it is a suggestive fact, that all these belong to types which abounded in Europe and India about the Miocene period. It is also known, from the prevalence of Tertiary deposits over the Sahara and much of Arabia, Persia, and Northern India, that during early Tertiary times a continuous sea from the Bay of Bengal to the British Isles completely cut off all land communication between Central and Southern Africa on the one side, and the great continent of the Eastern hemisphere on the other. When Africa was thus isolated, its fauna probably had a character somewhat analogous to that of South America at the same period. Most of the higher types of mammalian life were absent, while lemurs, Edentates, and Insectivora took their place. At this period Madagascar was no doubt united with Africa, and helped to form a great southern continent which must at one time have extended eastward as far as Southern India and Ceylon; and over the whole of this the lemurine type no doubt prevailed.

During some portion of this period, South Temperate Africa must have had a much greater extension, perhaps indicated by the numerous shoals and rocks to the south and east of the Cape of Good Hope, and by the Crozets and Kerguelen Islands further to the south-east. This would have afforded means for that intercommunion with Western Australia which is so clearly marked in the flora, and to some extent also in the insects of the two countries; and some such extension is absolutely required for the development of that wonderfully rich and peculiar temperate flora and fauna, which, now crowded into a narrow territory, is one of the greatest marvels of the organic world.

During this early period, when the great southern continents —South America, Africa, and Australia—were equally free from the incursions of the destructive felines of the north, the Struthious or ostrich type of birds was probably developed into its existing forms. It is not at all necessary to suppose that these three continents were at any time united, in order to account for the distribution of these great terrestrial birds; as this may have arisen by at least two other easily conceivable modes. The ancestral Struthious type may, like the Marsupial, have once spread over the larger portion of the globe; but as higher forms, especially of Carnivora, became developed, it would be exterminated everywhere but in those regions where it was free from their attacks. In each of these it would develope into special forms adapted to surrounding conditions; and the large size, great strength, and excessive speed of the ostrich, may have been a comparatively late development caused by its exposure to the attacks of enemies which rendered such modification necessary. This seems the most probable explanation of the distribution of Struthious birds, and it is rendered almost certain by the discovery of remains of this order in Europe in Eocene deposits, and by the occurrence of an ostrich among the fossils of the Siwalik hills; but it is just possible, also, that the ancestral type may have been a bird capable of flight, and that it spread from one of the three southern continents to the others at the period of their near approach, and more or less completely lost the power of flight owing to the long continued absence of enemies.

During the period we have been considering, the ancestors of existing apes and monkeys flourished (as we have seen in Chapter VI.) along the whole southern shores of the old Palæarctic continent; and it seems likely that they first entered Africa by means of a land connection indicated by the extensive and lofty plateaus of the Sahara, situated to the south-east of Tunis and reaching to a little north-west of Lake Tchad; and at the same time the elephant and rhinoceros type may have entered. This will account for the curious similarity between the higher faunas of West Africa and the Indo-Malay sub-region, for owing to the present distribution of land and sea and the narrowing of the tropical zone since Miocene times, these are now the only lowland, equatorial, forest-clad countries, which were in connection with the southern shores of the old Palæarctic continent at the time of its greatest luxuriance and development. This western connection did not probably last long, the junction that led to the greatest incursion of new forms, and the complete change in the character of the African fauna, having apparently been effected by way of Syria and the shores of the Red Sea at a somewhat later date. By this route the old South-Palæarctic fauna, indicated by the fossils of Pikermi and the Siwalik Hills, poured into Africa; and finding there a new and favourable country, almost wholly unoccupied by large Mammalia, increased to an enormous extent, developed into new forms, and finally overran the whole continent.

Before this occurred, however, a great change had taken place in the geography of Africa. It had gradually diminished on the south and east; Madagascar had been left isolated; while a number of small islands, banks, and coral reefs in the Indian Ocean alone remained to indicate the position of a once extensive equatorial land. The Mascarene Islands appear to represent the portion which separated earliest, before any carnivora had reached the country; and it was in consequence of this total exemption from danger, that several groups of birds altogether incapable of flight became developed here, culminating in the huge and unwieldy Dodo, and the more active Aphanapteryx. To the same cause may be attributed the development, in these islands, of gigantic land-tortoises, far surpassing any others now living on the globe. They appear to have formerly inhabited Mauritius, Bourbon, and Rodriguez, and perhaps all the other Mascarene islands, but having been recklessly destroyed, now only survive in the small uninhabited Aldabra islands north of the Seychelle group. The largest living specimen (5½ feet long) is now in our Zoological Gardens. The only other place where equally large tortoises (of an allied species) are found, is the Galapagos islands, where they were equally free from enemies till civilized man came upon the scene; who, partly by using them for food, partly by the introduction of pigs, which destroy the eggs, has greatly diminished their numbers and size, and will probably soon wholly exterminate them. It is a curious fact, ascertained by Dr. Günther, that the tortoises of the Galapagos are more nearly related to the extinct tortoises of Mauritius than is the living tortoise of Aldabra. This would imply that several distinct groups or sub-genera of *Testudo* have had a wide range over the globe, and that some of each have survived in very distant localities. This is rendered quite conceivable by the known antiquity of the genus *Testudo*, which dates back to at least the Eocene formation (in North America) with very little change of form. These sluggish reptiles, so long-lived and so tenacious of life, may have remained unchanged, while every higher animal type around them has become extinct and been replaced by very different forms; as in the case of the living *Emys tectum*, which is the sole survivor of the strange Siwalik fauna of the Miocene epoch. The ascertained history of the genus and the group, thus affords a satisfactory explanation of the close affinity of the gigantic tortoises of Mauritius and the Galapagos.

The great island of Madagascar seems to have remained longer united with Africa, till some of the smaller and more active

U

carnivora had reached it; and we consequently find there, no wholly terrestrial form of bird but the gigantic and powerful *Æpyornis*, well able to defend itself against such enemies. As already intimated, we refer the South American element in Madagascar, not to any special connection of the two countries independently of Africa, but to the preservation there of a number of forms, some derived from America through Africa, others of once almost cosmopolitan range, but which, owing to the severer competition, have become extinct on the African continent, while they have continued to exist under modified forms in the two other countries.

The depths of all the great oceans are now known to be so profound, that we cannot conceive the elevation of their beds above the surface without some corresponding depression elsewhere. And if, as is probable, these opposite motions of the earth's crust usually take place in parallel bands, and are to some extent dependent on each other, an elevation of the sea bed could hardly fail to lead to the submergence of large tracts of existing continents; and this is the more likely to occur on account of the great disproportion that we have seen exists between the mean height of the land and the mean depth of the ocean. Keeping this principle in view, we may, with some probability, suggest the successive stages by which the Ethiopian region assumed its present form, and acquired the striking peculiarities that characterise its several sub-regions. During the early period, when the rich and varied temperate flora of the Cape, and its hardly less peculiar forms of insects and of low type mammalia, were in process of development in an extensive south temperate land, we may be pretty sure that the whole of the east and much of the north of Africa was deep sea. At a later period, when this continent sank towards the south and east, the elevation may have occurred which connected Madagascar with Ceylon; and only at a still later epoch, when the Indian Ocean had again been formed, did central, eastern, and northern Africa gradually rise above the ocean, and effect a connection with the great northern continent by way of Abyssinia and Arabia. And if this last change took place with tolerable rapidity, or if the elevatory force acted from the north towards the south, there would be a new and unoccupied territory to be taken possession of by immigrants from the north, together with a few from the south and west. The more highly-organised types from the great northern continent, how-ever, would inévitably prevail; and we should thus have explained the curious uniformity in the fauna of so large an area, together with the absence from it of those peculiar Ethiopian types which so abundantly characterise the other three sub-regions.

We may now perhaps see the reason of the singular absence from tropical Africa of deer and bears; for these are both groups which live in fertile or well-wooded countries, whereas the line of immigration from Europe to Africa was probably always, as now, to a great extent a dry and desert tract, suited to antelopes and large felines, but almost impassable to deer and bears. We find, too, that whereas remains of antelopes and giraffes abound in the Miocene deposits of Greece, there were no deer (which are perhaps a somewhat later development); neither were there any bears, but numerous forms of Felidæ, Viverridæ, Mustelidæ, and ancestral forms of *Hyæna*, exactly suited to be the progenitors of the most prevalent types of modern African Zoology.

There appears to have been one other change in the geo-graphy of Africa and the Atlantic Ocean that requires notice. The rather numerous cases of close similarity in the insect forms of tropical Africa and America, seem to indicate some better means of transmission, at a not very remote epoch, than now exists. The vast depth of the Atlantic, and the absence of any corresponding likeness in the vertebrate fauna, entirely negative the idea of any union between the two countries; but a moderate extension of their shores towards each other is not improbable, and this, with large islands in the place of the Cape Verd group, St. Paul's Rocks, and Fernando Noronha, to afford resting places in the Atlantic, would probably suffice to explain the amount of similarity that actually exists.

Our knowledge of the geology and palæontology of Africa

being so scanty, it would be imprudent to attempt any more
detailed explanation of the peculiarities of its existing fauna.
The sketch now given is, it is believed, founded on a sufficient
basis of facts to render it not only a possible but a probable
account of what took place ; and it is something gained to be
able to show, that a large portion of the peculiarities and
anomalies of so remarkable a fauna as that of the Ethiopian
region, can be accounted for by a series of changes of physical
geography during the tertiary epoch, which can hardly be considered extreme, or in any way unlikely to have occurred.

TABLES OF DISTRIBUTION.

In drawing up these tables showing the distribution of various classes of animals in the Ethiopian Region, the following sources of information have been chiefly relied on, in addition to the general treatises, monographs, and catalogues, used for the Fourth Part of this work :—

Mammalia.—Blanford's Abyssinia ; Peters's Mozambique ; Heuglin and Schweinfurth for North East Africa ; Grandidier Schlegel, &c., for Madagascar ; the local lists given by Mr. Andrew Murray ; numerous papers by Fraser, Gray, Kirk, Mivart, Peters, Sclater, and Speke ; and a MS. list of Bovidæ from Sir Victor Brooke.

Birds.—Finsch and Hartlaub for East Africa ; Heuglin for North-East Africa ; Blanford for Abyssinia ; Layard for South Africa ; Hartlaub for West Africa ; Dohrn for Princes Island ; Andersson for Damaraland ; and papers by Gurney, Hartlaub, Kirk, Newton, Peters, Sharpe, Sclater, Schlegel, and Pollen and a MS. list of Madagascar Birds from Mr. Sharpe.

TABLE I.

FAMILIES OF ANIMALS INHABITING THE ETHIOPIAN REGION.

EXPLANATION.

Names in *italics* show families peculiar to the region.

Names inclosed thus (......) barely enter the region, and are not considered proper to belong to it.

Numbers are not consecutive, but correspond to those in Part IV.

Order and Family.	Sub-regions.				Range beyond the Region.
	East Africa.	West Africa.	South Africa.	Madagascar.	
MAMMALIA.					
PRIMATES.					
1. Simiidæ ...		—			Oriental
2. Semnopithecidæ	—	—			Oriental
3. Cynopithecidæ	—	—	—		Oriental, Palæarctic
6. Lemuridæ ...	—	—	—	—	Oriental
8. *Chiromyidæ* ...				—	
CHEIROPTERA.					
9. Pteropidæ ...	—	—	—	—	Oriental, Australian
11. Rhinolophidæ	—	—	—	—	The Eastern Hemisphere
12. Vespertilionidæ	—	—	—	—	Cosmopolite
13. Noctilionidæ...	—	—	—	—	All Tropical regions
INSECTIVORA.					
15. Macroscelididæ	—		—		South Palæarctic
17. Erinaceidæ ...			—		Palæarctic, Oriental
18. *Centetidæ* ...				—	Greater Antilles
19. *Potamogalidæ*		—			
20. *Chrysochloridæ*	—		—		
22. Soricidæ... ...	—	—	—	—	All regions but Australian and Neotropical
CARNIVORA.					
23. Felidæ	—	—	—	—	All regions but Australian
24. *Cryptoproctidæ*				—	
25. Viverridæ ...	—	—	—	—	Oriental, S. Palæarctic
26. *Protelidæ* ...			—		
27. Hyænidæ ..	—	—	—		S. Palæarctic, India
28. Canidæ	—	—	—		Almost cosmopolite
29. Mustelidæ ...	—	—	—		All regions but Australian
33. Otariidæ... ...			—		All temperate regions
CETACEA.					
36 to 41.					Oceanic
SIRENIA.					
42. Manatidæ ...	—	—			Neotropical, Oriental, Australian
UNGULATA.					
43. Equidæ	—	—		—	Palæarctic

Order and Family.	Sub-regions.				Range beyond the Region.
	East Africa.	West Africa.	South Africa.	Mada- gascar.	
45. Rhinocerotidæ	—	—	—		Oriental
46. *Hippopotamidæ*	—	—	—		
47. Suidæ	—	—	—	—	Cosmopolite ; excl. Australia
49. Tragulidæ ...		· —			Oriental
51. *Camelopardidæ*	—		—		
52. Bovidæ	—	—	—		All regions but Neotrop. and Australian
PROBOSCIDEA.					
53. Elephantidæ ...	—	—	—		Oriental
HYRACOIDEA.					
54. Hyracidæ ..	—	—	—	.	Syria
RODENTIA.					
55. Muridæ	—	—	—	—	Cosmopolite ; excl. Oceania
56. Spalacidæ ...	—	—	—		Palæarctic, Oriental
57. Dipodidæ ...	—	—	—	!	Palæarctic, Nearctic
58. Myoxidæ ..	—	—	—		Palæarctic
61. Sciuridæ... ...	—	—	—		All regions but Australian
64. Octodontidæ ...	—				N. Africa, Neotropical
65. Echimyidæ ...			—		Neotropical
67. Hystricidæ ...	—	—	—		S. Palæarctic, Oriental
70. Leporidæ ...	—		—		All regions but Australian
EDENTATA.					
72. Manididæ ...	—	—	—		Oriental
74. *Orycteropodidæ*	—		—		
BIRDS.					
PASSERES.					
1. Turdidæ... ...	—	—	—	—	Almost Cosmopolite
2. Sylviidæ... ...	—	—	—	—	Cosmopolite
3. Timaliidæ ...	—	—	—	—	Oriental, Australian
5. Cinclidæ? ...				—	Widely scattered
6. Troglodytidæ	—	—	—		Almost Cosmopolite
9. Sittidæ	—	—	—		Palæarctic, Oriental, Australian
10. Paridæ	—	—	—		All regions but Australian
13. Pycnonotidæ...	—	—	—	—	Oriental
14. Oriolidæ... ...	—	—	—	—	Oriental, Australian
15. Campephagidæ	—	—	—	—	Oriental, Australian
16. Dicruridæ ...	—	—	—	—	Oriental, Australian
17. Muscicapidæ...	—	—	—	—	The Eastern Hemisphere
19. Laniidæ	—	—	—	—	The Eastern Hemisphere and North America
20. Corvidæ	—	—	—	—	Cosmopolite
23. Nectariniidæ...	—	—	—	—	Oriental, Australian
24. Dicæidæ ...	—	—	—	—	Oriental, Australian
30. Hirundinidæ...	—	—	—	—	Cosmopolite
33. Fringillidæ ...	—	—	—	—	Cosmopolite, except Australian region
34. Ploceidæ ...	—	—	—	—	Oriental, Australian
35. Sturnidæ ...	—	—	—	—	Eastern Hemisphere
37. Alaudidæ ...	—	—	—	—	Eastern Hemisphere and North America

Order and Family.	Sub-regions.				Range beyond the Region.
	East Africa.	West Africa.	South Africa.	Mada-gascar.	
38. Motacillidæ ...	—	—	—	—	The Eastern Hemisphere
47. Pittidæ		—			Oriental, Australian
48. *Paictidæ* ...				—	
PICARIÆ.					
51. Picidæ	—	—	—		Cosmopolite, excl. Australian region
52. Yungidæ ...	—		—		Palæarctic
53. Indicatoridæ ...	—	—	—		Oriental
54. Megalæmidæ...	—	—	—		Oriental, Neotropical
56. *Musophagidæ*	—	—	—		
57. *Coliidæ*	—	—	—		
58. Cuculidæ ...	—	—	—	—	Cosmopolite
59. *Leptosomidæ* ...				—	
62. Coraciidæ ...	—	—	—	—	Oriental, Australian
63. Meropidæ ...	—	—	—	—	Oriental, Australian
66. Trogonidæ ...	—	—	—	—	Oriental, Neotropical
67. Alcedinidæ ...	—	—	—	—	Cosmopolite
68. Bucerotidæ ...	—	—	—		Oriental and to N. Guinea
69. Upupidæ ...	—	—	—	—	Palæarctic, Oriental
70. *Irrisoridæ* ...	—	—	—		
73. Caprimulgidæ ...	—	—	—	—	Cosmopolite
74. Cypselidæ ...	—	—	—	—	Almost Cosmopolite
PSITTACI.					
78. Palæornithidæ ...	—	—	—		Oriental
81. Psittacidæ ...	—	—	—	—	Neotropical
COLUMBÆ.					
84. Columbidæ ...	—	—	—	—	Cosmopolite
85. *Dididæ*				—	(Extinct)
GALLINÆ.					
86. Pteroclidæ ...	—	—	—		Palæarctic, Oriental
87. Tetraonidæ ...	—	—	—	—	Eastern Hemisphere and N. America
88. Phasianidæ ...	—	—	—	—	Old World and N. America
89. Turnicidæ ...	—	—	—	—	Eastern Hemisphere.
ACCIPITRES.					
94. Vulturidæ ...	—	—	—		All the continents but Australia
95. Falconidæ ...	—	—	—	—	Cosmopolite
96. *Serpentariidæ*	—	—	—		
97. Pandionidæ ...	—	—	—	—	Cosmopolite
98. Strigidæ ...	—	—	—	—	Cosmopolite
GRALLÆ.					
99. Rallidæ ...	—	—	—	—	Cosmopolite
100. Scolopacidæ...	—	—	—	—	Cosmopolite
103. Parridæ ...	—	—	—	—	Tropical
104. Glareolidæ ...	—	—	—	—	Eastern Hemisphere
105. Charadriidæ	—	—	—	—	Cosmopolite

Order and Family.	East Africa.	West Africa.	South Africa.	Mada-gascar.	Range beyond the Region.
106. Otididæ ...	—	—	—		Eastern Hemisphere
107. Gruidæ ...	—	-	—		All regions but Neotropical
113. Ardeidæ	—	—	—	—	Cosmopolite
114. Plataleidæ ...	—	—	—	—	Almost Cosmopolite
115. Ciconiidæ ...	—	—	—	—	Almost Cosmopolite
117. Phœnicopteridæ	—	—	—	—	Oriental and Neotropical
ANSERES.					
118. Anatidæ ...	—	—	—	—	Cosmopolite
119. Laridæ	—	—	—	—	Cosmopolite
120. Procellariidæ	—	—	—	—	Cosmopolite
121. Pelecanidæ ...	—	—	—	—	Cosmopolite
122. Spheniscidæ			—		South temperate regions
124. Podicipidæ ..	—	—	—	—	Cosmopolite
126. Struthionidæ	—		—		Temperate S. America
131. *Æpyornithidæ*				—	(Extinct)
REPTILIA.					
OPHIDIA.					
1. Typhlopidæ ...	—	—	—	—	All regions but Nearctic
5. Calamariidæ ...	—	—	—	—	Warms parts of all regions
7. Colubridæ ...	—	—	—	—	Almost Cosmopolite
8. Homalopsidæ		—			Oriental, and all other regions
9. Psammophidæ ...	—	—	—	—	Oriental and S. Palæarctic
10. *Rachiodontidæ*		—	—		
11. Dendrophidæ ...	—	—	—	—	Oriental, Australian, Neotropical
12. Dryiophidæ ...		—		—	Oriental, Neotropical
13. Dipsadidæ ...	—	—	—	—	Oriental, Australian, Neotropical
15. Lycodontidæ...	—	—	—		Oriental
17. Pythonidæ ...	—	—	—	—	All tropical regions
18. Erycidæ		—			Oriental, S. Palæarctic
20. Elapidæ	--	—	—		Tropical regions, S. U. States and Japan
21. *Dendraspididæ*	—	—			
22. *Atractaspididæ*		—	—		
23. Hydrophidæ ...			—		Oriental, Australian, Panama
25. Viperidæ ...	—	-	—	—	Oriental, Palæarctic
LACERTILIA.					
28. Amphisbænidæ	—	—			S. Europe, Neotropical
29. Lepidosternidæ		—	—		N. America
30. Varanidæ ...	—	—	—		Warm parts of E. Hemisphere
33. Lacertidæ ...	—	—	—		All continents but America
34. Zonuridæ ...	—	—	—	—	All America, N. India, S. Europe
40. *Chamæsauridæ*			—		
41. Gymnopthal- midæ ...		—		—	Palæarctic, Australian, Netropical
45. Scincidæ ...	—	—	—	—	Almost Cosmopolite
47. Sepidæ	—	—	—	—	South Palæarctic
48. Acontiadæ ...		—	—	—	Ceylon and Moluccas.
49. Geckotidæ ...	--	—	—	—	Almost cosmopolite

Order and Family.	Sub-regions.				Range beyond the Region.
	East Africa.	West Africa.	South Africa.	Madagascar.	
51. Agamidæ ...	—	—	—	.—	Oriental, Australian, S. Palæarctic
52. Chamæleonidæ	—	—	—	--	Oriental, S. Palæarctic
CROCODILIA.					
55. Crocodilidæ ...	—	—	—	—	Oriental, Neotropical
CHELONIA.					
57. Testudinidæ ...	—	—	—	—	All continents but Australia
58. Chelydidæ ...	—	—	—	—	Australia, S. America
59. Trionychidæ...	—	—	—		Oriental, Japan, E. United States
60. Cheloniidæ ...					Marine
AMPHIBIA.					
PSEUDOPHIDIA.					
1. Cæciliadæ ...		—			Oriental, Neotropical
ANOURA.					
7. Phryniscidæ ...	—	—			Neotropical, Australia, Java
9. Bufonidæ ...	—	—	—		All regions but Australian
11. Engystomidæ..		—	—		All regions but Palæarctic
14. Alytidæ... ...	—	--	—		All regions but Oriental
17. Polypedatidæ	—	—	—	—	All the regions
18. Ranidæ	—	—	—	—	Almost Cosmopolite
19. Discoglossidæ		—	—		All regions but Nearctic
21. *Dactylethridæ*	—	—			
FISHES (FRESH-WATER).					
ACANTHOPTERYGII.					
3. Percidæ	—				All regions but Australian
12. Scienidæ ...	—	—	—		All regions but Australian
35. Labyrinthici ...			—	—	Oriental, Moluccas
38. Mugillidæ ...	—	—	—	—	Australian, Neotropical
52. Chromidæ ...	—	—	—	—	Oriental, Neotropical
PHYSOSTOMI.					
59. Siluridæ... ...	—	—	—	—	All warm regions
60. Characinidæ ...	—	—			Neotropical
68. *Mormyridæ* ...	—	—			
69. *Gymnarchidæ*	—	—			
73. Cyprinodontidæ	—	—		—	Palæarctic, Oriental, American
75. Cyprinidæ ...	—	—	—	—	Absent from Australia and S. America
78. Osteoglossidæ	—	—			All tropical regions
82. Notopteridæ ...	—	—			Oriental
GANOIDEI.					
92. Sirenoidei ...	—	—			Neotropical, Australian
94. *Polypteridæ* ...	—	—			

Order and Family.	Sub-regions.				Range beyond the Region.
	East Africa.	West Africa.	South Africa.	Madagascar.	
INSECTS. LEPI-DOPTERA (PART).					
DIURNI (BUTTER-FLIES).					
1. Danaidæ ...	—	—	—	—.	All warm countries and Canada
2. Satyridæ ...	—	—	–	—	Cosmopolite
3. Elymniidæ ...		—			Oriental, Moluccas
6. Acræidæ... ...	—	—	—	—	All tropical regions
8. Nymphalidæ...		—	—	—	Cosmopolite
9. Libytheidæ ...		—		—	Absent from Australia only
10. Nemeobiidæ ..		—		—	Absent from Australia and Nearctic region
13. Lycænidæ ...	—	—	—	—	Cosmopolite
14. Pieridæ	—	—	—	—	Cosmopolite
15. Papilionidæ ...	—	–	—	—	Cosmopolite
16. Hesperidæ ...	—	—	—	—	Cosmopolite
SPHINGIDEA.					
17. Zygænidæ ...	—	—	—	—	Cosmopolite
19. Agaristidæ ...	—	—	—	—	Australian, Oriental
20. Uraniidæ ...		—		—	All tropical regions
22. Ægeriidæ ...	—	—	—	—	Cosmopolite, excl. Australia
23. Sphingidæ ...	—	—	—	—	Cosmopolite

TABLE II.

LIST OF GENERA OF TERRESTRIAL MAMMALIA AND BIRDS INHABITING THE ETHIOPIAN REGION.

EXPLANATION.

Names in *italics* show genera peculiar to the region.
Names inclosed thus (...) show genera which just enter the region, but are not consider properly to belong to it.
Genera which undoubtedly belong to the region are numbered consecutively.

MAMMALIA.

Order, Family, and Genus.	No. of Species	Range within the Region.	Range beyond the Region.
PRIMATES.			
SIMIIDÆ.			
1. *Troglodytes* ...	2	W. Africa to Western Nile Sources	
SEMNOPITHECIDÆ.			
2. *Colobus*	11	Abyssinia to West Africa	
CYNOPITHECIDÆ.			
3. *Myiopithecus* ...	1	West Africa	
4. *Cercopithecus* ...	24	Tropical Africa	
5. *Cercocebus* ...	5	West Africa	
6. *Theropithecus* ...	2	North-east Africa, Arabia,	Palestine
7. *Cynocephalus* ...	10	Nubia to Cape, W. Africa, Arabia	
(Sub-Order) *LEMUROIDEA.*			
LEMURIDÆ.			
8. *Indris*	6	Madagascar	
9. *Lemur*	15	Madagascar	
10. *Hapalemur* ...	2	Madagascar	
11. *Microcebus* ...	4	Madagascar	
12. *Chirogaleus* ...	5	Madagascar	
13. *Lepilemur* ...	2	Madagascar	
14. *Perodicticus* ...	1	Sierra Leone	
15. *Arctocebus* ...	1	Old Calabar	
16. *Galago*	14	Tropical and S. Africa	
CHIROMYIDÆ.			
17. *Chiromys*	1	Madagascar	
CHIROPTERA.			
PTEROPIDÆ.			
18. Pteropus	7	Africa and Madagascar	Tropics of Eastern Hemisphere
19. Xantharpya ...	1	All Africa	Oriental, Austro-Malayan

Order, Family, and Genus.	No. of Species	Range within the Region.	Range beyond the Region.
20. Cynopterus ...	1	Tropical Africa	Oriental
21. *Epomophorus* ...	6	Tropical Africa and Abyssinia	
22. *Hypsignathus* ...	1	W. Africa	
RHINOLOPHIDÆ.			
23. Rhinolophus ...	6	Africa and Madagascar	Warmer parts of Eastern Hemisphere
24. *Macronycterys* ...	1	W. Africa	
25. Phyllorhina ...	4	Tropical Africa,	Indo-Malaya, Austro-Malaya
26. Asellia	1	Nubia	Indo-Malaya, Austro-Malaya
27. Megaderma ...	1	Senegal, Upper Nile	Oriental, Moluccas
28. Nycteris	3	All Africa	Java
VESPERTILIONIDÆ.			
29. Vespertilio ...	14	Africa and Madagascar	Cosmopolite
30. Kerivoula ...	1	S. Africa	Oriental
31. Miniopteris ...	1	S. Africa	Indo-Malaya
32. Nycticejus ...	7	Tropical Africa	India
33. Taphozous ...	2	Africa and Madagascar	Oriental, Austro-Malayan, Neotropical
NOCTILIONIDÆ.			
34. Nyctinomus ...	1	Madagascar	Oriental, American, S. Palæarctic
35. Molossus	3	Africa, Bourbon	Neotropical, S. Palæarctic
INSECTIVORA.			
MACROSCELIDIDÆ.			
36. *Macroscelides* ...	2	South and East Africa	N. Africa.
37. *Petrodromus* ..	1	Mozambique	
38. *Rhynchocyon* ...	1	Mozambique	
ERINACEIDÆ.			
39. Erinaceus... ...	2	Cen. and South Africa	Palæarctic, N. India
CENTETIDÆ.			
40. *Centetes*	2	Madagascar and Mauritius	
41. *Hemicentetes* ...	2	Madagasear	
42. *Ericulus* ...	2	Madagascar	
43. *Oryzorictes* ...	1	Madagascar	
44. *Echinops*	3	Madagascar	
POTAMOGALIDÆ.			
45. *Potamogale* ...	1	Old Calabar	
CHRYSOCHLORIDÆ.			
46. *Chrysochloris* ...	3	Cape to Mozambique	
SORICIDÆ.			
47. Sorex	15	All Africa and Madagascar	Palæarc., Nearc., Ori

Order, Family, and Genus.	No. of Species.	Range within the Region.	Range beyond the Region.
CARNIVORA.			
FELIDÆ.			
48. Felis...	8	All Africa	All reg. but Australian
49. Lynx [?]	1	N. and S. Africa	Palæarctic and Nearctic
50. Cynælurus ...	1	Cape of Good Hope	Persia, India
CRYPTOPROCTIDÆ.			
51. *Cryptoprocta* ...	1	Madagascar	
VIVERRIDÆ.			
52. Viverra	1	Tropical Africa	Oriental
53. Genetta	4	Tropical and S. Africa	S. Palæarctic
54. *Fossa*	2	Madagascar	
55. *Poiana*	1	W. Africa	
56. *Galidia*	3	Madagascar	
57. *Nandinia* ...	1	W. Africa	
58. *Galidictis* ...	2	Madagascar	
59. Herpestes ...	13	All Africa	S. Europe, Oriental
60. *Athylax*	3	S. and E. Africa (?) Madagascar	
61. Calogale	9	Tropical and S. Africa	Oriental
62. *Galerella*	1	E. Africa	
63. *Ariela*	1	S. Africa	
64. *Ichneumia* ...	4	E. Africa, Senegal, S. Africa	
65. Bdeogale	3	Tropical Africa	
66. *Helogale*	2	E. and S. Africa	
67. *Cynictis*	3	S. Africa	
68. *Rhinogale* ...	1	E. Africa	
69. Mungos	3	Tropical and S. Africa	
70. *Crossarchus* ...	1	Tropical Africa	
71. *Eupleres*	1	Madagascar	
72. *Suricata*	1	S. Africa	
PROTELIDÆ.			
73. *Proteles*	1	S. Africa	
HYÆNIDÆ.			
74. Hyæna	3	All Africa	S. Palæartic, India
CANIDÆ.			
75. *Lycaon*	1	S., Central, and E. Africa	
76. Canis	5	All Africa	Almost Cosmopolitan
77. *Megalotis*	1	S. Africa	
MUSTELIDÆ.			
78. Mustela	1	Angola	Palæarctic, Nearctic
79 Gymnopus [?] ...	1	S. Africa	Oriental
80. Aonyx	1	S. and W. Africa	Oriental
81. *Hydrogale* ...	1	S. Africa	
82. Mellivora ..	2	South and Tropical Africa.	India
83. *Ictonyx*	2	Tropical and S. Africa	
OTARIIDÆ.			
84. Arctocephalus	1	Cape of Good Hope	South Temperate Zone

Order, Family, and Genus.	No of Species.	Range within the Region.	Range beyond the Region.
SIRENIA.			
MANATIDÆ.			
85. Manatus ...	1	W. Africa	Tropical America
86. Halicore ...	1	E. Africa	Oriental and Australian
UNGULATA.			
EQUIDÆ.			
87. Equus	3	Tropical and S. Africa	Palæarctic
RHINOCEROTIDÆ.			
88. Rhinoceros ...	4	All Tropical and S. Africa	Oriental
HIPPOPOTAMIDÆ.			
89. *Hippopotamus*	2	Great Rivers of Africa	
SUIDÆ.			
90. *Potamochœrus*	3	Tropical Africa and Madg.	
91. *Phacochœrus* ...	2	Abyssinia to Caffraria	
TRAGULIDÆ.			
92. *Hyomoschus* ...	1	W. Africa	
CAMELOPARDALIDÆ.			
93. *Camelopardalis*	1	All open country	
BOVIDÆ.			
94. Bubalus... ...	3	Trop. and S. Africa	India
95. *Oreas*	2	Africa S. of Sahara	
96. *Tragelaphus* ...	8	Africa S. of Sahara	
97. *Oryx*	3	Arabian and African deserts	S. Palæarctic
98. *Gazella*	12	Africa N. of Equator and S. Africa	Palæarctic Deserts
99. *Æpyceros* ...	1	S. E. Africa	
100. *Cervicapra* ...	4	All Tropical Africa	
101 *Kobus*	6	Pastures of all Africa	
102. *Pelea*	1	South Africa	
103. *Nanotragus* ..	9	Africa S. of Sahara	
104. *Neotragus* ...	1	Abyssinia and N. E. Africa	
105. *Cephalophus* ...	22	All tropical Africa	
106. *Hippotragus* ...	3	Gambia, Central Africa to Cape	
107. *Alcephalus* ...	9	All Africa	
108. *Catoblepas* ...	2	Africa S. of Equator	
(Capra	1	Abyssinia, high)	Palæarctic genus
PROBOSCIDEA.			
ELEPHANTIDÆ.			
109. Elephas... ...	1	Tropical and S. Africa	Oriental

Order, Family, and Genus.	No. of Species.	Range within the Region.	Range beyond the Region.
HYRACOIDEA.			
HYRACIDÆ.			
110. Hyrax	10	Tropical and S. Africa	Syria
RODENTIA.			
MURIDÆ.			
111. Mus	26	All Africa	E. Hemis. excl. Oceania
112. *Lasiomys* .	1	W. Africa	
113. Acanthomys ...	4	Tropical Africa	India
114. *Cricetomys* ...	1	Tropical Africa	
115. *Saccostomus* ...	2	Mozambique	
116. *Dendromys* ...	2	S. Africa	
117. *Nesomys*... ...	1	Madagascar	
118. *Steatomys* ...	2	East and S. Africa	
119. *Pelomys*	1	Mozambique	
120. *Otomys*	6	S. and E. Africa	
121. Meriones ...	14	Africa	Palæarctic, India
122. *Malacothrix* ...	2	S. Africa	
123. *Mystromys* ...	1	S. Africa	
124. *Brachytarsomys*	1	Madagascar	
125. *Hypogeomys* ...	1	Madagascar	
126. *Lophiomys* ...	1	S. Arabia and N. E. Africa	
SPALACIDÆ.			
127. Rhizomys ...	4	Abyssinia	Oriental to Malacca
128. *Bathyerges* ..	1	S. Africa	
129. *Georychus* ...	6	E. Central, and S. Africa	
130. *Heliophobius*...	1	Mozambique	
DIPODIDÆ.			
131. Dipus	7	N. and Central Africa	Central Palæarctic
132. *Pedetes*	1	S. Af. to Mozambique and Angel	
MYOXIDÆ.			
133. Myoxus ...	1	Africa to Cape	Palæarctic
SCIURIDÆ.			
134. Sciurus	18	All woody districts of Africa	All regions but Australia
135. *Anomalurus* ...	5	W. Africa and Fernando Po.	
OCTODONTIDÆ.			
136. *Pectinator* ...	1	Abyssinia	
ECHIMYIDÆ.			
137. *Petromys* ...	1	S. Africa	
138. *Aulacodes* ...	1	W., E., and S. Africa	
HYSTRICIDÆ.			
139. Hystrix	1	Africa to Cape	S. Palæarctic Oriental
140. Atherura ...	1	W. Africa	Palæarctic

Order, Family, and Genus.	No. of Species.	Range within the Region.	Range beyond the Region.
LEPORIDÆ.			
141. Lepus	5	East and South Africa	All regions but Australian
EDENTATA.			
MANIDIDÆ.			
142. Manis	4	Sennaar to W. Africa and Cape	Oriental
ORYCTEROPODIDÆ.			
143. *Orycteropus* ...	2	N. E. Africa to Nile Sources, and S. Africa	

BIRDS.

Order, Family, and Genus.	No. of Species.	Range within the Region.	Range beyond the Region.
PASSERES.			
TURDIDÆ.			
1. Turdus	13	The whole reg. (excl. Madagas.)	Almost Cosmopolite
2. Monticola ...	2	S. Africa	Palæarctic and Oriental
3. *Chætops*	3	S. Africa	
4. *Bessonornis* ...	15	The whole region	Palestine
SYLVIIDÆ.			
5. *Drymœca*... ...	70	The whole region	Palestine
6. Cisticola	13	The whole region	Palæarc.,Orien., Austral.
7. Sphenœacus ...	1	S. Africa	Australian
8. *Camaroptera* ...	5	Africa	
9. Acrocephalus ...	8	The whole region	Palæarc.,Orien., Austral.
10. Bradyptetus ...	8	Abyssinia and S. Africa	S. Europe, Palestine
11. *Catriscus*	3	All Africa	
12. *Bernieria*... ...	1	Madagascar	
13. *Ellisia*	1	Madagascar	
14. *Mystacornis* ...	1	Madagascar	
15. Phylloscopus ...	1	S. Africa	Palæarctic, Oriental
16. *Eremomela* ...	16	All Africa	
17. *Eroessa*	1	Madagascar	
18. Hypolais	2	S. Africa	Palæarctic, Oriental
19. Aedon	8	E. and S. Africa	Palæarctic
20. Sylvia	3	N. E. Africa, Gambia, Cape Verd Ids.	Palæarctic, Oriental
21. Curruca	2	S. Africa	Palæarctic
22. Ruticilla	2	Abyssinia and Senegal	Palæarctic, Oriental
23. Cyanecula ...	2	N. E. Africa	Palæarctic
24. Copsychus ...	2	Madagascar and Seychelle Ids.	Oriental
25. Thamnobia ...	7	All Africa	Oriental
26. *Cercotrichas* ...	2	W. and N. E. Africa	
27. *Pœoptera*... ...	1	W. Africa	
28. *Gervasia*	2	Madagascar and Seychelle Ids.	
29. Dromolæa ...	13	All Africa	S. Palæarctic, India
30. Saxicola	14	Central, E. and S. Africa	Palæarctic, India
31. Cercomela ...	3	N. E. Africa	Palestine, N. W. India
32. Pratincola ...	7	Africa and Madagascar	Palæarctic, Oriental

X

Order, Family, and Genus.	No. of Species.	Range within the Region.	Range beyond the Region.
TIMALIIDÆ.			
33. Chatarrhæa ...	1	Abyssinia	Oriental, Palestine
34. *Crateropus* ...	17	All Africa	N. Africa, Persia
35. *Hypergerus* ...	1	W. Africa	
36. *Cichladusa* ...	3	W. and E. Africa	
37. *Alethe*	4	W. Africa	
38. *Oxylabes*	2	Madagascar	
CINCLIDÆ. [?]			
39. *Mesites*	1	Madagascar	
TROGLODYTIDÆ.[?]			
40. Sylvietta... ...	2	Central, E. and S. Africa	
SITTIDÆ.			
41. *Hypherpes* ...	1	Madagascar	
PARIDÆ.			
42. Parus	5	All Africa	Palæarc., Orien., Nearc.
43. *Parisoma*... ...	5	All Africa	
44. Ægithalus ...	4	W., Central, and S. Africa	Palæarctic
45. *Parinia*	1	W. Africa, Prince's Island	
PYCNONOTIDÆ.			
46. Pycnonotus ...	8	All Africa	S. Palæarctic, Oriental
47. *Phyllastrephus*	4	W. and S. Africa	
48. Hypsipetes .	4	Madagascar and Mascarene Ids.	Oriental
49. *Tylas*	1	Madagascar	
50. Criniger	14	W. and S. Africa	Oriental
51. *Ixonotus*	8	W. Africa	
52. *Andropadus* ...	9	Africa and Madagascar	
53. *Lioptilus*... ...	1	S. Africa	
ORIOLIDÆ.			
54. Oriolus	10	All Africa	Palæarctic, Oriental
55. *Artamia* [?] ...	3	Madagascar	
56. *Cyanolanius* [?]	1	Madagascar	
CAMPEPHAGIDÆ.			
57. *Lanicterus* ...	5	All Africa	
58. *Oxynotus*... ...	2	Mauritius and Bourbon	
59. Campephaga ...	5	The whole region	Celebes to New Caledonia
DICRURIDÆ			
60. Dicrurus	11	The whole region	Oriental, Australian
MUSCICAPIDÆ.			
61. Butalis	3	All Africa	Palæarctic, N. Oriental
62. Muscicapa ...	10	All tropical Africa	Palæarctic
63. Alseonax	4	S. Africa	Oriental
64. *Newtonia*... ...	1	Madagascar	

Order, Family, and Genus.	No. of Species.	Range within the Region.	Range beyond the Region.
65. *Hyliota*	2	W. Africa	
66. *Erythrocercus* ...	2	Tropical Africa	
67. *Artomyias* ...	2	W. Africa	
68. *Pseudobias* ..	1	Madagascar	
69. *Smithornis* ...	2	W. and S. Africa	
70. *Megabias*	1	W. Africa	
71. *Cassinia*	2	W. Africa	
72. *Bias*	1	Tropical Africa	
73. *Elminia*	2	Tropical Africa	
74. *Platystira* ...	12	All Africa	
75. *Tchitrea*	18	The whole region	Oriental
76. *Pogonocichla* ...	1	S. Africa	
77. *Bradyornis* ...	7	All Africa	
LANIIDÆ.			
78. *Parmoptila* [?]...	1	W. Africa	
79. *Calicalicus* ...	1	Madagascar	
80. Lanius	15	All Africa	Palæarc., Orien , Nearc.
81. *Hypocolius* ...	1	Abyssinia	
82. *Corvinella* ...	1	S. and W. Africa	
83. *Urolestes*	1	S. Africa	
84. *Fraseria*	2	W. Africa	
85. *Hypodes*	1	W. Africa	
86. *Cuphoterus* ...	1	Prince's Island	
87. *Nilaus*	1	All Africa	
88. *Prionops*	9	All Africa	
89. *Eurocephalus* ...	2	N. E. and S. Africa	
90. *Chaunonotus* ...	1	W. Africa	
91. *Vanga*	4	Madagascar	
92. *Laniarius* ...	38	All Africa, Madagascar [?]	
93. *Meristes*	2	W. and S. E. Africa	
94. *Nicator*	1	E. Africa	
95. *Telephonus* ...	10	All Africa	N. Africa
CORVIDÆ.			
96. *Ptilostomus* ...	2	W. and E. Africa	
97. *Corvus*	7	All Africa and Madagascar	Cosmop., excl. S. Amer.
98. *Corvultur*	2	N. E. to S. Africa	
99. *Picathartes* ...		W. Africa	
(*Fregilus*	1	Abyssinia)	Palæarctic genus
NECTARINIIDÆ.			
100. *Nectarinia* ...	55	The whole region	
101. *Promerops* ...	1	S. Africa	
102. *Cinnyricinclus*	4	W. Africa	
103. *Neodrepanis* ..	1	Madagascar	
DICÆIDÆ.			
104. Zosterops ...	23	The whole region	Oriental and Australian
105. *Pholidornis* ...	1	W. Africa	
HIRUNDINIDÆ.			
106. Hirundo... ...	17	The whole region	Cosmopolite

x 2

Order, Family, and Genus.	No. of Species.	Range within the Region.	Range beyond the Region.
107. *Psalidoprogne*	10	The whole region	
108. *Phedina*	2	Madagascar and Mauritius	
109. Petrochelidon	1	S. Africa	Neotropical
110. Chelidon ...	1	Bogos-land	Palæarctic, Oriental
111. Cotyle	6	All Africa	Palæarctic, Oriental
112. *Waldenia* ...	1	W. Africa	
FRINGILLIDÆ.			
113. *Dryospiza* ...	8	All Africa	S. Palæarctic
114. *Chlorospiza* ...	4	Abyssinia to Cape	Palæarctic
115. Passer	18	All Africa	Palæarctic, Oriental
116. *Crithagra* ...	12	All Africa	N. Africa, Syria
117. *Ligurnus* ...	2	W. Africa	
(Erythrospiza	1	Nubia, Arabia)	S. Palæarctic genus
118. Pinicola [?] ...	1	Cameroons, W. Africa	N. Temperate genus
119. *Fringillaria* ...	9	All Africa	South Palæarctic
PLOCEIDÆ.			
120. *Textor*	5	All Africa	
121. *Hyphantornis*	32	Tropical and S. Africa	
122. *Symplectes* ...	8	Tropical and S. Africa	
123. *Malimbus* ...	9	W. and E. Africa	
124. Ploceus	2	W. and E. Africa	Oriental
125. *Nelicurvius* ...	1	Madagascar	
126. *Foudia*	11	Tropical Africa, Madagascar, &c.	
127. *Sporopipes* .	1	Tropical and S. Africa	
128. *Pyromelana* ...	12	Tropical and S Africa	
129. *Philetærus* ...	1	S. Africa	
130. *Nigrita*	7	W. and N. E. Africa	
131. *Plocepasser* ...	4	E. and S. Africa	
132. *Vidua*	6	Tropical and S. Africa	
133. *Colliuspasser*...	9	Tropical and S. Africa	
134. *Chera*	1	S. Africa	
135. *Spermospiza* ...	2	W. Africa	
136. *Pyrenestes* ...	6	Tropical and S. Africa	
137. Estrilda	16	Tropical and S. Africa	Oriental
138. *Pytelia*	20	Tropical and S. Africa	
139. *Hypargos* ...	2	E. Africa, Madagascar	
140. *Amadina* ...	6	Tropical and S. Africa	
141. *Spermestes* ...	7	The whole region	
142. *Amauresthes*...	1	E. and W. Africa	
143. *Hypochera* ...	2	Tropical and S. Africa	
STURNIDÆ.			
144. *Dilophus* ...	1	S. Africa, Loanda, Sennaar	
145. *Buphaga* ..	2	Trop. and S. Africa ([?] a family)	
146. *Euryceros* ...	1	Madagascar ([?] a family)	
147. *Juida*	5	Tropical and S. Africa	
148. *Lamprocolius*	16	Tropical and S. Africa	
149. *Cinnyricinclus*	2	Tropical and S. Africa	
150. *Onychognathus*	2	W. Africa	
151. *Spreo*	5	Tropical and S. Africa	
152. *Amydrus* ...	5	N. E. Africa	Palestine
153. *Hartlaubius* ...	1	Madagascar	

Order, Family, and Genus.	No. of Species.	Range within the Region.	Range beyond the Region.
154. *Falculia* ...	1	Madagascar	
155. *Fregilupus* ...	1	Bourbon	
ALAUDIDÆ.			
156. Alauda	3	Abyssinia and S. W. Africa	Palæarctic, Indian
157. *Spizocorys* ...	1	South Africa	
158. Galerida... ...	4	North of tropical Africa	Palæarctic, Indian
159. *Calendula* ...	2	Abyssinia, S. Africa	
(Melanocorypha	1	Abyssinia)	Palæarctic genus
160. Certhilauda ...	8	South Africa	S. Europe
161. Alaemon ...	3	South Africa	S. Palæarctic
162. *Heterocorys* ...	1	South Africa	
163. Mirafra	10	South Africa, Madagascar	Oriental, Australian
164. Ammomanes ..	4	African deserts	S. Palæarctic, Indian
165. *Megalophonus*	5	Tropical and S. Africa	
166. *Tephrocorys* ...	2	S. Africa	
167. Pyrrhulauda ..	6	Tropical and S. Africa	Oriental, Canary Islands
MOTACILLIDÆ.			
168. Motacilla ...	8	The whole region	Palæarctic, Oriental, Australian
169. Anthus	10	Tropical and S. Africa	All regions, exc. Australia
170. *Macronyx* ...	4	Tropical and S. Africa	
PITTIDÆ.			
171. Pitta	1	W. Africa	Oriental, Australian
PAICTIDÆ.			
172. *Philepitta* ...	2	Madagascar	
PICARIÆ.			
PICIDÆ.			
173. *Verreauxia* ...	1	W. Africa	
174. *Dendropicus* ...	14	Tropical and S. Africa	
175. *Campethera* ...	14	Tropical and S. Africa	
176. *Geocolaptes* ...	1	South Africa	
YUNGIDÆ.			
177. Yunx	1	N. E. Africa, S. Africa	Palæarctic
INDICATORIDÆ.			
178. Indicator ...	8	Tropical and S. Africa	Oriental
MEGALÆMIDÆ.			
179. *Pogonorhynchus*	14	Tropical and S. Africa	
180. *Buccanodon* ...	1	West Africa	
181. *Stactolœma* ...	1	West Africa	
182. *Barbatula* ...	9	West and South Africa	
183. *Xylobucco* ...	3	West and South Africa	
184. *Gymnobucco* ...	3	West Africa	
185. *Trachyphonus*	6	Tropical and South Africa	
MUSOPHAGIDÆ.			
186. *Musophaga* ...	2	West Africa	

Order, Family, and Genus.	No. of Species.	Range within the Region.	Range beyond the Region.
187. *Turacus..* ...	10	Tropical and S. Africa	
188. *Schizorhis* ...	6	Tropical and S. Africa	
COLIIDÆ.			
189. *Colius*	7	Tropical and S. Africa	
CUCULIDÆ.			
190. *Ceuthmocharcs*	2	Africa and Madagascar	
191. *Coua*	9	Madagascar	
192. *Cochlothraustes*	1	Madagascar	
193. Centropus ...	8	Africa and Madagascar	Oriental, Australian
194. Cuculus	10	Africa and Madagascar	Palæarc., Orien., Austral.
195. Chrysococcyx	7	Tropical and S. Africa	Oriental, Australian
196. Coccystes ...	6	Tropical and S. Africa	S. Palæarctic, Oriental
LEPTOSOMIDÆ.			
197. *Leptosomus* ..	1	Madagascar	
CORACIIDÆ.			
198. Coracias... ...	5	Africa and Madagascar	S. Palæarctic, Oriental
199. Eurystomus ...	3	Africa and Madagascar	Oriental, Australian
200. *Atelornis* ...	2	Madagascar	
201. *Brachypteracias*	1	Madagascar	
202. *Geobiastes* ...	1	Madagascar	
MEROPIDÆ.			
203. Merops	11	Africa and Madagascar	S.Palæar., Orien.,Austral.
204. *Melittophagus*	5	Tropical and S. Africa	
TROGONIDÆ.			
205. *Apaloderma* ...	2	Tropical and S. Africa	
ALCEDINIDÆ.			
206. Alcedo	2	W. Africa, Abyssinia, Natal	Palæar., Orien., Austral.
207. *Corythornis* ...	3	Africa and Madagascar	
208. Ceryle	1	W. Africa, Abyssinia, Natal	American, Palæarctic
209. *Myioceyx* ...	2	West Africa	
210. *Ispidina*... ...	4	Africa and Madagascar	
211. Halcyon... ...	10	Africa, Prince's Is., St. Thomé	S.Palæar., Orien.Austral.
BUCEROTIDÆ.			
212. Berenicornis ...	1	West Africa	Malaya
213. *Tockus*	12	Tropical and S. Africa	
214. *Bycanistes* ...	6	Tropical and S. Africa	
215. *Bucorvus* ...	2	Tropical and S. Africa	
UPUPIDÆ.			
216. Upupa	3	Africa and Madagascar	S. Palæarctic, Oriental
IRRISORIDÆ.			
217. *Irrisor*	12	Africa and Madagascar	

Order, Family, and Genus.	No. of Species	Range within the Region.	Range beyond the Region.
CAPRIMULGIDÆ.			
218. Caprimulgus ...	18	Africa and Madagascar	Palæarc.,Orien.,Austral.
219. *Scortornis* ...	3	Tropical Africa	
220. *Macrodipteryx*	2	W. Africa to Abyssinia	
221. *Cosmetornis* ...	1	Tropical Africa to the Zambesi	
CYPSELIDÆ.			
222. Cypselus	6	The whole region	Palæarctic, Oriental
223. Collocalia ...	1	Mascarene Ids., Madagascar	Oriental, Australian
224. Chætura	4	Tropical Africa and Madagascar	Cosmop., exc. W. Palæarctic
PSITTACI.			
PALÆORNITHIDÆ.			
225. Palæornis ...	3	W. Africa to Abys. & Mauritius	Oriental
PSITTACIDÆ.			
226. *Coracopsis* ..	5	Madagascar and Seychelle Ids.	
227. *Psittacus* ..	2	W. Africa	
228. *Pæocephalus* ...	9	Tropical and S. Africa	
229. *Agapornis* ...	4	Tropical and S. Africa	
230. *Poliopsitta* ...	2	Trop. Africa and Madagascar	
COLUMBÆ.			
COLUMBIDÆ.			
231. Treron	6	Africa and Madagascar	Oriental
232. *Alectrœnas* ...	5	Madagascar and Masc. Ids. (extct in Mauritius and Rodriguez)	
233. Columba ...	12	Africa and Madagascar	Palæarctic, Oriental
234. *Œna*	1	Tropical and S. Africa	
235. Turtur	10	Africa, Madagascar, Comoroand Seychelle Islands	Palæarctic, Oriental
236. *Aplopelia*	4	Abyssinia, S. Africa and West African Islands	
237. *Chalcopelia* ...	3	Tropical and S. Africa	
DIDIDÆ (extinct)			
238. *Didus*	5	Mascarene Islands	
GALLINÆ.			
PTEROCLIDÆ.			
239. Pterocles ...	9	Africa and Madagascar	S. Palæarctic, Indian
TETRAONIDÆ.			
240. *Ptilopachus* ...	1	West Africa	
241. *Francolinus* ...	30	Africa and Madagascar	S. Palæarctic, Indian
242. *Peliperdix* ...	1	West Africa	
243. *Margaroperdix*	1	Madagascar	
244. Coturnix ..	2	Tropical and S. Africa	Palæar., Orient.,Austral.
(Caccabis ..	1	Abyssinia)	Palæarctic genus
PHASIANIDÆ.			
245. *Phasidus* ..	1	West Africa	

Order, Family, and Genus.	No. of Species.	Range within the Region.	Range beyond the Region.
246. *Agelastes* ...	1	West Africa	
247. *Acryllium* ...	1	West Africa	
248. *Numida* ...	9	Africa to Natal and Madagascar	
TURNICIDÆ.			
249. Turnix	4	S. Africa and Madagascar	Palæarc.,Orient., Austrl.
250. *Ortyxelos* ...	1	Africa	
ACCIPITRES.			
VULTURIDÆ.			
251. Gyps	2	Africa, except W. sub-region	Palæarctic, Oriental
252. Pseudogyps ...	1	N. E. Africa to Senegal	Oriental
253. Otogyps... ...	1	N. E. and S. Africa	Palæarctic, Oriental
254. *Lophogyps* ...	1	N. E. and S. Africa and Senegal	
255. Neophron ...	2	Africa, excl. west coast	S. Palæarctic, Oriental
FALCONIDÆ.			
256. *Polyboroides* ...	2	Africa and Madagascar	
257. Circus	4	Africa and Madagascar	Almost Cosmopolite
258. *Urotriorchis* ...	1	W. Africa	
259. *Melierax* ...	5	Africa, excl. west coast	
260. Astur	5	Africa and Madagascar	Almost Cosmopolite
261. Nisoides... ...	1	Madagascar	
262. *Eutriorchis* ...	1	Madagascar	
263. Accipiter ...	8	Africa and Madagascar	Almost Cosmopolite
264. Buteo	5	Africa and Madagascar	Cosmop., excl. Austral.
265. Gypaëtus ...	1	N. E. and S. Africa	S. Palæarctic
266. Aquila	5	All Africa	Nearc., Palæarc., Indian
267. Nisaëtus... ...	1	W. Africa	S. Palæarctic, Oriental, Australia
268. Spizaëtus ...	3	All Africa	Neotropical, Oriental to N. Guinea
269. *Lophoœtus* ...	1	All Africa	
270. *Asturinula* ...	1	Tropical Africa	
271. *Dryotriorchis*	1	W. Africa	
272. Circaëtus ...	5	All Africa	Palæarctic, Oriental
273. Butastur ...	1	N. E. Africa	Oriental to New Guinea
274. *Helotarsus* ...	2	Tropical and S. Africa	
275. Haliæetus ...	2	The whole region	Cosmopolite, excl. Neotropical region
276. *Gypohierax* ...	1	West and East Africa	
277. *Elanoides* ...	1	West and N.E. Africa	
278. Milvus	1	The whole region	The Eastern Hemisphere
279. Elanus	1	Africa	India to Australia
280. Machærhamphus	1	S. W. Africa and Madagascar	Malacca
281. Pernis	1	S. Africa and Madagascar	Palæarctic, Oriental
282. Baza	3	Africa and Madagascar	India to N. Australia
283. Poliohierax ...	1	East Africa	Burmah
284. Falco	4	All Africa	Almost Cosmopolite
285. Cerchneis ...	8	The whole region	Almost Cosmopolite
SERPENTARIIDÆ.			
286. *Serpentarius* ...	1	The greater part of Africa	

Order, Family, and Genus.	No. of Species.	Range within the Region.	Range beyond the Region.
PANDIONIDÆ.			
287. Pandion... ...	1	All Africa	Cosmopolite
STRIGIDÆ.			
288. Athene	5	Africa and Madagascar, Rodriquez (extinct)	Palæarctic, Oriental, Australian
289. Bubo	8	Africa and Madagascar	Cosmopolite
290. *Scotopelia* ...	2	West and S. Africa to Zambesi	
291. Scops	3	W. and S. Africa, Madagascar, Comoro Islands	Almost Cosmopolite
292. Syrnium ...	2	Africa	Palæarctic, Oriental, American
293. Asio	1	N. E. and S. Africa	Cosmopolite
294. Strix	4	Africa and Madagascar	Cosmopolite

Peculiar or very Characteristic Genera of Wading or Swimming Birds.

GRALLÆ.			
RALLIDÆ.			
Himantornis ...	1	West Africa	
Podica... ..	3	Africa	Burmah
GLAREOLIDÆ.			
Cursorius ...	8	All Africa	S. Europe, India
OTIDIDÆ.			
Eupodotis ...	16	All Africa	India, Australia
GRUIDÆ.			
Balearica. ...	2	All Africa	
ARDEIDÆ.			
Balœniceps ...	1	Upper Nile	
PLATALEIDÆ.			
Scopus ...	1	Tropical and S. Africa	
ANSERES.			
ANATIDÆ.			
Thalassornis...	1	South Africa	
STRUTHIONES.			
STRUTHIONIDÆ.			
295. *Struthio* ...	2	All Africa	Syria
ÆPYORNITHIDÆ.		(Extinct)	
296. *Æpyornis* ...	3 [?]	Madagascar	

CHAPTER XII.

THE ORIENTAL REGION.

This region is of comparatively small extent, but it has a very diversified surface, and is proportionately very rich. The deserts on the north-west of India are the debatable land that separates it from the Palæarctic and Ethiopian regions. The great triangular plateau which forms the peninsula of India is the poorest portion of the region, owing in part to its arid climate and in part to its isolated position ; for there can be little doubt that in the later Tertiary period it was an island, separated by an arm of the sea (now forming the valleys of the Ganges and Indus) from the luxuriant Himalayan and Burmese countries. Its southern extremity, with Ceylon, has a moister climate and more luxuriant vegetation, and exhibits indications of a former extension southwards, with a richer and more peculiar fauna, partly Malayan and partly Mascarene in its character. The whole southern slopes of the Himalayas, with Burmah, Siam and Western China, as well as the Malay peninsula and the Indo-Malay islands, are almost everywhere covered with tropical forests of the most luxuriant character, which abound in varied and peculiar forms of vegetable and animal life. The flora and fauna of this extensive district are essentially of one type throughout; yet it may be usefully divided into the Indo-Chinese and the Malayan sub-regions, as each possesses a number of peculiar or characteristic animals. The former sub-region, besides having many tropical and sub-tropical types of its own, also possesses a large number of peculiarly modified temperate forms on the mountain ranges of its northern

ORIENTAL REGION

Scale 1 inch=1,000 miles

EXPLANATION

Terrestrial Contours

From Sea level to 1,000 feet   White
1,000 feet to 2,500
2,500 " 5,000
5,000 " 10,000
10,000 " 20,000
Above 20,000 feet

The Marine Contour of 1,000 feet
is shown by a dotted line

Pasture lands shown thus
Forest
Desert

The boundaries and reference numbers
of the Sub-regions are shown in Red

boundary, which are wholly wanting in the Malayan sub-region. The Philippine islands are best classed with the Indo-Malay group, although they are strikingly deficient in many Malayan types, and exhibit an approach to the Celebesian division of the Austro-Malay sub-region.

Zoological Characteristics of the Oriental Region.—The Oriental Region possesses examples of 35 families of Mammalia, 71 of Birds, 35 of Reptiles, 9 of Amphibia, and 13 of Fresh-water Fishes. Of these 163 families, 12 are peculiar to the region; namely, Tarsiidæ, Galeopithecidæ, and Tupaiidæ among Mammalia, while Æluridæ, though confined to the higher Himalayas, may perhaps with more justice be claimed by the Palæarctic region; Liotrichidæ, Phyllornithidæ, and Eurylæmidæ among birds; Xenopeltidæ (extending, however, to Celebes), Uropeltidæ, and Acrochordidæ among reptiles; Luciocephalidæ, Ophiocephalidæ and Mastacembelidæ among fresh-water fishes. A number of other families are abundant, and characteristic of the region; and it possesses many peculiar and characteristic genera, which must be referred to somewhat more in detail.

Mammalia.—The Oriental region is rich in quadrumana, and is especially remarkable for its orang-utans and long-armed apes (*Simia, Hylobates,* and *Siamanga*); its abundance of monkeys of the genera *Presbytes* and *Macacus ;* its extraordinary long-nosed monkey (*Presbytes nasalis*); its Lemuridæ (*Nycticebus* and *Loris*); and its curious genus *Tarsius,* forming a distinct family of lemurs. All these quadrumanous genera are confined to it, except *Tarsius* which extends as far as Celebes. It possesses more than 30 genera of bats, which are enumerated in the lists given at the end of this chapter. In Insectivora it is very rich, and possesses several remarkable forms, such as the flying lemur (*Galeopithecus*); the squirrel-like Tupaiidæ consisting of three genera; and the curious *Gymnura* allied to the hedge-hogs. In Carnivora, it is especially rich in many forms of civets (Viverridæ), possessing 10 peculiar genera, among which *Prionodon* and *Cynogale* are remarkable; numerous Mustelidæ, of which *Gymnopus, Mydaus, Aonyx* and *Helictis* are the most conspicuous; *Ælurus,* a curious animal, cat-like in appearance but

more allied to the bears, forming a distinct family of Carnivora,
and confined to the high forest-districts of the Eastern Himalayas and East Thibet; *Melursus* and *Helarctos*, peculiar forms of
bears ; *Platanista*, a dolphin peculiar to the Ganges and Indus.
Among Ruminants it has the beautiful chevrotain, forming
the genus *Tragulus* in the family Tragulidæ; with one peculiar
genus and three peculiar sub-genera of true deer. The Antilo-
pinæ and Caprinæ are few, confined to limited districts and not
characteristic of the region ; but there are everywhere wild cattle
of the genera *Bibos* and *Bubalus*, which, with species of *Rhinoceros*
and *Elephas*, form a prominent feature in the fauna. The Rodents
are less developed than in the Ethiopian region, but several forms
of squirrels everywhere abound, together with some species of
porcupine; and the Edentata are represented by the scaly manis.

Birds.—The families and genera of birds which give a character to Oriental lands, are so numerous and varied, that we
can here only notice the more prominent and more remarkable.
The Timaliidæ, represented by the babblers (*Garrulax, Pomatorhinus, Timalia*, &c.), are almost everywhere to be met with, and
no less than 21 genera are peculiar to the region; the elegant
fork-tailed *Enicurus* and rich blue *Myiophonus*, though comparatively scarce, are characteristic of the Malayan and Indo-
Chinese faunas; the elegant little "hill-tits" (Liotrichidæ)
abound in the same part of the region; the green bulbuls (*Phyllornis*) are found everywhere ; as are various forms of Pycnonotidæ, the black and crimson "minivets" (*Pericrocotus*), and the
glossy "king-crows" (*Dicrurus*); *Urocissa, Platylophus* and *Dendrocitta* are some of the interesting and characteristic forms
of the crow family; sun-birds (Netariniidæ) of at least three
genera are found throughout the region, as are the beautiful little
flower-peckers (Dicæidæ), and some peculiar forms of weaver-
birds (*Ploceus* and *Munia*). Of the starling family, the most
conspicuous are the glossy mynahs (*Eulabes*). The swallow-
shrikes (*Artamus*) are very peculiar, as are the exquisitely
coloured pittas (Pittidæ), and the gaudy broad-bills (Eurylæmidæ). Leaving the true Passeres, we find woodpeckers,
barbets, and cuckoos everywhere, often of peculiar and re-

markable forms; among the bee-eaters we have the exquisite *Nyctiornis* with its pendent neck-plumes of blue or scarlet; brilliant kingfishers and strangely formed hornbills abound everywhere; while brown-backed trogons with red and orange breasts, though far less frequent, are equally a feature of the Ornithology. Next we have the frog-mouthed goatsuckers (*Battrachostomus*), and the whiskered swifts (*Dendrochelidon*), both wide-spread, remarkable, and characteristic groups of the Oriental region. Coming to the parrot tribe, we have only the long-tailed *Palæornis* and the exquisite little *Loriculus*, as characteristic genera. We now come to the pigeons, among which the fruit-eating genera *Treron* and *Carpophaga* are the most conspicuous. The gallinaceous birds offer us some grand forms, such as the peacocks (*Pavo*); the argus pheasants (*Argusianus*); the fire-backed pheasants (*Euplocamus*); and the jungle-fowl (*Gallus*), all strikingly characteristic; and with these we may close our sketch, since the birds of prey and the two Orders comprising the waders and swimmers offer nothing sufficiently remarkable to be worthy of enumeration here.

Reptiles.—Only the more abundant and characteristic groups will here be noticed. In the serpent tribe, the Oligodontidæ, a small family of ground-snakes; the Homalopsidæ, or fresh-water snakes; the Dendrophidæ, or tree-snakes; the Dryiophidæ, or whip-snakes; the Dipsadidæ, or nocturnal tree-snakes; the Lycodontidæ or fanged ground-snakes; the Pythonidæ, or rock-snakes; the Elapidæ, or venomous colubrine snakes (including the "cobras"); and the Crotalidæ, or pit-vipers, are all abundant and characteristic, ranging over nearly the whole region, and pre-senting a great variety of genera and species. Among lizards, the Varanidæ or water-lizards; the Scincidæ or "scinks;" the Gecko-tidæ, or geckoes; and the Agamidæ, or eastern iguanas; are the most universal and characteristic groups. Among crocodiles the genus *Crocodilus* is widely spread, *Gavialis* being characteristic of the Ganges. Among Chelonia, or shielded reptiles, forms of fresh-water Testudinidæ and Trionychidæ (soft tortoises) are tolerably abundant.

Amphibia —The only abundant and characteristic groups of

this class are toads of the family Engystomidæ; tree-frogs of the family Polypedatidæ; and several genera of true frogs, Ranidæ.

Fresh-water Fishes.—The more remarkable and characteristic fishes inhabiting the fresh waters of the Oriental region belong to the following families: Nandidæ, Labyrinthici, Ophiocephalidæ, Siluridæ, and Cyprinidæ; the last being specially abundant.

The sketch here very briefly given, must be supplemented by an examination of the tables of distribution of the genera of all the Mammalia and Birds inhabiting the region. We will now briefly summarize the results.

Summary of the Oriental Vertebrata.—The Oriental region possesses examples of 163 families of Vertebrata of which 12 are peculiar, a proportion of a little more than one-fourteenth of the whole.

Out of 118 genera of Mammalia 54 seem to be peculiar to the region, equal to a proportion of $\frac{9}{20}$ or a little less than half. Of Land-Birds there are 342 genera of which 165 are peculiar, bringing the proportion very close to a half.

In the Ethiopian region the proportion of peculiar forms both of Mammalia and Birds is greater; a fact which is not surprising when we consider the long continued isolation of the latter region—an isolation which is even now very complete, owing to the vast extent of deserts intervening between it and the Palæarctic region; while the Oriental and Palæarctic were, during much of the Tertiary epoch, hardly separable.

Insects.

Lepidoptera.—We can only glance hastily at the more prominent features of the wonderfully rich and varied butterfly-fauna of the Oriental region. In the first family Danaidæ, the genera *Danais* and *Euplœa* are everywhere abundant, and the latter especially forms a conspicuous feature in the entomo-logical aspect of the country; the large "spectre-butterflies" (*Hestia*) are equally characteristic of the Malayan sub-region. Satyridæ, though abundant are not very remarkable, *Debis, Melanitis, Mycalesis,* and *Ypthima* being the most characteristic genera. Morphidæ are well represented by the genera *Amathusia, Zeuxidia, Discophora*, and *Thaumantis*, some of the species of which almost equal the grand South American Morphos. The Nymphalidæ furnish us with a host of characteristic genera, among the most remarkable of which are, *Terinos, Adolias, Cethosia, Cyrestis, Limenitis*, and *Nymphalis*, all abounding in beautiful species. Among the Lycænidæ are a number of fine groups, among which we may mention *Ilerda, Myrina, Deudoryx, Aphneus, Iolaus*, and *Amblypodia*, as characteristic examples. The Pieridæ furnish many fine forms, such as *Thyca, Iphias, Thestias, Eronia, Prioneris*, and *Dercas*, the last two being peculiar. The Papilionidæ are unsurpassed in the world, presenting such grand genera as *Teinopalpus* and *Bhutanitis;* the yellow-marked *Ornithopteræ;* the superb "Brookiana;" the elegant *Leptocercus;* and *Papilios* of the "Coon," "Philoxenus," "Memnon," "Protenor," and especially the 'green-and-gold-dusted' "Paris" groups.

The Moths call for no special observations, except to notice the existence in Northern India of a number of forms which resemble in a striking manner some of the most remarkable of the above mentioned groups of the genus *Papilio*, especially the "Protenor" group, which there is reason to believe is protected by a peculiar smell or taste like the *Heliconias* and Danaidæ.

Coleoptera.—The most characteristic Oriental form of the Cicindelidæ or tiger beetles, is undoubtedly the elegant genus *Collyris*, which is found over the whole region and is almost confined to it. Less abundant, but equally characteristic, is the wingless ant-like *Tricondyla*. Two small genera *Apteroessa* and *Dromicidia* are confined to the Indian Peninsula, while *Therates* only occurs in the Malayan sub-region.

The Carabidæ, or ground carnivorous beetles, are so numerous that we can only notice a few of the more remarkable and characteristic forms. The wonderful *Mormolyce* of the Indo-Malay sub-region, stands pre-eminent for singularity in the entire family. *Thyreopterus, Orthogonius, Catascopus*, and *Pericallus* are very characteristic forms, as well as *Planetes* and

Distrigus, the latter having a single species in Madagascar. There are 80 genera of this family peculiar to the region, 10 of which have only been found in Ceylon.

Among the Lucanidæ, or stag-beetles, *Lucanus, Odontolabris*, and *Cladognathus* are the most characteristic forms. Sixteen genera inhabit the region, of which 7 are altogether peculiar, while three others only extend eastward to the Austro-Malayan sub-region.

The beautiful Cetoniidæ, or rose-chafers, are well represented by *Rhomborhina, Heterorhina, Clinteria, Macronota, Agestrata, Chalcothea* and many fine species of *Cetonia*. There are 17 peculiar genera, of which *Mycteristes, Phœdimus, Plectrone*, and *Rhagopteryx*, are Malayan; while *Narycius, Clerota, Bombodes*, and *Chiloloba* are Indian.

In Buprestidæ—those elongate metallic-coloured beetles whose elytra are used as ornaments in many parts of the world—this region stands pre-eminent, in its gigantic *Catoxantha*, its fine *Chrysochroa*, its Indian *Sternocera*, its Malayan *Chalcophora* and *Belionota*, as well as many other beautiful forms. It possesses 41 genera, of which 14 are peculiar to it, the rest being generally of wide range or common to the Ethiopian and Australian regions.

In the extensive and elegant group of Longicorns, the Oriental region is only inferior to the Neotropical. It possesses 360 genera, 25 of which are Prionidæ, 117 Cerambicidæ, and 218 Lamiidæ;—about 70 per cent. of the whole being peculiar. The most characteristic genera are *Rhaphidopodus* and *Ægosoma* among Prionidæ; *Neocerambyx, Euryarthrum, Pachyteria, Acrocyrta, Tetraommatus, Chloridolum*, and *Polyzonus* among Cerambycidæ; and *Cœlosterna, Rhytidophora, Batocera, Agelasta*, and *Astathes* among Lamiidæ.

Of remarkable forms in other families, we may mention the gigantic horned *Chalcosoma* among Scarabæidæ; the metallic *Campsosternus* among Elateridæ; the handsome but anomalous *Trictenotoma* forming a distinct family; the gorgeous *Pachyrhynchi* of the Philippine Islands among Curculionidæ; *Diurus* among Brenthidæ; with an immense number and variety of Anthotribidæ, Heteromera, Malacoderma, and Phytophaga.

THE ORIENTAL SUB-REGIONS.

The four sub-regions into which we have divided the Oriental region, are very unequal in extent, and perhaps more so in productiveness, but they each have well-marked special features, and serve well to exhibit the main zoological characteristics of the region. As they are all tolerably well defined and their faunas comparatively well-known, their characteristics will be given with rather more than usual detail.

I. Hindostan, or Indian Sub-region.

This includes the whole peninsula of India from the foot of the Himalayas on the north to somewhere near Seringapatam on the south, the boundary of the Ceylonese sub-region being unsettled. The deltas of the Ganges and Brahmaputra mark its eastern limits, and it probably reaches to about Cashmere in the north-west, and perhaps to the valley of the Indus further south; but the great desert tract to the east of the Indus forms a transition to the south Palæarctic sub-region. Perhaps on the whole the Indus may be taken as a convenient boundary. Many Indian naturalists, especially Mr. Blyth and Mr. Blanford, are impressed with the relations of the greater part of this sub-region to the Ethiopian region, and have proposed to divide it into several zoological districts dependent on differences of climate and vegetation, and characterized by possessing faunas more or less allied either to the Himalayan or the Ethiopian type. But these sub-divisions appear far too complex to be useful to the general student, and even were they proved to be natural, would be beyond the scope of this work. I agree, however, with Mr. Elwes in thinking that they really belong to local rather than to geographical distribution, and confound "station" with "habitat." Wherever there is a marked diversity of surface and vegetation the productions of a country will correspondingly differ; the groups peculiar to forests, for example, will be absent from open

Y

plains or arid deserts.   It happens that the three great Old
World regions are separated from each other by a debatable land
which is chiefly of a desert character; hence we must expect to
find a resemblance between the inhabitants of such districts in
each region. We also find a great resemblance between the aquatic
birds of the three regions; and as we generally give little weight
to these in our estimate of the degree of affinity of the faunas of
different countries, so we should not count the desert fauna as of
equal weight with the more restricted and peculiar types which
are found in the fertile tracts,—in the mountains and valleys, and
especially in the primeval forests.   The supposed preponderance
of exclusively Ethiopian groups of Mammalia and Birds in this
sub-region, deserves however a close examination, in order to
ascertain how far the facts really warrant such an opinion.

Mammalia.—The following list of the more important genera
of Mammalia which range over the larger part of this sub-region
will enable naturalists to form an independent judgment as to
the preponderance of Ethiopian, or of Oriental and Palæarctic
types, in this, the most important of all the classes of animals
for geographical distribution.

RANGE OF THE GENERA OF MAMMALIA WHICH INHABIT THE SUB-REGION
OF HINDOSTAN.

1. Presbytes   ... Oriental only.
2. Macacus   ... Oriental only.
3. Erinaceus   ... Palæarctic genus.
4. Sorex ...   ... Widely distributed.
5. Felis ...   ... Almost Cosmopolitan.
6. Cynælurus   ... Ethiopian and S. Palæarctic.
7. Viverra   ... Ethiopian and Oriental to China and Malaya.
8. Viverricula   ... Oriental only.
9. Paradoxurus ... Oriental only.
10. Herpestes   ... Ethiopian, S. Palæarctic, and Oriental to Malaya.
11. Calogale   ... Ethiopian, Oriental to Cambodja.
12. Tæniogale   ... Oriental.
13. Hyæna   ... Palæarctic and Ethiopian (a Palæarctic species.)
14. Canis ...   ... Palæarctic and Oriental to Malaya.
15. Cuon ...   ... Oriental to Malaya.
16. Vulpes   ... Very wide range.
17. Lutra ...   ... Oriental and Palæarctic.
18. Mellivora   ... Ethiopian.
19. Melursus   ... Oriental only; family not Ethiopian.
20. Sus   ...   ... Palæarctic and Oriental, not Ethiopian.
21. Tragulus   ... Oriental.

22. Cervus	...	Oriental and Palæarctic ; family not Ethiopian.
23. Cervulus	...	Oriental ; family not Ethiopian.
24. Bibos ...	...	Palæarctic and Oriental.
25. Portax	...	Oriental.
26. Gazella	...	Palæarctic and Ethiopian.
27. Antilope	...	Oriental.
28. Tetraceros	...	Oriental.
29. Elephas	...	Oriental species.
30. Mus ...	...	Cosmopolite nearly.
31. Platacanthomys		Oriental.
32. Meriones	...	Very wide range.
33. Spalacomys	...	Oriental.
34. Sciurus	...	Almost Cosmopolite.
35. Pteromys	...	Palæarctic and Oriental to China and Malaya.
36. Hystrix	...	Wide range.
37. Lepus ...	...	Wide range.
38. Manis ...	...	Ethiopian and Oriental to Malaya.

Out of the above 38 genera, 8 have so wide a distribution as to give no special geographical indications. Of the remaining 30, whose geographical position we have noted, 14 are Oriental only ; 5 have as much right to be considered Oriental as Ethiopian, extending as they do over the greater part of the Oriental region ; 2 (the hyæna and gazelle) show Palæarctic rather than Ethiopian affinity ; 7 are Palæarctic and Oriental but not Ethiopian ; and only 2 (*Cynœlurus* and *Mellivora*) can be considered as especially Ethiopian. We must also give due weight to the fact that we have here Ursidæ and Cervidæ, two families entirely absent from the Ethiopian region, and we shall then be forced to conclude that the affinities of the Indian peninsula are not only clearly Oriental, but that the Ethiopian element is really present in a far less degree than the Palæarctic.

Birds.—The naturalists who have adopted the " Ethiopian theory " of the fauna of Hindostan, have always supported their views by an appeal to the class of birds ; maintaining, that not only are almost all the characteristic Himalayan and Malayan genera absent, but that their place is to a great extent supplied by others which are characteristic of the Ethiopian region. After a careful examination of the subject, Mr. Elwes, in a paper read before the Zoological Society (June 1873) came to the conclusion, that this view was an erroneous one, founded on the fact that the birds of the plains are the more abundant and more open to observation; and that these are often of wide-spread types, and some few almost exclusively African. The facts he adduced do not, however, seem to have satisfied the objectors; and as the subject is an important one, I will here give lists of all the genera of Passeres, Picariæ, Psittaci, Columbæ, and Gallinæ, which inhabit the sub-region, leaving out those which only just enter within its boundaries from adjacent sub-regions. These are arranged under four heads:—1. Oriental genera; which are either wholly confined to, or strikingly prevalent in, the Oriental region beyond the limits of the Indian peninsula. 2. Genera of Wide Range; which are fully as much entitled to be considered Oriental or Palæarctic as Ethiopian, and cannot be held to prove any Ethiopian affinity. 3. Palæarctic genera; which are altogether or almost absent from the Ethiopian region. 4. Ethiopian genera; which are confined to, or very prevalent in, the Ethiopian region, whence they extend into the Indian peninsula but not over the whole Oriental region. The last are the only ones which can be fairly balanced against those of the first list, in order to determine the character of the fauna.

1. ORIENTAL GENERA IN CENTRAL INDIA.

Geocichla, Orthotomus, Prinia, Megalurus, Abrornis, Larvivora, Copsychus, Kittacincla, Pomatorhinus, Malacocercus, Chatarrhœa, Layardia, Garrulax, Trochalopteron, Pellorneum, Dumetia, Pyctoris, Alcippe, Myiophonus, Sitta, Dendrophila, Phyllornis, Iora, Hypsipetes, Pericrocotus, Graucalus, Volvocivora, Chibia, Chaptia, Irena, Erythrosterna, Hemipus, Hemichelidon, Niltava, Cyornis, Eumyias, Hypothymis, Myialestes, Tephrodornis, Dendrocitta, Arachnechthra, Nectarophila, Arachnothera, Dicæum, Piprisoma, Munia, Eulabes, Pastor, Acridotheres, Sturnia, Sturnopastor, Artamus, Nemoricola, Pitta, Yungipicus, Chrysocolaptes, Hemicircus, Gecinus, Mulleripicus, Brachypternus, Tiga, Micropternus, Megalæma, Xantholæma, Rhopodytes, Taccocoua, Surniculus, Hierococcyx, Eudynamnis, Nyctiornis, Harpactes, Pelargopsis, Ceyx, Hydrocissa, Meniceros, Batrachostomus, Dendrochelidon, Collocalia, Palæornis, Treron, Carpophaga, Chalcophaps, Ortygornis, Perdix, Pavo, Gallus, Galloperdix;—87 genera; and one peculiar genus, *Salpornis*, whose affinities are Palæarctic or Oriental.

2. GENERA OF WIDE RANGE OCCURRING IN CENTRAL INDIA.

Tardus, Monticola, Drymœca, Cisticola, Acrocephalus, Phylloscopus, Pratincola, Parus, Pycnonotus, Criniger, Oriolus, Dicrurus, Tchitrea, Lanius, Corvus, Zosterops, Hirundo, Cotyle, Passer, Ploceus, Estrilda, Alauda, Calandrella, Mirafra, Ammomanes, Motacilla, Anthus, Picus, Yunx, Centropus, Cuculus, Chrysoccocyx, Coccystes, Coracias, Eurystomus, Merops, Alcedo, Ceryle, Halcyon, Upupa, Caprimulgus, Cypselus, Chætura, Columba, Turtur, Pterocles, Coturnix, Turnix ;—48 genera.

3. PALÆARCTIC GENERA OCCURRING IN CENTRAL INDIA.

Hypolais, Sylvia, Curruca, Cyanecula, Calliope, Chelidon, Euspiza, Emberiza, Galerita, Calobates, Corydalla ;—11 genera.

4. ETHIOPIAN GENERA OCCURRING IN CENTRAL INDIA.

Thamnobia, Pyrrhulauda, Pterocles, Francolinus ;—4 genera.

A consideration of the above lists shows us, that the Hindostan sub-region is by no means so poor in forms of bird-life as is generally supposed (and as I had myself anticipated, it would prove to be), possessing, as it does, 151 genera of land-birds, without counting the Accipitres. It must also set at rest the question of the zoological affinities of the district, since a preponderance of 88 genera, against 4, cannot be held to be insufficient, and cannot be materially altered by any corrections in details that may be proposed or substantiated. Even of these four, only the first two are exclusively Ethiopian, *Pterocles* and *Francolinus* both being Palæarctic also. It is a question, indeed, whether anywhere in the world an outlying sub-region can be found, exhibiting less zoological affinity for the adjacent regions ; and we have here a striking illustration of the necessity of deciding all such cases, not by *examples*, which may be so chosen as to support any view, but by carefully weighing and contrasting the whole of the facts on which the solution of the problem admittedly depends. It will, perhaps, be said that a great many of the 88 genera above given are very scarce and very local; but this is certainly not the case with the majority of them; and even where it is so, that does not in any degree affect their value as indicating zoo-geographical affinities. It is the *presence* of a type in a region, not its abundance or scarcity, that is the important fact; and when we have to do, as we have here, with many groups whose habits and mode of life necessarily seclude them from observation, their supposed scarcity may not even be a fact.

Reptiles and Amphibia.—Reptiles entirely agree with Mammalia and Birds in the main features of their distribution. Out of 17 families of snakes inhabiting Hindostan, 16 range over the greater part of the entire region, and only two can be supposed to show any Ethiopian affinity. These are the Psammophidæ and Erycidæ, both desert-haunting groups, and almost as much South Palæarctic as African. The genus *Tropidococcyx* is peculiar to the sub-region, and *Aspidura, Passerita* and *Cynophis* to the peninsula and Ceylon; while a large number of the most characteristic genera, as *Dipsas, Simotes, Bungarus, Naja, Trimeresurus, Lycodon* and *Python*, are characteristically Oriental.

Of the six families of lizards all have a wide range The genera *Eumeces, Pentadactylus, Gecko, Eublepharis,* and *Draco,* are characteristically or wholly Oriental; *Ophiops* and *Uromastix* are Palæarctic; while *Chamœleon* is the solitary case of decided Ethiopian affinity.

Of the Amphibia not a single family exhibits special Ethiopian affinities.

II. Sub-region of Ceylon and South-India.

The Island of Ceylon is characterised by such striking peculiarities in its animal productions, as to render necessary its separation from the peninsula of India as a sub-region; but it is found that most of these special features extend to the Neilgherries and the whole southern mountainous portion of India, and that the two must be united in any zoo-geographical province. The main features of this division are,—the appearance of numerous animals allied to forms only found again in the Himalayas or in the Malayan sub-region, the possession of several peculiar generic types, and an unusual number of peculiar species.

Mammalia.—Among Mammalia the most remarkable form is *Loris*, a genus of Lemurs altogether peculiar to the sub-region; several peculiar monkeys of the genus *Presbytes;* the Malayan genus *Tupaia;* and *Platacanthomys*, a peculiar genus of Muridæ.

Birds.—Among birds it has *Ochromela*, a peculiar genus of flycatchers ; *Phœnicophaës* (Cuculidæ) and *Drymocataphus* (Timaliidæ), both Malayan forms ; a species of *Myiophonus* whose nearest ally is in Java; *Trochalopteron, Brachypteryx, Buceros* and *Loriculus*, which are only found elsewhere in the Himalayas and Malayana. It also possesses about 80 peculiar species of birds, including a large jungle fowl, one owl and two hornbills.

Reptiles.—It is however by its Reptiles, even more than by its higher vertebrates, that this sub-region is clearly characterised. Among snakes it possesses an entire family, Uropeltidæ, consisting of 5 genera and 18 species altogether confined to it,—*Rhinophis* and *Uropeltis* in Ceylon, *Silybura, Plecturus* and *Melanophidium* in Southern India. Four other genera of snakes, *Haplocercus, Cercaspis, Peltopelor,* and *Hypnale* are also peculiar; *Chersydrus* is only found elsewhere in Malaya; while *Aspidura, Passerita,* and *Cynophis,* only extend to Hindostan; species of *Eryx, Echis,* and *Psammophis* show an affinity with Ethiopian and Palæarctic forms. Among lizards several genera of *Agamidæ* are peculiar, such as *Otocryptis, Lyricoephalus, Ceratophora, Cophotis, Salea, Sitana* and *Charasia.* In the family Acontiadæ, *Nessia* is peculiar to Ceylon, while a species of the African genus *Acontias* shows an affinity for the Ethiopian region.

Amphibia.—The genera of Amphibians that occur here are generally of wide range, but *Nannophrys, Haplobatrachus,* and *Cacopus* are confined to the sub-region; while *Megalophrys* is Malayan, and the species found in Ceylon also inhabit Java.

Insects.—The insects of Ceylon also furnish some curious examples of its distinctness from Hindostan, and its affinity with Malaya. Among its butterflies we find *Papilio jophon,* closely allied to *P. antiphus* of Malaya. The remarkable genus *Hestia,* so characteristic of the Malay archipelago, only occurs elsewhere on the mountains of Ceylon; while its *Cynthia* and *Parthenos* are closely allied to, if not identical with, Malayan species. Among Coleoptera we have yet more striking examples. The highly characteristic Malayan genus *Tricondyla* is represented in Ceylon by no less than 10 species; and among Longicorns we find the genera *Tetraommatus, Thranius, Cacia, Praonetha, Ropica,* and *Serixia,* all exclusively Malayan or only just entering the Indo-Chinese peninsula, yet all represented in Ceylon, while not a single species occurs in any part of India or the Himalayas.

The Past History of Ceylon and South-India as indicated by its Fauna.—In our account of the Ethiopian region we have already had occasion to refer to an ancient connection between this sub-region and Madagascar, in order to explain the distribution of the Lemurine type, and some other curious affinities between the two countries. This view is supported by the geology of India, which shows us Ceylon and South India consisting mainly of granitic and old metamorphic rocks, while the greater part of the peninsula, forming our first sub-region, is of tertiary formation, with a few isolated patches of secondary rocks. It is evident therefore, that during much of the tertiary period, Ceylon and South India were bounded on the north by a considerable extent of sea, and probably formed part of an extensive southern continent or great island. The very numerous and remarkable cases of affinity with Malaya, require however some closer approximation to these islands, which probably occurred at a later period. When, still later, the great plains and table-lands of Hindostan were formed, and a permanent land communication effected with the rich and highly developed Himalo-Chinese fauna, a rapid immigration of new types took place, and many of the less specialised forms of mammalia and birds (particularly those of ancient Ethiopian type) became extinct. Among reptiles and insects the competition was less severe, or the older forms were too well adapted to local conditions to be expelled; so that it is among these groups alone that we find any considerable number, of what are probably the remains of the ancient fauna of a now submerged southern continent.

III. Himalayan or Indo-Chinese Sub-region.

This, which is probably the richest of all the sub-regions, and perhaps one of the richest of all tracts of equal extent on the face of the globe, is essentially a forest-covered, mountainous country, mostly within the tropics, but on its northern margin extending some degrees beyond it, and rising in a continuous mountain range till it meets and intercalates with the Manchurian sub-division of the Palæarctic region. The peculiar mammalia, birds and insects of this sub-region begin to appear at the very foot of the Himalayas, but Dr. Gunther has shown that many of the reptiles characteristic of the plains of India are found to a height of from 2,000 to 4,000 feet.

In Sikhim, which may be taken as a typical example of the Himalayan portion of the sub-region, it seems to extend to an altitude of little less than 10,000 feet, that being the limit of the characteristic Timaliidæ or babbling thrushes; while the equally characteristic Pycnonotidæ, or bulbuls, and Treronidæ, or thick-billed fruit-pigeons, do not, according to Mr. Blanford, reach quite so high. We may perhaps take 9,000 feet as a good approximation over a large part of the Himalayan range; but it is evidently not possible to define the line with any great precision. Westward, the sub-region extends in diminishing breadth, till it terminates in or near Cashmere, where the fauna of the plains of India almost meets that of the Palæarctic region, at a moderate elevation. Eastward, it reaches into East Thibet and North-west China, where Père David has found a large number of the peculiar types of the Eastern Himalayas. A fauna, in general features identical, extends over Burmah and Siam to South China; mingling with the Palæarctic fauna in the mountains south of the Yang-tse-kiang river, and with that of Indo-Malaya in Tenasserim, and to a lesser extent in Southern Siam and Cochin China.

Zoological Characteristics of the Himalayan Sub-region.— Taking this sub-region as a whole, we find it to be characterised by 3 genera of mammalia (without counting bats), and 44 genera of land-birds, which are altogether peculiar to it; and by 13 genera of mammalia and 36 of birds, which it possesses in common with the Malayan sub-region; and besides these it has almost all the genera before enumerated as "Oriental," and several others of wide range, more especially a number of Palæarctic genera which appear in the higher Himalayas. The names of the more characteristic genera are as follows:—

PECULIAR HIMALO-CHINESE GENERA.

Mammalia.—*Urva, Arctonyx, Ælurus.*

Birds. — *Suya, Horites, Chæmarrhornis, Tarsiger, Oreicola, Acanthoptila, Grammatoptila, Trochalopteron, Actinodura, Sibia, Suthora, Paradoxornis, Chlenasicus, Tesia, Rimator, Ægithaliscus, Cephalopyrus, Liothrix, Siva, Minla, Proparus, Cutia, Yuhina, Ixulus, Myzornis, Erpornis, Hemixus, Chibia, Niltava, Anthipes, Chelidorhynx, Urocissa, Pachyglossa, Heterura, Hæmatospiza, Ampeliceps, Saroglossa, Psarisomus, Serilophus, Vivia, Hyopicus, Gecinulus, Aceros, Ceriornis.*

GENERA COMMON TO THE HIMALO-CHINESE AND MALAYAN SUB-REGIONS.

Mammalia. — *Hylobates, Nycticebus, Viverricula, Prionodon, Arctitis, Paguma, Arctogale, Cuon, Gymnopus, Aonyx, Helictis, Rhinoceros, Nemorhedus, Rhizomys.*

Birds.—*Oreocincla, Notodela, Janthocincla, Timalia, Stachyris, Mixornis, Trichastoma, Enicurus, Pnœpyga, Melanochlora, Allotrius, Microscelis, Iole, Analcipus, Cochoa, Bhringa, Xanthopygia, Hylocharis, Cissa, Temnurus, Crypsirhina, Chalcostetha, Anthreptes, Chalcoparia, Cymbirhynchus, Hydrornis, Sasia, Venilia, Indicator, Carcineutes, Lyncornis, Macropygia, Argusianus Polyplectron, Euplocamus, Phodilus.*

Plate VII. Scene in Nepal, with Characteristic Himalayan Animals.—Our illustration contains figures of two mammals

PLATE VII.

SCENE IN NEPAUL, WITH CHARACTERISTIC ANIMALS.

and two birds, characteristic of the higher woody region of the Himalayas. The lower figure on the left is the *Helictis nepalensis*, confined to the Eastern Himalayas, and belonging to a genus of the weasel family which is exclusively Oriental. It is marked with white on a grey-brown ground. Above it is the remarkable Panda (*Ælurus fulgens*), a beautiful animal with a glossy fur of a reddish colour, darker feet, and a white somewhat cat-like face. It is distantly allied to the bears, and more nearly to the American racoons, yet with sufficient differences to constitute it a distinct family. The large bird on the tree, is the horned Tragopan (*Ceriornis satyra*), one of the fine Himalayan pheasants, magnificently spotted with red and white, and ornamented with fleshy erectile wattles and horns, of vivid blue and red colours. The bird in the foreground is the *Ibidorhynchus struthersii*, a rare and curious wader, allied to the curlews and sandpipers but having the bill and feet red. It frequents the river-beds in the higher Himalayas, but has also been found in Thibet.

Reptiles.—Very few genera of reptiles are peculiar to this sub-region, all the more important ranging into the Malay islands. Of snakes the following are the more characteristic genera :—*Typhline, Cylindrophis, Xenopeltis, Calamaria, Xenelaphis, Hypsirhina, Fordonia*, several small genera of Homalopsidæ (*Herpeton* and *Hipistes* being characteristic of Burmah and Siam) *Psammodynastes, Gonyosoma, Chrysopelea, Tragops, Dipsas, Pareas, Python, Bungarus, Naja, Callophis*, and *Trimeresurus. Naja* reaches 8,000 feet elevation in the Himalayas, *Tropidonotus* 9,000 feet, *Ablabes* 10,000 feet, and *Simotes* 15,000 feet.

Of lizards, *Pseudopus* has one species in the Khasya hills while the other inhabits South-east Europe; and there are two small genera of Agamidæ peculiar to the Himalayas, while *Draco* and *Calotes* have a wide range and *Acanthosaura, Dilophyrus, Physignathus*, and *Liolepis* are found chiefly in the Indo-Chinese peninsula. There are several genera of Scincidæ; and the extensive genus of wall-lizards, *Gecko*, ranges over the whole region.

Of Amphibia, the peculiar forms are not numerous. *Icthyopsis* a genus of Ceciliadæ, is peculiar to the Khasya Hills; *Tylo-tritron* (Salamandridæ) to Yunan in Western China, and perhaps belongs to the Palæarctic region.

Of the tail-less Batrachians, *Glyphoglossus* is found in Pegu; *Xenophys* in the Eastern Himalayas; while *Callula, Ixalus, Rhacophorus, Hylurana, Oxyglossus,* and *Phrynoglossus,* are common to the Himalo-Chinese and Malayan sub-regions.

Of the lizards, *Colotes, Barycephalus,* and *Hinulia,*—and of the Batrachia, *Bufo,*—are found at above 11,000 feet elevation in the Himalayas.

Insects.—So little has been done in working out the insect faunas of the separate sub-regions, that they cannot be treated in detail, and the reader is referred to the chapter on the distribution of insects in the part of this work devoted to Geographical Zoology. A few particulars may, however, be given as to the butterflies, which have been more systematically collected in tropical countries than any other order of insects. The Himalayan butterflies, especially in the eastern portions of the range—in Assam and the Khasya Hills—are remarkably fine and very abundant; yet all the larger groups extend into the Malayan sub-region, many to Ceylon, and a considerable proportion even to Africa and Austro-Malaya. There are a large number of peculiar types, but most of them consist of few or single species. Such are *Neope, Orenoma,* and *Rhaphicera,* genera of Satyridæ; *Enispe* (Morphidæ); *Hestina, Penthema,* and *Abrota* (Nymphalidæ); *Dodona* (Erycinidæ); *Ilerda* (Lycænidæ); *Calinaga, Teinopalpus,* and *Bhutanitis* (Papilionidæ). Its more prominent features are, however, derived from what may be termed Malayan, or even Old World types, such as *Euplœa,* among Danaidæ; *Amathusia, Clerome,* and *Thaumantis,* among Morphidæ; *Euripus, Diadema, Athyma, Limenitis,* and *Adolias,* among Nymphalidæ, *Zemeros* and *Taxila* among Erycinidæ; *Amblypodia, Miletus, Ilerda,* and *Myrina,* among Lycænidæ; *Thyca, Prioneris, Dercas, Iphias,* and *Thestias* among Pieridæ; and Papilios of the "*Amphrisius*," "*Coon*," "*Philoxenus*," "*Protenor*," "*Paris*," and "*Sarpedon*" groups. In the Himalayas there is an unusual abundance of large and gorgeous species of the genus *Papilio,* and of large and showy Nymphalidæ, Morphidæ, and Danaidæ, which render it, in favoured localities, only second to South America for a display of this form of beauty and variety in insect life.

Among the other orders of insects in which the Himalayas are remarkably rich, we may mention large and brilliant Cetoniidæ, chiefly of the genus *Rhomborhima ;* a magnificent Lamellicorn, *Euchirus macleayii*, allied to the gigantic long-armed beetle (*E. longimanus*) of Amboyna; superb moths of the families Agaristidæ and Sesiidæ; elegant and remarkable Fulgoridæ, and strange forms of the gigantic Phasmidæ; most of which appear to be of larger size or of more brilliant colours than their Malayan allies.

Islands of the Indo-Chinese Sub-region.—A few important islands belong to this sub-region, the Andamans, Formosa, and Hainan being the most interesting.

Andamans.—The only mammalia are a few rats and mice, a *Paradoxurus*, and a pig supposed to be a hybrid race,—all of which may have been introduced by man's agency. The birds of the Andaman Islands have been largely collected, no less than 155 species having been obtained; and of these 17, (all land-birds) are peculiar. The genera are all found on the continent, and are mostly characteristic of the Indo-Chinese fauna, to which most of the species belong. Reptiles are also tolerably abundant; about 20 species are known, the majority being found also on the continent, while a few are peculiar. There are also a few Batrachia, and some fresh-water fishes, closely resembling those of Burmah. The absence of such mammalia as monkeys and squirrels, which abound on the mainland, and which are easily carried over straits or narrow seas by floating trees, is sufficient proof that these islands have not recently formed part of the continent. The birds are mostly such as may have reached the islands while in their present geographical position; and the occurrence of reptiles and fresh-water fishes, said to be identical in species with those of Burmah, must be due to the facilities, which some of these animals undoubtedly possess, for passing over a considerable width of sea. We must conclude, therefore, that these islands do not owe their existing fauna to an actual union with the mainland; but it is probable that they may have been formerly more extensive, and have then been less distant from the continent than at the present time.

The Nicobar Islands, usually associated with the Andamans, are less known, but present somewhat similar phenomena. They are, however, more Malayan in their fauna, and seem properly to belong to the Indo-Malay sub-region.

Formosa.—This island has been carefully examined by Mr. Swinhoe, who found 144 species of birds, of which 34 are peculiar. There is one peculiar genus, but the rest are all Indo-Chinese, though some of the species are more allied to Malayan than to Chinese or Himalayan forms. About 30 species of mammalia were found in Formosa, of which 11 are peculiar species, the rest being either Chinese or Himalayan. The peculiar species belong to the genera *Talpa, Helictis, Sciuropterus, Pteromys, Mus, Sus, Cervus,* and *Capricornis.* A few lizards and snakes of continental species have also been found. These facts clearly indicate the former connection of Formosa with China and Malaya, a connection which is rendered the more probable by the shallow sea which still connects all these countries.

Hainan.—The island of Hainan, on the south coast of China, is not so well known in proportion, though Mr. Swinhoe collected 172 species of birds, of which 130 were land-birds. Of these about 20 were peculiar species; the remainder being either Chinese, Himalayan, or Indo-Malayan. Mr. Swinhoe also obtained 24 species of mammalia, all being Chinese, Himalayan, or Indo-Malayan species except a hare, which is peculiar. This assemblage of animals would imply that Hainan, as might be anticipated from its position, has been more recently separated from the continent than the more distant island of Formosa.

IV. Indo-Malaya, or the Malayan Sub-region.

This sub-region, which is almost wholly insular (including only the Malayan peninsula on the continent of Asia), is equal, if not superior, in the variety and beauty of its productions, to that which we have just been considering. Like Indo-China, it is a region of forests, but it is more exclusively tropical ; and it is therefore deficient in many of those curious forms of the temperate zone of the Himalayas, which seem to have been developed from Palæarctic rather than from Oriental types.   Here alone, in the Oriental region, are found the most typical equatorial forms of life—organisms adapted to a climate characterised by uniform but not excessive heat, abundant moisture, and no marked departure from the average meteorological state, throughout the year. These favourable conditions of life only occur in three widely separated districts of the globe—the Malay archipelago, Western Africa, and equatorial South America.   Hence perhaps it is, that the tapir and the trogons of Malacca should so closely resemble those of South America ; and that the great anthropoid apes and crested hornbills of Western Africa, should find their nearest allies in Borneo and Sumatra.

Although the islands which go to form this sub-region are often separated from each other by a considerable expanse of sea, yet their productions in general offer no greater differences than those of portions of the Indo-Chinese sub-region separated by an equal extent of dry land. The explanation is easy, however, when we find that the sea which separates them is a very shallow one, so shallow that an elevation of only 300 feet would unite Sumatra, Java, and Borneo into one great South-eastern prolongation of the Asiatic continent. As we know that our own country has been elevated and depressed to a greater amount than this, at least twice in recent geological times, we can have no difficulty in admitting similar changes of level in the Malay archipelago, where the subterranean forces which bring about such changes are still at work, as manifested by the great chain of active volcanoes in Sumatra and Java.   Proofs of somewhat earlier changes of level are to be seen in the Tertiary coal formations of Borneo, which demonstrate a succession of elevations and subsidences, with as much certainty as if we had historical record of them.

It is not necessary to suppose, nor is it probable, that all these great islands were recently united to the continent, and that their separation took place by one general subsidence of the whole. It is more consonant with what we know of such matters, that the elevations and depressions were partial, varying in their points of action and often recurring; sometimes extending one part of an island, sometimes another; now joining an island to the main.land, now bringing two islands into closer proximity. There is reason to believe that sometimes an intervening island has sunk or receded and allowed others which it before separated to effect a partial union independently of it. If we recognise the probability that such varied and often-renewed changes of level have occurred, we shall be better able to understand how certain anomalies of distribution in these islands may have been brought about. We will now endeavour to sketch the general features of the zoology of this interesting district, and then proceed to discuss some of the relations of the islands to each other.

Mammalia.—We have seen that the Indo-Chinese sub-region possesses 13 species of mammalia in common with the Indo-Malay sub-region, and 4 others peculiar to itself, besides one Ethiopian and several Oriental and Palæarctic forms of wide range. Of this latter class the Malay islands have comparatively few, but they possess no less than 14 peculiar genera, viz· *Simia, Siamanga, Tarsius, Galeopithecus, Hylomys, Ptilocerus, Gymnura, Cynogale, Hemigalea, Arctogale, Barangia, Mydaus, Helarctos,* and *Tapirus.* The islands also possess tigers, deer, wild pigs, wild cattle, elephants, the scaly ant-eater, and most of the usual Oriental genera; so that they are on the whole fully as rich as, if not richer than, any part of Asia; a fact very unusual in island faunas, and very suggestive of their really continental nature.

Plate VIII. Scene in Borneo with Characteristic Malayan Quadrupeds.—The Malayan fauna is so rich and peculiar that we devote two plates to illustrate it. We have here a group of mammalia, such as might be seen together in the vast forests of Borneo. In the foreground we have the beautiful deer-like Chevrotain (*Tragulus javanicus*). These are delicate little

PLATE VIII.

A FOREST IN BORNEO, WITH CHARACTERISTIC MAMMALIA.

animals whose body is not larger than a rabbit's, thence often called "mouse-deer." They were formerly classed with the "musk-deer," owing to their similar tusk-like upper canines; but their anatomy shows them to form quite a distinct family, having more resemblance to the camels. On the branch above is the curious feather-tailed Tree-Shrew (*Ptilocerus lowii*), a small insectivorous animal altogether peculiar to Borneo. Above this is the strange little Tarsier (*Tarsius spectrum*), one of the lemurs confined to the Malay islands, but so distinct from all others as to constitute a separate family. The other small animals are the Flying Lemurs (*Galæopithecus volans*) formerly classed with the lemurs, but now considered to belong to the Insectivora. They have a very large expansion of the skin connecting the fore and hind limbs and tail, and are able to take long flights from one tree to another, and even to rise over obstacles in their course by the elevatory power of the tail-membrane. They feed chiefly on leaves, and have a very soft and beautifully marbled fur.

In the distance is the Malayan tapir (*Tapirus indicus*), a representative of a group of animals now confined to the larger Malay islands and tropical America, but which once ranged over the greater part of temperate Europe.

Birds.—Owing to several of the families consisting of very obscure and closely allied species, which have never been critically examined and compared by a competent ornithologist, the number of birds inhabiting this sub-region is uncertain. From the best available materials there appear to be somewhat less than 650 species of land-birds actually known, or excluding the Philippine Islands somewhat less than 600. The larger part of these are peculiar species, but mostly allied to those of Indo-China; 36 of the genera, as already stated, being common to these two sub-regions. There are, however, no less than 46 genera which are peculiarly or wholly Indo Malayan and, in many cases, have no close affinity with other Oriental groups. These peculiar genera are as follows:—*Timalia, Malacopteron, Macronus, Napothera, Turdinus,* and *Trichixos*—genera

z

of Timaliidæ ; *Eupetes*, a most remarkable form, perhaps allied to *Enicurus*, and *Cinclus; Rhabdornis* (Certhiidæ) found only in the Philippines; *Psaltria*, a diminutive bird of doubtful affinities, provisionally classed among the tits (Paridæ); *Setornis* (Pycnonotidæ); *Lalage* (Campephagidæ) extending eastward to the Pacific Islands ; *Pycnosphrys, Philentoma* (Muscicapidæ) ; *Laniellus*, a beautiful bird doubtfully classed with the shrikes (Laniidæ); *Platylophus* and *Pityriasis*, the latter a most anomalous form— perhaps a distinct family, at present classed with the jays, in Corvidæ; *Prionochilus*, a curious form classed with Dicæidæ; *Erythrura* (Ploceidæ), extending eastwards to the Fiji Islands ; *Gymnops, Calornis*, (Sturnidæ); *Eurylæmus, Corydon*, and *Calyptomena* (Eurylæmidæ) ; *Eucichla*, the longest tailed and most elegantly marked of the Pittidæ ; *Reinwardtipicus* and *Miglyptes* (Picidæ); *Psilopogon* and *Calorhamphus*, (Megalæmidæ); *Rhinococcyx, Dasylophus, Lepidogrammus, Carpococcyx, Zanclostomus, Poliococcyx, Rhinortha*, (Cuculidæ) ; *Berenicornis, Caldo, Cranorhinus, Penelopides, Rhinoplax*, (Bucerotidæ) ; *Psittinus*, (Psittacidæ); *Ptilopus, Phapitreron*, (Columbidæ); *Rollulus*, (Treronidæ); *Muchœrhamphus*, (Falconidæ). Many of these genera are abundant and wide-spread, while some of the most characteristic Himalayan genera, such as *Larvivora, Garrulax, Hypsipetes, Pomatorhinus*, and *Dendrocitta*, are here represented by only a few species.

Among the groups that are characteristic of the Malayan sub-region, the Timaliidæ and Pycnonotidæ stand pre-eminent; the former represented chiefly by the genera *Timalia, Malacopteron, Macronus*, and *Trichastoma*, the latter by *Criniger, Microscelis*, and many forms of *Pycnonotus*. The Muscicapidæ, Dicruridæ, Campephagidæ, Ploceidæ, and Nectariniidæ are also well developed ; as well as the Pittidæ, and the Eurylæmidæ, the limited number of species of the latter being compensated by a tolerable abundance of individuals. Among the Picariæ are many conspicuous groups; as, woodpeckers (Picidæ); barbets (Megalæmidæ); trogons (Trogonidæ); kingfishers (Alcedinidæ); and hornbills (Bucerotidæ) ; five families which are perhaps the most conspicuous in the whole fauna. Lastly come the pigeons

(Columbidæ), and the pheasants (Phasianidæ), which are fairly represented by such fine genera as *Treron, Ptilopus, Euplocamus*, and *Argusianus*. A few forms whose affinities are Australian rather than Oriental, help to give a character to the ornithology, though none of them are numerous. The swallow-shrikes (*Artamus*); the wag-tail fly-catchers (*Rhipidura*); the green fruit-doves (*Ptilopus*); and the mound-makers (*Megapodius*), are the chief of these.

There are a few curious examples of remote geographical alliances that may be noted. First, we have a direct African connection in *Machærhamphus*, a genus of hawks, and *Berenicornis*, a genus of hornbills; the only close allies being, in the former case in South, and in the latter in West Africa. Then we have a curious Neotropical affinity, indicated by *Carpococcyx*, a large Bornean ground-cuckoo, whose nearest ally is the genus *Neomorphus* of South America; and by the lovely green-coloured *Calyptomena* which seems unmistakably allied to the orange-coloured *Rupicola*, or " Cock of the rock," in general structure and in the remarkable form of crest, a resemblance which has been noticed by many writers.

In the preceding enumeration of Malayan genera several are included which extend into the Austro-Malay Islands, our object, at present, being to show the differences and relations of the two chief Oriental sub-regions.

Plate IX. A Malayan Forest with some of its peculiar Birds.— Our second illustration of the Malayan fauna is devoted to its bird-life; and for this purpose we place our scene in the Malay peninsula, where birds are perhaps more abundant and more interesting, than in any other part of the sub-region. Conspicuous in the foreground is the huge Rhinoceros Hornbill (*Buceros rhinoceros*), one of the most characteristic birds of the Malayan forests, the flapping of whose wings, as it violently beats the air to support its heavy body, may be heard a mile off. On the ground behind, is the Argus pheasant (*Argusianus giganteus*) whose beautifully ocellated wings have been the subject of a most interesting description in Mr. Darwin's *Descent of Man*. The wing-feathers are here so enormously

z 2

developed for display (as shown in our figure) that they
become almost, if not quite, useless for their original purpose of
flight; yet the colours are so sober, harmonizing completely
with the surrounding vegetation, and the bird is so wary, that
in the forests where it abounds an old hunter assured me he had
never been able to see a specimen till it was caught in his
snares.   It is interesting to note, that during the display of the
plumage the bird's head is concealed by the wings from a
spectator in front, and, contrary to what usually obtains among
pheasants, the head is entirely unadorned, having neither crest
nor a particle of vivid colour,—a remarkable confirmation of
Mr. Darwin's views, that gayly coloured plumes are developed
in the male bird for the purpose of attractive display in
the breeding season.   The long-tailed bird on the right is
one of the Drongo-shrikes (*Edolius remifer*), whose long bare
tail-feathers, with an oar-like web at the end, and blue-
black glossy plumage, render it a very attractive object as it
flies after its insect prey.   On the left is another singular bird
the great Broad-bill (*Corydon sumatranus*), with dull and sombre
plumage, but with a beak more like that of a boat-bill than of a
fruit-eating passerine bird.   Over all, the white-handed Gibbon
(*Hylobates lar*) swings and gambols among the topmost branches
of the forest.

Reptiles and Amphibia.—These are not sufficiently known to
be of much use for our present purpose.   Most of the genera
belong to the continental parts of the Oriental region, or have a
wide range.   Of snakes *Rhabdosoma*, *Typhlocalamus*, *Tetragono-
soma*, *Acrochordus*, and *Atropos*, are the most peculiar, and there
are several peculiar genera of Homalopsidæ.   Of Oriental genera,
Cylindrophis, *Xenopeltis*, *Calamaria*, *Hypsirhina*, *Psammodynastes*, *Gonyosoma*, *Tragops*, *Dipsas*, *Pareas*, *Python*, *Bungarus*,
Naja, and *Callophis* are abundant; as well as *Simotes*, *Ablabes*,
Tropidonotus, and *Dendrophis*, which are widely distributed.
Among lizards *Hydrosaurus* and *Gecko* are common; there are
many isolated groups of Scincidæ; while *Draco*, *Calotes*, and
many forms of Agamidæ, some of which are peculiar, abound.

Among the Amphibia, toads and frogs of the genera *Micrhyla*,

PLATE IX.

A MALAYAN FOREST, WITH ITS CHARACTERISTIC BIRDS.

Kalophrynus, Ansonia, and *Pseudobufo,* are peculiar : while the Oriental *Megalophrys, Ixalus, Rhacophorus,* and *Hylorana* are abundant and characteristic.

Fishes.—The fresh-water fishes of the Malay archipelago have been so well collected and examined by the Dutch naturalists, that they offer valuable indications of zoo-geographical affinity ; and they particularly well exhibit the sharply defined limits of the region, a large number of Oriental and even Ethiopian genera extending eastward as far as Java and Borneo, but very rarely indeed sending a single species further east, to Celebes or the Moluccas.   Thirteen families of fresh-water fishes are found in the Indo-Malay sub-region.   Of these the Sciœnidæ and Symbranchidæ have mostly a wide range in the tropics.   Ophiocephalidæ are exclusively Oriental, reaching Borneo and the Philippine islands.   The Mastacembelidæ are also Oriental, but one species is found as far as Ceram. Of the Nandidæ, 3 genera range over the whole region.   The Labyrinthici extend from Africa through the Oriental region to Amboyna.   The single species constituting the family Luciocephalidæ is confined to Borneo and the small islands of Biliton and Banca.   Of the extensive family Siluridæ 17 genera are Oriental and Malayan, and 11 are Malayan exclusively ; and not one of these appears to pass beyond the limits of the sub-region.   The Cyprinidæ offer an equally striking example, 23 genera ranging eastward to Java and Borneo and not one beyond ; 14 of these being exclusively Malayan.   It must be remembered that this is not from any want of knowledge of the countries farther east, as extensive collections have also been made in Celebes, the Moluccas, and Timor ; so that the facts of distribution of fresh-water fishes come, most unexpectedly, to fortify that division of the archipelago into two primary regions, which was founded on a consideration of mammalia and birds only.

Insects.—Few countries in the world can present a richer and more varied series of insects than the Indo-Malay islands, and we can only here notice a few of their more striking peculiarities and more salient features.

The butterflies of this sub-region, according to the best estimate that can be formed, amount to about 650 described species, a number that will yet, no doubt, be very considerably increased. The genera which appear to be peculiar to it are *Erites* (Satyridæ) ; *Zeuxidia* (Morphidæ) ; *Amnosia, Xanthotænia,* and *Tanæcia* (Nymphalidæ). The groups which are most characteristic of the region, either from their abundance in individuals or species, or from their size and beauty, are—the rich dark-coloured *Euplœa ;* the large semi-transparent *Hestia ;* the plain-coloured *Mycalesis,* which replace our meadow-brown butterflies (*Hipparchia*) ; the curious *Elymnias,* which often closely resemble Euplœas ; the large and handsome *Thamantis* and *Zeuxidia,* which take the place of the giant Morphos of South America ; the *Cethosia,* of the brightest red, and marked with a curious zigzag pattern ; the velvety and blue-glossed *Terinos ;* the pale and delicately-streaked *Cyrestis ;* the thick-bodied and boldly coloured *Adolias ;* the small wine-coloured *Taxila ;* the fine blue *Amblypodia ;* the beautiful *Thyca,* elegantly marked underneath with red and yellow, which represent our common white butterflies and are almost equally abundant ; the pale blue *Eronia,* and the large red-tipped *Iphias.* The genus *Papilio* is represented by a variety of fine groups ; the large *Ornithoptera,* with satiny yellow under wings ; the superb green-marked "*brookeana ;* " the "*paradoxa*" group, often closely resembling the Euplœas that abound in the same district ; the "*paris*" group richly dusted with golden-green specks ; the "*helenus*" group with wide-spreading black and white wings ; the black and crimson "*polydorus*" group ; the "*memnon*" group, of the largest size and richly-varied colours ; and the "*eurypilus*" group, elegantly banded or spotted with blue or green : all these are so abundant that some of them are met with in every walk, and are a constant delight to the naturalist who has the privilege of observing them in their native haunts.

The Coleoptera are far less prominent and require to be carefully sought after ; but they then well repay the collector. As affording some measure of the productiveness of the tropics in insect life it will not be out of place to give a few notes of the number of species collected by myself in some of the best localities. At Singapore 300 species of Coleoptera were collected in 15 days, and in a month the number had increased to 520; of which 100 were Longicorns and 140 Rhyncophora. At Sarawak in Borneo I obtained 400 species in 15 days, and 600 in a month. In two months this number had increased to about 850, and in three months to 1,000 species. This was the most prolific spot I ever collected in, especially for Longicorns which formed about one-fifth of all the species of beetles. In the Aru Islands in one month, I obtained only 235 species of Coleoptera, and about 600 species of insects of all orders; and this may be taken as a fair average, in localities where no specially favourable conditions existed. On the average 40 to 60 species of Coleoptera would be a good day's collecting; 70 exceptionally good; while the largest number ever obtained in one day was 95, and the majority of these would be very minute insects. It must be remembered, however, that many very common species were passed over, yet had every species met with been collected, not much more than 100 species would ever have been obtained in one day's collecting of four or five hours. These details may afford an interesting standard of comparison for collectors in other parts of the world.

Of Cicindelidæ the most peculiarly Malayan form is *Therates*, found always on leaves in the forests in the same localities as the more widely spread *Collyris*. Five genera of this family are Indo-Malayan.

The Carabidæ, though sufficiently plentiful, are mostly of small size, and not conspicuous in any way. But there is one striking exception in the purely Malayan genus *Mormolyce*, the largest and most remarkable of the whole family. It is nocturnal, resting during the days on the under side of large *boleti* in the virgin forest. *Pericallus* and *Catascopus* are among the few genera which are at all brilliantly coloured.

Buprestidæ are abundant, and very gay; the genus *Belionota* being perhaps one of the most conspicuous and characteristic. The giant *Catoxantha* is, however, the most peculiar, though comparatively scarce. *Chrysochroa* and *Chalcophora* are also abundant and characteristic. Out of the 41 Oriental genera 21 are Malayan, and 10 of these are not found in the other sub-regions.

In Lucanidæ the Malay islands are rich, 14 out of the 16 Oriental genera occurring there, and 3 being peculiar. There are many fine species of *Odontolabris*, which may be considered the characteristic genus of the sub-region.

The Cetoniidæ are well represented by 16 genera and about 120 species. The genera *Mycteristes*, *Phœdimus*, *Plectrone*, *Euremina*, *Rhagopteryx* and *Centrognathus* are peculiar, while *Agestrata*, *Chalcothea*, and *Macronota* are abundant and characteristic.

The Longicorns, as in all continental forest regions near the equator, are very abundant and in endlessly varied forms. No less than 55 genera containing about 200 species are peculiar to this sub-region, the Cerambycidæ being much the most numerous. *Euryarthrum*, *Cœlosterna*, *Agelasta*, and *Astathes* may be considered as most characteristic; but to name the curious and interesting forms would be to give a list of half the genera. For the relations of the Longicorns of the Indo-Malay, and those of the Austro-Malay region, the reader is referred to the chapter on the distribution of insects in the succeeding part of this work.

Terrestrial Mollusca.—The Philippine islands are celebrated as being one of the richest parts of the world for land shells, about 400 species being known. The other islands of the sub-region are far less rich, not more than about 100 species having yet been described from the whole of them. *Helix* and *Bulimus* both abound in species in the Philippines, whereas the latter genus is very scarce in Borneo and Java. Ten genera of Helicidæ inhabit the sub-region; *Pfeifferia* is found in the Philippines and Moluccas, while the large genus *Cochlostyla* is almost peculiar to the Philippines. Of the Operculata there are representatives of 20 genera, of which *Dermatoma* and *Pupinella* are peculiar, while *Registoma* and *Callia* extend to the Australian region. *Cyclophorus*, *Leptopoma*, and *Pupina* are perhaps the most characteristic genera.

The Zoological Relations of the Several Islands of the Indo-Malay Sub-region.

Although we have grouped the Philippine islands with the Indo-Malay sub-region, to which, as we shall see, they undoubtedly belong, yet most of the zoological characteristics we have just sketched out, apply more especially to the other groups of islands and the Malay peninsula. The Philippine islands stand, to Malaya proper, in the same relation that Madagascar does to Africa or the Antilles to South America; that is, they are remarkable for the absence of whole families and genera which everywhere characterise the remainder of the district. They are, in fact, truly insular, while the other islands are really continental in all the essential features of their natural history. Before, therefore, we can conveniently compare the separate islands of Malaya[1] with each other, we must first deal with the Philippine group, showing in what its speciality consists, and why it must be considered apart from the sub-region to which it belongs.

Mammals of the Philippine Islands.—The only mammalia recorded as inhabiting the Philippine Islands are the following :—

QUADRUMANA.	1. Macacus cynomolgus.	
	2. Cynopithecus niger.	Dr. Semper doubts this being a Philippine species.
LEMUROIDEA.	3. Tarsius spectrum.	
INSECTIVORA.	4. Galeopithecus philippinensis.	
	5. Tupaia (species).	On Dr. Semper's authority.
CARNIVORA.	6. Viverra tangalunga.	
	7. Paradoxurus philippensis.	
UNGULATA.	8. Sus (species).	On Dr. Semper's authority.
	9. Cervus mariannus.	
	10. Cervus philippensis.	
	11. Cervus alfredi.	
	12. Bos (species).	Wild cattle ; perhaps introduced.
RODENTIA.	13. Phlæomys cummingii.	
	14. Scuirus philippinensis.	

Also 24 species, belonging to 17 genera, of bats.

---

[1] As so many typical Malay groups are absent only from the Philippines, I have adopted the term "Malaya," to show the distribution of these, using the term " Indo-Malaya " when the range of the group includes the Philippines. This must be remembered when consulting the tables of distribution at the end of this chapter.

The foregoing list, although small, contains an assemblage of species which are wholly Oriental in character, and several of which (*Tarsius, Galeopithecus, Tupaia*) are characteristic and highly peculiar Malayan forms. At the same time these islands are completely separated from the rest of Malaya by the total absence of *Semnopithecus, Hylobates, Felis, Helarctos, Rhinoceros, Manis*, and other groups constantly found in the great Indo-Malay islands and peninsula of Malacca. We find apparently two sets of animals : a more ancient series, represented by the deer, *Galeopithecus*, and squirrel, in which the species are distinct from any others; and a more recent series, represented by *Macacus cynomolgus*, and *Viverra tangalunga*, identical with common Malayan animals. The former indicate the earliest period when these volcanic islands were connected with some part of the Malayan sub-region, and they show that this was not geologically remote, since no peculiar generic types have been preserved or differentiated. The latter may indicate either the termination of the period of union, or merely the effects of introduction by man. The reason why a larger number of mammalian forms were not introduced and established, was probably because the union was effected only with some small islands, and from these communicated to other parts of the archipelago ; or it may well be that later subsidences extinguished some of the forms that had established themselves.

Birds of the Philippine Islands.—These have been carefully investigated by Viscount Walden, in a paper read before the Zoological Society of London in 1873, and we are thus furnished with ample information on the relations of this important portion of the fauna.

The total number of birds known to inhabit the Philippines is 219, of which 106 are peculiar. If, however, following our usual plan, we take only the land-birds, we find the numbers to be 159 species, of which 100 are peculiar ; an unusually large proportion for a group of islands so comparatively near to various parts of the Oriental and Australian regions. The families of birds which are more especially characteristic of the Indo-Malay sub-region are about 28 in number and examples

of all these are found in the Philippines except four, viz., Cinclidæ, Phyllornithidæ, Eurylæmidæ, and Podargidæ. The only
Philippine families which are, otherwise, exclusively Austro-
Malayan are, Cacatuidæ and Megapodiidæ. Yet although the
birds are unmistakably Malayan, as a whole, there are, as in
the mammalia (though in a less degree), marked deficiencies of
most characteristic Malayan forms. Lord Walden gives a list
of no less than 69 genera thus absent; but it will be sufficient
here to mention such wide-spread and specially Indo-Malay
groups as,—*Eurylœmus, Nyctiornis, Arachnothera, Geocichla,
Malacopteron, Timalia, Pomatorhinus, Phyllornis, Iora, Criniger,
Enicurus, Chaptia, Tchitrea, Dendrocitta, Eulabes, Palæornis,
Miglyptes, Tiga,* and *Euplocamus.* These deficiencies plainly
show the isolated character of the Philippine group, and imply
that it has never formed a part of that Indo-Malayan extension
of the continent which almost certainly existed when the peculiar Malayan fauna was developed ; or that, if it has been so
united, it has been subsequently submerged and broken up to
such an extent, as to cause the extinction of many of the absent
types.

It appears from Lord Walden's careful analysis, that 31 of the
Philippine species occur in the Papuan sub-region, and 47 in
Celebes ; 69 occur also in India, and 75 in Java. This last fact
is curious, since Java is the most remote of the Malayan islands,
but it is found to arise almost wholly from the birds of that
island being better known, since only one species, *Xantholœma
rosea,* is confined to the Philippine Islands and Java.

The wading and swimming birds are mostly of wide-spread
forms, only 6 out of the 60 species being peculiar to the Philippine
archipelago. Confining ourselves to the land-birds, and combining several of the minutely subdivided genera of Lord Walden's paper so as to agree with the arrangement adopted in this
work, we find that there are 112 genera of land-birds represented in the islands. Of these, 50 are either cosmopolitan, of
wide range, or common to the Oriental and Australian regions,
and may be put aside as affording few indications of geographical
affinity. Of the remaining 62 no less than 40 are exclusively

or mainly Oriental, and most of them are genera which range widely over the region, only two (*Philentoma* and *Rollulus*) being exclusively Malayan, and two others (*Megalurus* and *Malacocircus*) more especially Indian or continental. Five other genera, though having a wide range, are typically Palæarctic, and have reached the islands through North China. They are, *Monticola*, *Acrocephalus*, *Phylloscopus*, *Calliope*, and *Passer;* the two first having extended their range southward into the Moluccas. The peculiarly Australian genera are only 12, the majority being characteristic Papuan and Moluccan forms; such as—*Campephaga*, *Alcyone*, *Cacatua*, *Tanygnathus*, *Ptilopus*, *Janthœnas*, *Phlogœnas*, and *Megapodius*. One is peculiar to Celebes (*Prioniturus*); one to the Papuan group (*Cyclopsitta*) ; and one is chiefly Australian (*Gerygone*). The beautiful little parroquets forming the genus *Loriculus*, are characteristic of the Philippines, which possess 5 species, a larger number than occurs in any other group of islands, though they range from India to Flores. There remain six peculiar genera—*Rhabdornis*, an isolated form of creepers (Certhiidæ): *Gymnops*, a remarkable bareheaded bird belonging to the starlings (Sturnidæ); *Dasylophus*, and *Lepidogrammus*, remarkable genera of cuckoos (Cuculidæ) ; *Penelopides,* a peculiar hornbill, and *Phapitreron*, a genus of pigeons. Besides these there are four other types (here classed as sub-genera, but considered to be distinct by Lord Walden) which are peculiar to the Philippines. These are *Pseudoptynx*, an owl of the genus *Athene; Pseudolalage*, a sub-genus of *Lalage; Zeocephus*, a sub-genus of *Tchitrea ;* and *Ptilocolpa*, included under *Carpophaga*.

When we look at the position of the Philippine group, connected by the Bashee islands with Formosa, by Palawan and the Sooloo archipelago with Borneo, and by the Tulour and other islets with the Moluccas and Celebes, we have little difficulty in accounting for the peculiarities of its bird fauna. The absence of a large number of Malayan groups would indicate that the actual connection with Borneo, which seems necessary for the introduction of the Malay types of mammalia, was not of long duration ; while the large proportion of wide-spread continental genera of birds would seem to imply that greater facilities had once existed for immigration from Southern China, perhaps by a land connection through Formosa, at which time the ancestors of the peculiar forms of deer entered the country. It may indeed be objected that our knowledge of these islands is far too imperfect to arrive at any satisfactory conclusions as to their former history ; but although many more species no doubt remain to be discovered, experience shows that the broad characters of a fauna are always determined by a series of collections made by different persons, at various localities, and at different times, even when more imperfect than those of the Philippine birds really are. The isolated position, and the volcanic structure of the group, would lead us to expect them to be somewhat less productive than the Moluccas, close to the rich and varied Papuan district,—or than Celebes, with its numerous indications of an extensive area and great antiquity; and taking into account the excessive poverty of its mammalian fauna, which is certain to be pretty well known, I am inclined to believe that no future discoveries will materially alter the character of Philippine ornithology, as determined from the materials already at our command.

Java.—Following the same plan as we have adopted in first discussing the Philippine islands, and separating them from the body of the sub-region on account of special peculiarities, we must next take Java, as possessing marked individuality, and as being to some extent more isolated in its productions than the remaining great islands.

Java is well supplied with indigenous mammalia, possessing as nearly as can be ascertained 55 genera and 90 species. None of these genera are peculiar, and only about 5 of the species,—3 quadrumana, a deer and a wild pig. So far then there is nothing remarkable in its fauna, but on comparing it with that of the other great islands, viz., Borneo and Sumatra, and the Malay peninsula, we find an unmistakable deficiency of characteristic forms, the same in kind as that we have just commented on in the case of the Philippines, though much less in degree. First, taking genera which are found in all three of the above-named localities and which must therefore be held to be typical Malayan groups, the following are absent from Java: *Viverra, Gymnopus, Lutra, Helarctos, Tapirus, Elephas,* and *Gymnura;* while of those *known* to occur in two, and which, owing to our imperfect knowledge, may very probably one day be discovered in the third, the following are equally wanting: *Simia, Siamanga, Hemigalea, Paguma, Rhinosciurus,* and *Rhizomys.* It may be said this is only negative evidence, but in the case of Java it is much more, because this island is not only the best known of any in the archipelago, but there is perhaps no portion of British India of equal extent so well known. It is one of the oldest of the Dutch possessions and the seat of their colonial government; good roads traverse it in every direction, and experienced naturalists have been resident in various parts of it for years together, and have visited every mountain and every forest, aided by bands of diligent native collectors. We should be almost as likely to find new species of mammalia in Central Europe as in Java; and therefore the absence of such animals as the Malay bear, the elephant, tapir, gymnura, and even less conspicuous forms, must be accepted as a positive fact.

In the other islands there are still vast tracts of forest in the hands of natives and utterly unexplored, and any similar absence in their case will prove little; yet on making the same comparison in the case of Borneo, the most peculiar and the least known of the other portions of the sub-region, we find only 2 genera absent which are found in the three other divisions, and only 3 which are found in two others. A fact to be noted also is, that the only genus found in Java but not in other parts of the sub-region (*Helictis*) occurs again in North India; and that some Javan *species,* as *Rhinoceros javanicus,* and *Lepus kurgosa* occur again in the Indo-Chinese sub-region, but not in the Malayan.

Among the birds we meet with facts of a similar import; and though the absence of certain types from Java is not quite so certain as among the mammalia, this is more than balanced by the increased number of such deficiencies, so that if a few should be proved to be erroneous, the main result will remain unaltered.

Java possesses about 270 species of land birds, of which about 40 are peculiar to it. There are, however, very few peculiar genera, *Laniellus*, a beautiful spotted shrike, being the most distinct, while *Cochoa* and *Psaltria* are perhaps not different from their Indian allies. The island has however a marked individuality in two ways—in the absence of characteristic Malayan types, and in the presence of a number of forms not yet found in any of the other Malay islands, but having their nearest allies in various parts of the Indo-Chinese sub-region. The following 16 genera are all found in Malacca, Sumatra, and Borneo, but are absent from Java: *Setornis, Temnurus, Dendrocitta, Corydon, Calyptomena, Venilia, Reinwardtipicus, Caloramphus, Rhinortha, Nyctiornis, Cranorhinus, Psittinus, Polyplectron, Argusianus, Euplocamus,* and *Rollulus.* The following 9 are known from *two* of the above localities, and will very probably be found in the third, but are absent from, and not likely to occur in, Java: *Trichixos, Eupetes, Melanochlora, Chaptia, Pityriasis, Lyncornis, Carpococcyx, Poliococcyx,* and *Rhinoplax.* We have thus 25 typically Malayan genera which are not known to occur in Java.

The following genera, on the other hand, do not occur in any of the Malayan sub-divisions except Java, and they all occur again, or under closely allied forms, in the Indo-Chinese sub-region: *Brachypteryx* (allied species in Himalayas); *Zoothera* allied species in Aracan); *Notodela* (allied species in Pegu); *Pnoëpyga* (allied species in Himalayas); *Allotrius* (allied species in the Himalayas); *Cochoa* (allied species in the Himalayas); *Crypsirhina* (allied species in Burmah); *Estrilda* (allied species in India); *Psaltria* (allied genus—*Ægithaliscus*—in Himalayas); *Pavo muticus* and *Harpactes oreskios* (same species in Siam and Burmah); *Cecropis striolata* (same species in Java and Formosa, and allied species in India).

Here we have 12 instances of very remarkable distribution, and considering that there are nearly as many birds known from Sumatra and Borneo as from Java, and considerably more from the Malay peninsula, it is not likely that many of these well marked forms will be discovered in these countries.

There are also a considerable number of species of birds common to Malacca, Sumatra, and Borneo, but represented in Java by distinct though closely allied species. Such are,—

Venilia malaccensis	(represented in Java by)	*V. miniata.*
Drymocataphus nigrocapitatus	„	„	*D. capistratus.*
Malacopteron coronatum	„	„	*M. rufifrons.*
Irena cyanea	„	„	*I. turcosa.*
Ploceus baya	„	„	*P. hypoxantha.*
Loriculus galgulus	„	„	*L. pusillus.*
Ptilopus jambu	„	„	*P. porphyreus.*

Now if we look at our map of the region, and consider the position of Java with regard to Borneo, Sumatra, and the Indo-Chinese peninsula, the facts just pointed out appear most anomalous and perplexing. First, we have Java and Sumatra forming one continuous line of volcanoes, separated by a very narrow strait, and with all the appearance of having formed one continuous land; yet their productions differ considerably, and those of Sumatra show the closest resemblance to those of Borneo, an island ten times further off than Java and differing widely in the absence of volcanoes or any continuous range of lofty mountains. Then again, not only does Java differ from these two, but it agrees with a country beyond them both—a country from which they seem to have a much better chance to have been supplied by immigration than Java has, and to have (almost necessarily) participated, even more largely, in the benefits of any means of transmission capable of reaching the latter island. Yet more; whatever changes have occurred to bring about the anomalous state of things that exists must have been, zoologically and geologically, recent; for the strange cross-affinities between Java and the Indo-Chinese continent (in which Sumatra and Borneo have not participated), as well as that between Malacca, Sumatra, and Borneo (in which Java has not participated) are exhibited, in many cases by community of *species*, in others by the presence of very closely allied forms of the same *genera*, of mammalia and birds. Now we know that these higher animals become replaced by allied species much more rapidly than the mollusca; and it is also pretty certain that the modification by which this replacement is effected takes place more rapidly when the two sets of individuals are isolated from each other, and especially when they are restricted to islands, where they are necessarily subject to distinct and pretty constant conditions, both physical and organic. It becomes therefore almost a certainty, that Siam and Java on the one hand, and Sumatra, Borneo, and Malacca on the other must have been brought into some close connexion, not earlier than the newer Pliocene period; but while the one set of countries were having their meeting, the other must have been by some means got out of the way. Before attempting to indicate the mode by which this might have been effected in accordance with what we know of the physical geography, geology, and vegetation of the several islands, it will be as well to complete our sketch of their zoological relations to each other, so as ascertain with some precision, what are the facts of distribution which we have to explain.

Malacca, Sumatra, and Borneo.—After having set apart the Philippine Islands and Java, we have remaining two great islands and a peninsula, which, though separated by considerable arms of the sea, possess a fauna of wonderful uniformity having all the typical Malayan features in their full development. Their unity is indeed so complete, that we can find hardly any groups of sufficient importance by which to differentiate them from each other; and we feel no confidence that future discoveries may not take away what speciality they possess. One after another, species or genera once peculiar to Borneo or Sumatra have been found elsewhere; and this has gone to such an extent in birds, that hardly a peculiar genus and very few peculiar species are left in either island. Borneo however is undoubtedly the most peculiar. It possesses three genera of Mammalia not found elsewhere: *Cynogale,* a curious carnivore allied to the otters; with *Dendrogale* and *Ptilocerus,* small insectivora allied to *Tupaia.* It has *Simia,* the Orang-

A A

utan, and *Paguma*, one of the Viverridæ, in common with
Sumatra ; as well as *Rhinosciurus*, a peculiar form of squirrel, and
Hemigalea, one of the Viverridæ, in common with Malacca.
Sumatra has only one genus not found in any other Malayan
district—*Nemorhedus*, a form of antelope which occurs again
in North India. It also has *Siamanga* in common with Malacca,
Mydaus with Java, and *Rhizomys* with India. The Malay Peninsula seems to have no peculiar forms of Mammalia, though
it is rich in all the characteristic Malay types.

The bats of the various islands have been very unequally
collected, 36 species being recorded from Java, 23 from Sumatra,
but only 16 each from Borneo and Malacca. Leaving these out
of consideration, and taking into account the terrestrial mammals only, we find that Java is the poorest in species, while
Borneo, Sumatra, and Malacca are tolerably equal; the numbers
being 55, 62, 66, and 65 respectively. Of these we find that
the species confined to each island or district are (in the same
order) 6, 16, 5, and 6. It thus appears that Borneo is, in its
mammalia, the most isolated and peculiar ; next comes Sumatra,
and then Malacca and Java, as shown by the following table.

	Peculiar Genera.	Peculiar Species.
Borneo ...	4	16
Sumatra	1	5
Malacca	0	6
Java ...	0	6

This result differs from that which we have arrived at by the
more detailed consideration of the fauna of Java; and it serves
to show that the estimate of a country by the number of its
peculiar genera and species alone, may not always represent its
true zoological importance or its most marked features. Java,
as we have seen, is differentiated from the other three districts
by the absence of numerous types common to them all, and by
its independent continental relations. Borneo is also well distinguished by its peculiar genera and specific types, yet it is at
the same time more closely related to Sumatra and Malacca
than is Java. The two islands have evidently had a very
different history, which a detailed knowledge of their geology

would alone enable us to trace. Should we ever arrive at a fair knowledge of the physical changes that have resulted in the present condition, we shall almost certainly find that many of the differences and anomalies of their existing fauna and flora will be accounted for.

In Birds we hardly find anything to differentiate Borneo and Sumatra in any clear manner. *Pityriasis* and *Carpococcyx*, once thought peculiar to the former, are now found also in the latter; and we have not a single genus left to characterize Borneo except *Schwaneria* a peculiar fly-catcher, and *Indicator*, an African and Indian group not known to occur elsewhere in the Malay sub-region. Sumatra as yet alone possesses *Psilopogon*, a remarkable form of barbet, but we may well expect that it will be soon found in the interior of Borneo or Malacca; it also has *Berenicornis*, an African form of hornbill. The Malay Peninsula appears to have no genus peculiar to it, but it possesses some Chinese and Indian forms which do not pass into the islands. As to the species, our knowledge of them is at present very imperfect. The Malay Peninsula is perhaps the best known, but it is probable that both Sumatra and Borneo are quite as rich in species. With the exception of the genera noted above, and two or three others as yet found in two islands only, the three districts we are now considering may be said to have an almost identical bird-fauna, consisting largely of the same species and almost wholly of these together with closely allied species of the same genera. There are no well-marked groups which especially characterise one of these islands rather than the other, so that even the amount of speciality which Borneo undoubtedly exhibits as regards mammalia, is only faintly shown by its birds. The Pittidæ may perhaps be named as the most characteristic Bornean group, that island possessing six species, three of which are peculiar to it and are among the most beautiful birds of an unusually beautiful family. Yet Sumatra possesses two peculiar, and hardly less remarkable species.

In other classes of vertebrates, in insects, and in land-shells, our knowledge is far too imperfect to allow of our making any useful comparison between the faunas.

Banca.—We must, however note the fact of peculiar species occurring in Banca, a small island close to Sumatra, and thus offering another problem in distribution. A squirrel (*Sciurus bangkanus*) is allied to three species found in Malacca, Sumatra, and Borneo respectively, but quite as distinct from them all as they are from each other.   More curious are the two species of *Pitta* peculiar to Banca ; one, *Pitta megarhynchus*, is allied to the *P. brachyurus*, which inhabits the whole sub-region and extends to Siam and China, but differs from it in its very large bill and differently coloured head ; the other, *P. bangkanus*, is allied to *P. cucullatus*, which extends from Nepal to Malacca, and to *P. sordidus*, which inhabits both Borneo and Sumatra as well as the Philippines.

We have here, on a small scale, a somewhat similar problem to that of Java, and as this is comparatively easy of solution we will consider it first.   Although, on the map, Banca is so very close to Sumatra, the observer on the spot at once sees that the proximity has been recently brought about.   The whole southeast coast of Sumatra is a great alluvial plain, hardly yet raised above the sea level, and half flooded in the wet season.   It is plainly a recent formation, caused by the washing down into a shallow sea of the *débris* from the grand range of volcanic mountains 150 miles distant.   Banca, on the other hand is, though low, a rugged and hilly island, formed almost wholly of ancient rocks of apparently volcanic origin, and closely resembling parts of the Malay Peninsula and the intervening chain of small islands.   There is every appearance that Banca once formed the extremity of the Peninsula, at which time it would probably have been separated from Sumatra by 50 or 100 miles of sea.   Its productions should, therefore, most resemble those of Singapore and Malacca, and the few peculiar species it possesses will be due to their isolation in a small tract of country, surrounded by a limited number of animal and vegetable forms, and subject to the influence of a peculiar soil and climate.   The parent species existing in such large tracts as Borneo or Sumatra, subjected to more varied conditions of soil, climate, vegetation, food, and enemies, would preserve, almost or quite unchanged, the characteristics which had been developed under nearly identical conditions when the great island formed part of the continent. Geology teaches us that similar changes in the forms of the higher vertebrates have taken place during the Post-Tertiary epoch ; and there are other reasons for believing that, under such conditions of isolation as in Banca, the change may have required but a very moderate period, even reckoned in years. We will now return to the more difficult problem presented by the peculiar continental relations of Java, as already detailed.

Probable Recent Geographical Changes in the Indo-Malay Islands.—Although Borneo is by far the largest of the Indo-Malay islands, yet its physical conformation is such that, were a depression to occur of one or two thousand feet, it would be reduced to a smaller continuous area than either Sumatra or Java. Except in its northern portion it possesses no lofty mountains, while alluvial valleys of great extent penetrate far into its interior. A very moderate depression, of perhaps 500 feet, would convert it into an island shaped something like Cele-bes ; and its mountains are of so small an average elevation, and consist so much of isolated hills and detached ranges, that a depression of 2,000 feet would almost certainly break it up into a group of small islands, with a somewhat larger one to the north. Sumatra (and to a less extent Java) consists of an almost continuous range of lofty mountains, connected by plateaus from 3,000 to 4,000 feet high ; so that although a depression of 2,000 feet would greatly diminish their size, it would probably leave the former a single island, while the latter would be separated into two principal islands of still considerable extent. The en-ormous amount of volcanic action in these two islands, and the great number of conical mountains which must have been slowly raised, chiefly by ejected matter, to the height of 10,000 and 12,000 feet, and whose shape indicates that they have been for-med above water, renders it almost certain that for long periods they have not undergone submersion to any considerable extent. In Borneo, however, we have no such evidences. No volcano, active or extinct, is known in its entire area; while extensive beds of coal of tertiary age, in every part of it, prove that it has been subject to repeated submersions, at no distant date geologically. An indication, if not a proof, of still more recent submersion is to be found in the great alluvial valleys which on the south and south-west extend fully 200 miles inland, while they are to a less degree a characteristic feature all round the island. These swampy plains have been formed by the combined action of rivers and tides; and they point clearly to an immediately preceding state of things, when that which is even now barely raised above the ocean, was more or less sunk below it.

These various indications enable us to claim, as an admissible and even probable supposition, that at some epoch during the Pliocene period of geology, Borneo, as we now know it, did not exist; but was represented by a mountainous island at its present northern extremity, with perhaps a few smaller islets to the south. We thus have a clear opening from Java to the Siamese Peninsula; and as the whole of that sea is less than 100 fathoms deep, there is no difficulty in supposing an elevation of land connecting the two together, quite independent of Borneo on the one hand and Sumatra on the other. This union did not probably last long; but it was sufficient to allow of the introduction into Java of the *Rhinoceros javanicus*, and that group of Indo-Chinese and Himalayan species of mammalia and birds which it alone possesses. When this ridge had disappeared by subsidence, the next elevation occurred a little more to the east, and produced the union of many islets which, aided by subaerial denudation, formed the present island of Borneo. It is probable that this elevation was sufficiently extensive to unite Borneo for a time with the Malay Peninsula and Sumatra, thus helping to produce that close resemblance of genera and even of species, which these countries exhibit, and obliterating much of their former speciality, of which, however, we have still some traces in the long-nosed monkey and *Ptilocerus* of Borneo, and the considerable number of genera both of mammalia and birds confined to two only out of the three divisions of typical Malaya. The subsidence which again divided these countries by arms of the sea rather wider than at present, might have left Banca isolated, as already referred to, with its proportion of the common fauna to be, in a few instances subsequently modified.

Thus we are enabled to understand how the special relations of the *species* of these islands to each other may have been brought about. To account for their more deep-seated and general zoological features, we must go farther back.

Probable Origin of the Malayan Fauna.—The typical Malayan fauna is essentially an equatorial one, and must have been elaborated in an extensive equatorial area. This ancient land almost certainly extended northward over the shallow sea as far as the island of Palawan, the Paracels shoals and even Hainan. To the east, it may at one time have included the Philippines and Celebes, but not the Moluccas. To the south it was limited by the deep sea beyond Java. It included all Sumatra and the Nicobar islands, and there is every reason to believe that it stretched out also to the west so as to include the central peak of Ceylon, the Maldive isles, and the Cocos islands west of Sumatra. We should then have an area as extensive as South America to 15° south latitude, and well calculated to develop that luxuriant fauna and flora which has since spread to the Himalayas. The submergence of the western half of this area (leaving only a fragment in Ceylon) would greatly diminish the number of animals and perhaps extinguish some peculiar types; but the remaining portion would still form a compact and extensive district, twice as large as the peninsula of India, over the whole of which a uniform Malayan fauna would prevail. The first important change would be the separation of Celebes ; and this was probably effected by a great subsidence, forming the deep strait that now divides that island from Borneo. During the process Celebes itself was no doubt greatly submerged, leaving only a few islands in which were preserved that remnant of the ancient Malayan fauna that now constitutes one of its most striking and anomalous features. The Philippine area would next be separated, and perhaps be almost wholly submerged ; or broken up into many small volcanic islets in which a limited number of Malayan types alone survived. Such a condition of things will account for the very small variety of mammalia compared with the tolerably numerous genera of birds, that now characterise its fauna; while both here and in Celebes we find some of the old Malayan types preserved, which, in the extended area of the Sunda Isles have been replaced by more dominant forms.

The next important change would be the separation of Java; and here also no doubt a considerable submergence occurred, rendering the island an unsuitable habitation for the various Malay types whose absence forms one of its conspicuous features. It has since remained permanently separated from the other islands, and has no doubt developed some peculiar species, while it may have preserved some ancient forms which in the larger area have become changed. From the fact that a number of its species are confined either to the western or the eastern half of the island, it is probable that it long continued as two islands, which have become united at a comparatively recent period. It has also been subjected to the immigration of Indo-Chinese forms, as already referred to in the earlier part of this sketch.

We have thus shown how the main zoological features of the several sub-divisions of the Malayan sub-region may be accounted for, by means of a series of suppositions as to past changes which, though for the most part purely hypothetical, are always in accordance with what we know both of the physical geography and the zoology of the districts in question and those which surround them. It may also be remarked, that we know, with a degree of certainty which may be called absolute, that alternate elevation and subsidence is the normal state of things all over the globe; that it was the rule in the earliest geological epochs, and that it has continued down to the historical era. We know too, that the *amount* of elevation and subsidence that can be proved to have occurred again and again in the same area, is often much greater than is required for the changes here speculated on,—while the *time* required for such changes is certainly less than that necessitated by the changes of specific and generic forms which have coincided with, and been to a large extent dependent on them. We have, therefore, true causes at work, and our only suppositions have been as to how those causes could have brought about the results which we see; and however complex and unlikely some of the supposed changes may seem to the reader, the geologist who has made a study of such changes, as recorded in the crust of the earth, will not only admit them to be probable, but will be inclined to believe that they have really been far more complex and more unexpected than any supposition we can make about them.

There is one other external relation of the Malayan fauna about which it may be necessary to say a few words. I have supposed the greatest westward extension of the Malayan area to be indicated by the Maldive islands, but some naturalists would extend it to include Madagascar in order to account for the range of the Lemuridæ. Such an extension would, however, render it difficult to explain the very small amount of correspondence with a pervading diversity, between the Malayan and Malagasy faunas. It seems more reasonable to suppose an approximation of the two areas, without actual union having ever occurred. This approximation would have allowed the interchange of certain genera of birds, which are common to the Oriental Region and the Mascarene islands, but it would have been too recent to account for the diffusion of the lemurs which belong to distinct genera and even distinct families. This probably dates back to a much earlier period, when the lemurine type had a wide range over the northern hemisphere. Subjected to the competition of higher forms, these imperfectly developed groups have mostly died out, except a few isolated examples, chiefly found in islands, and a few groups in Africa.

In our discussion of the origin of the Ethiopian fauna, we have supposed that a close connection once existed between Madagascar and Ceylon. This was during a very early tertiary epoch; and if, long after it had ceased and the fauna of Ceylon and South India had assumed somewhat more of their present character, we suppose the approximation or union of Ceylon and Malaya to have taken place, we shall perhaps be able to account for most of the special affinities they present, with the least amount of simultaneous elevation of the ocean bed; which it must always be remembered, requires a corresponding depression elsewhere to balance it.

Concluding Remarks on the Oriental Region.—We have already so fully discussed the internal and external relations of the several sub-regions, that little more need be said. The rich and varied fauna which inhabited Europe at the dawn of the tertiary period,—as shown by the abundant remains of mammalia wherever suitable deposits of Eocene age have been discovered,— proves, that an extensive Palæarctic continent then existed; and the character of the flora and fauna of the Eocene deposits is so completely tropical, that we may be sure there was then no barrier of climate between it and the Oriental region. At that early period the northern plains of Asia were probably under water, while the great Thibetan plateau and the Himalayan range, had not risen to more than a moderate height, and would have supported a luxuriant sub-tropical flora and fauna. The Upper Miocene deposits of northern and central India, and Burmah, agree in their mammalian remains with those of central and southern Europe, while closely allied forms of elephant, hyæna, tapir, rhinoceros, and *Chalicotherium* have occurred in North China; leading us to conclude that one great fauna then extended over much of the Oriental and Palæarctic regions. Perim island at the mouth of the Red Sea, where similar remains are found, probably shows the southern boundary of this part of the old Palæarctic region in the Miocene period. Towards the equator there would, of course, be some peculiar groups; but we can hardly doubt, that, in that wonderful time when even the lands that stretched out furthest towards the pole, supported a luxuriant forest vegetation, substantially one fauna ranged over the whole of the great eastern continent of the northern hemisphere. During the Pliocene period, however, a progressive change went on which resulted in the complete differentiation of the Oriental and Palæarctic faunas. The causes of this change were of two kinds. There was a great geographical and physical revolution effected by the elevation of the Himalayas and the Thibetan plateau, and, probably at the same time, the northward extension of the great Siberian plains. This alone would produce an enormous change of climate in all the extra-tropical part of Asia, and inevitably lead to a segregation of the old fauna into tropical and temperate, and a modification of the latter so as to enable it to support a climate far more severe than it had previously known. But it is almost certain that, concurrently with this, there was a change going on of a cosmical nature, leading to an alteration of the climate of the northern hemisphere from equable to extreme, and culminating in that period of excessive cold which drove the last remnants of the old sub-tropical fauna beyond the limits of the Palæarctic region. From that time, the Oriental and the Ethiopian regions alone contained the descendants of many of the most remarkable types which had previously flourished over all Europe and Asia; but the early history of these two regions, and the peculiar equatorial types developed in each, sufficiently separate them, as we have already shown. The Malayan sub-region is that in which characteristic Oriental types are now best developed, and where the fundamental contrast of the Oriental, as compared with the Ethiopian and Palæarctic regions, is most distinctly visible.

TABLES OF DISTRIBUTION.

In constructing these tables, showing the distribution of various classes of animals in the Oriental region, the following sources of information have been chiefly relied on, in addition to the general treatises, monographs, catalogues, &c., used for the compilation of the Fourth Part of this work.

Mammalia.—Jerdon's Indian Mammalia; Kelaart's Fauna of Ceylon; Horsfield and Moore's Catalogue of the East India Museum; Swinhoe's Catalogue of Chinese Mammalia; S. Müller's Zoology of the Indian Archipelago; Dr. J. E. Gray's list of Mammalia of the Malay Archipelago (Voyage of Samarang); and papers by Anderson, Blyth, Cantor, Gray, Peters, Swinhoe, &c.

Birds.—Jerdon's Birds of India; Horsfield and Moore's Catalogue; Holdsworth's list of Ceylon Birds; Schlegel's Catalogue of the Leyden Museum; Swinhoe on the Birds of China, Formosa, and Hainan; Salvadori on the Birds of Borneo; Lord Walden on the Birds of the Philippine Islands; and papers by Blyth, Blanford, Elwes, Elliot, Stoliczka, Sclater, Sharpe, Swinhoe, Verreaux, and Lord Walden.

Reptiles.—Günther's Reptiles of British India; papers by same author, and by Dr. Stoliczka.

TABLE I.

FAMILIES OF ANIMALS INHABITING THE ORIENTAL REGION.

EXPLANATION.

Names in *italics* show families peculiar to the region.
Numbers correspond with those in Part IV.
Names enclosed thus (......) barely enter the region, and are not considered really to belong to it.

Order and Family.	Sub-regions.				Range beyond the Region.
	Hindo-stan.	Ceylon.	Indo-China.	Indo-Malaya.	
MAMMALIA.					
PRIMATES.					
1. Simiidæ			—	...	W. Africa
2. Semnopithecidæ	—	—	—	—	Tropical Africa
3. Cynopithecidæ	—	—	—	—	All Africa, S. Palæarctic
6. Lemuridæ ...		—	—	—	Ethiopian
7. *Tarsiidæ...* ...				—	Celebes
CHIROPTERA.					
9. Pteropidæ ...	—	—	—	—	Ethiopian, Australian
11. Rhinolophidæ	—	—	—	—	The Eastern Hemisphere
12. Vespertilionidæ	—	—	—	--	Cosmopolite
13. Noctilionidæ ...	—	—	—	—	Tropical regions
INSECTIVORA.					
14. *Galeopithecidæ*				—	
16. *Tupaiidæ* ...		—	—	--	
17. Erinaceidæ ...	—	—		—	Palæarctic, S. Africa
21. Talpidæ			—		Palæarctic, Nearctic
22. Soricidæ	—	—	—	—	Palæarctic, Ethiopian, N. America
CARNIVORA.					
23. Felidæ	—	—	—	—	All regions but Australian
25. Viverridæ ...	--	—	—	—	Ethiopian, S. Palæarctic
27. Hyænidæ ..	—				Ethiopian, S. Palæarctic
28. Canidæ	—	—	—	—	All regions but Australian [?]
29. Mustelidæ ...	—	—	—	—	All regions but Australian
31. Æluridæ			—		Palæarctic
32. Ursidæ	—	—	—	—	Palæarctic, Nearctic, Chili
CETACEA.					Oceanic
SIRENIA.					
42. Manatidæ ...	—	—	—	--	Ethiopian, N. Pacific
UNGULATA.					
3. (Equidæ)... ...	—				Palæarctic, Ethiopian

Order and Family.	Sub-regions.				Range beyond the Region.
	Hindo-stan.	Ceylon.	Indo-China.	Indo-Malayn.	
44. Tapiridæ... ..				—	Neotropical
45. Rinocerotidæ ...			—	—	Ethiopian
47. Suidæ	—	—	—	—	Palæarctic, Ethiopian, Neotropical
49. Tragulidæ ...	—	—	—	—	W. Africa
50. Cervidæ	—	—	—	—	All regions but Ethiopian and Australian
52. Bovidæ	—	—	—	—	All regions but Australian and Neotropical
53. Elephantidæ ...	—	—	—	—	Ethiopian
RODENTIA.					
55. Muridæ	—	—	—	—	Cosmopolite, excl. Oceania
56. Spalacidæ ...			—	—	Palæarctic, Ethiopian
61. Sciuridæ... ...	—	—	—	—	All regions but Australian
67. Hystricidæ ...	—	—	—	—	S. Palæarctic, Ethiopian
70. Leporidæ ...	—	—	—	—	All regions but Australian
EDENTATA.					
72. Manididæ ...	—	—	—	—	Ethiopian
BIRDS.					
PASSERES.					
1. Turdidæ	—	—	—	—	Almost Cosmopolite
2. Sylviidæ	—	—	—	—	Almost Cosmopolite
3. Timaliidæ ...	—	—	—	—	Ethiopian, Australian
4. Panuridæ ..				—	Palæarctic
5. Cinclidæ... ...	—	—	—	—	Not Ethiopian or Australian
6. Troglodytidæ...			—	—	American and Palæarctic
8. Certhiidæ ...	—		—	—	Palæarctic, Nearctic, Australian
9. Sittidæ ...	—	—	—	—	Palæarctic, Nearctic, Australian, Madagascar
10. Paridæ	—	—	—	—	The Eastern Hemisphere and North America
11. Liotrichidæ ...			—	—	
12. Phyllornithidæ	—	—	—	—	
13. Pycnonotidæ ...	—	—	—	—	Ethiopian, Moluccas
14. Oriolidæ... ...	—	—	—	—	The Eastern Hemisphere
15. Campephagidæ	—	—	—	—	Ethiopian, Australian
16. Dicruridæ ...	—	—	—	—	Ethiopian, Australian
17. Muscicapidæ ...	—	—	—	—	The Eastern Hemisphere
18. Pachycephalidæ			—	—	Australian
19. Laniidæ	—	—	—	—	The Eastern Hemisphere and North America
20. Corvidæ	—	—	—	—	Cosmopolite
23. Nectariniidæ ...	—	—	—	—	Ethiopian, Australian
24. Dicæidæ... ...	—	—	—	—	Ethiopian, Australian
30. Hirundinidæ ...	—	—	—	—	Cosmopolite
33. Fringillidæ ...	—	—	—	—	All regions but Australian
34. Ploceidæ ...	—	—	—	—	Ethiopian, Australian
35. Sturnidæ ...	—	—	—	—	The Eastern Hemisphere
36. Artamidæ ...	—	—	—	—	Australian
37. Alaudidæ ...	—	—	—	—	All regions but Neotropical
38. Motacillidæ ...	—	—	—	—	Cosmopolite
43. Eurylæmidæ ...			—	—	
47. Pittidæ	—	—	—	—	Ethiopian, Australian

Order and Family.	Sub-regions.				Range beyond the Region.
	Hindostan.	Ceylon.	Indo-China.	Indo-Malaya.	
PICARIÆ.					
51. Picidæ	--	...	—	—	All regions but Australian
52. Yungidæ ...	—				Palæarctic
53. Indicatoridæ			—	—	Ethiopian
54. Megalæmidæ	--	—	—	—	Ethiopian, Neotropical
58. Cuculidæ ...	—	—	—	—	Cosmopolite
62 Coraciidæ ...	--	—	—	--	Ethiopian, Australian
63. Meropidæ ...	—	—	...	...	Ethiopian, Australian
66. Trogonidæ ...	—	—	—	—	Neotropical, Ethiopian
67. Alcedinidæ ...	--	—	—	—	Cosmopolite
68. Bucerotidæ ...	—	—	—	—	Ethiopian, Austro-Malayan
69. Upupidæ ...	—	-	—		Ethiopian, S. Palæarctic
71. Podargidæ ...	—	—	—	—	Australian
73. Caprimulgidæ	--	—	—	—	Cosmopolite
74. Cypselidæ ...	—	—	—	—	Cosmopolite
PSITTACI.					
76. (Cacatuidæ) ...			—		Australian
78. Palæornithidæ	—	...	—	—	Ethiopian, Austro-Malayan
COLUMBÆ.					
84. Columbidæ ...	—	—	—	—	Cosmopolite
GALLINÆ.					
86. Pteroclidæ ...	—				Ethiopian, Palæarctic
87. Tetraonidæ ...	—	—	—	—	Eastern Hemisphere and North America
88. Phasianidæ	—	—	—	—	Ethiopian, Palæarctic, North America
89. Turnicidæ ...	—	—	—	—	Ethiopian, Australian, S. Palæarctic
90. Megapodiidæ				—	Australian
ACCIPITRES.					
94. Vulturidæ ..	—	—	—		All regions but Australian
96. Falconidæ ...	—	—	--	—	Cosmopolite
97. Pandionidæ ...	—	—	--	—	Cosmopolite
98. Strigidæ	—	—	—	—	Cosmopolite
GRALLÆ.					
99. Rallidæ ...	—	—	—	—	Cosmopolite
100. Scolopacidæ...	—	—	—	—	Cosmopolite
103. Parridæ... ...	—	—	—	—	Tropical regions
104. Glareolidæ ...	—	—	—	—	Eastern Hemisphere
105. Charadriidæ...	—	—	—	—	Cosmopolite
106. Otididæ... ...	—	—	—		Eastern Hemisphere
107. Gruidæ	—	—	—		All regions but Neotropical
113. Ardeidæ ...	—	—	—	—	Cosmopolite
114. Plataleidæ ...	—	—	—	—	Almost Cosmopolite
115. Ciconiidæ ...	—	—	--	—	Almost Cosmopolite
117. Phænicopteridæ	--	—			Ethiopian, Neotropical, S. Palæarctic

Order and Family.	Sub-regions.				Range beyond the Region.
	Hindo-stan.	Ceylon.	Indo-China.	Indo-Malaya.	
ANSERES.					
118. Anatidæ ...	—	—	—	—	Cosmopolite
119. Laridæ	—	—	—	--	Cosmopolite
120. Procellariidæ	—	—	—	—	Cosmopolite
121. Pelecanidæ ...	—	—	—	—	Cosmopolite
124. Podicipidæ ...	—	—	—	—	Cosmopolite
REPTILIA.					
OPHIDIA.					
1. Typhlopidæ ...	—	—	—	—	All regions but Nearctic
2. Tortricidæ... ...	—	—	—	—	Austro-Malaya, S. America
3. *Xenopeltidæ* ...			—	—	Celebes
4. *Uropeltidæ* ...		—			
5. Calamariidæ ...	—	—	—	—	All the warmer regions
6. Oligodontidæ ...	—	—	—	—	S. America, Japan
7. Colubridæ ...	—	—	—	—	Almost Cosmopolite
8. Homalopsidæ...	—	—	—	—	All the regions
9. Psammophidæ	—		—	—	Ethiopian, S. Palæarctic
11. Dendrophidæ...	—		—	—	Ethiopian, Australian, Neotropical
12. Dryiophidæ ...	—		—	—	Ethiopian, Neotropical
13. Dipsadidæ ...	—		—	—	Ethiopian, Australian, Neotropical
14. Scytalidæ ...				—	Tropical America
15. Lycodontidæ ...	—		—	—	Ethiopian
16. Amblycephalidæ			—	—	Neotropical
17. Pythonidæ ...	—		—	—	The tropical regions, and California
18. Erycidæ	—		—		Ethiopian, S. Palæarctic
19. *Acrochordidæ*...		—			
20. Elapidæ	—		—	—	Tropical regions, Japan, S. Carolina
23. Hydrophidæ ...	—	—	—	—	Australian, Panama, Madagascar
24. Crotalidæ ...	—	—	—	—	America, E. Palæarctic
25. Viperidæ ...	—	—	—	—	Ethiopian, Palæarctic
LACERTILIA.					
30. Varanidæ ...	—	—	—	—	Africa, Australia
33. Lacertidæ ...	—	—	—	—	The Eastern Hemisphere
34. Zonuridæ ...			—		America, S. Europe, Ethiopian
45. Scincidæ... ...	—	—	—	—	Almost Cosmopolite
48. Acontiadæ ...		—			Ethiopian, Moluccas
49. Geckotidæ ..	—	—	—	—	Almost Cosmopolite
51. Agamidæ ...	—	—	—	—	The Eastern Hemisphere
52. Chamæleonidæ	—	—			Ethiopian
CROCODILIA.					
54. Gavialidæ ...	—		—		N. Australia
55. Crocodilidæ ...	—	—	—	—	Ethiopian, Neotropical, N. Australia
CHELONIA.					
57. Testudinidæ ...	—	—	—	—	All continents but Australia
59. Trionychidæ ...	—	—	—	—	Japan, E. of N. America, Africa
60. Cheloniidæ ...					Marine

Order and Family.	Sub-regions.				Range beyond the Region.
	Hindo-stan.	Ceylon.	Indo-China.	Indo-Malaya.	
AMPHIBIA.					
PSEUDOPHIDIA.					
1. Cæciliadæ	—	—	—		Ethiopian, Neotropical
URODELA.					
5. Salamandridæ		—			North temperate zone
ANOURA.					
7. Phryniscidæ ...				—	Ethiopian, Australian, Neotropical
9. Bufonidæ ..	—	—	—	—	All continents but Australia
11. Engystomidæ...	—	—	—	—	All regions but Palæarctic
16. Hylidæ			—		All regions but Ethiopian
17. Polypedatidæ...	—	—	—	—	Neotropical and all other regions
18. Ranidæ	—	—	—	—	Almost Cosmopolite
19. Discoglossidæ		—	—	—	All regions but Nearctic
FISHES.					
(FRESHWATER).					
ACANTHOPTERYGII.					
3. Percidæ	—	—	—	—	All regions but Australian
12. Scienidæ ...	—	—	—	—	All regions but Australian
33. Nandidæ ...	—	—	—	—	Neotropical
35. Labyrinthici ...	—	—	—	—	S. Africa, Moluccas
36. *Luciocephalidæ*				—	
39. *Ophiocephalidæ*	—	—	—	—	
46. *Mastacembelidæ*	—	--	—	—	
52. Chromidæ ...		—			Ethiopian, Neotropical
PHYSOSTOMI.					
59. Siluridæ	—	—	—	—	All warm regions
73. Cyprinodontidæ	—	—	—	—	S. Palæarctic, Ethiopian, American
75. Cyprinidæ ...	—	—	—	—	Not in S. America and Australia
78. Osteoglossidæ...				—	All tropical regions
82. Notopteridæ ...	—	—	—	—	W. Africa
85. Symbranchidæ	—		—	—	Australian (? Marine) Neotropical
INSECTS.					
LEPIDOPTERA (PART).					
DIURNI (BUTTERFLIES.)					
1. Danaidæ... ...	—	—	—	—	All warm regions and to Canada
2. Satyridæ... ...	—	—	—	--	Cosmopolite
3. Elymniidæ ...			—	—	Ethiopian, Moluccas
4. Morphidæ ...			—	—	Neotropical, Moluccas, and Polynesia
6. Acræidæ	—	—	—	—	All tropical regions
8. Nymphalidæ ...	--	—	—	—	Cosmopolite

B B

Order and Family.	Sub-regions.				Range beyond the Region.
	Hindo-stan.	Ceylon.	Indo-China.	Indo-Malaya.	
9. Libytheidæ ...	—	—	—	—	Absent from Australia
10. Nemeobeidæ ..			—	—	Not in Australia or Nearctic regions
13. Lycænidæ ...	—	—	—	—	Cosmopolite
14. Pieridæ	—	—	—	—	Cosmopolite
15. Papilionidæ ...	—	—	—	—	Cosmopolite
16. Hesperidæ ...	—	—	—	—	Cosmopolite
SPHINGIDEA.					
17. Zygænidæ ...	—	—	—	—	Cosmopolite
19. Agaristidæ ...	—	—	—	—	Australian, Ethiopian
20. Uraniidæ ...	—	—	—	—	All tropical regions
22. Ægeriidæ ...	—	—	—	—	Absent from Australia
23. Sphingidæ ...	—	—	—	—	Cosmopolite

TABLE II.

GENERA OF TERRESTRIAL MAMMALIA AND BIRDS INHABITING THE ORIENTAL REGION.

EXPLANATION.

Names in *italics* show genera peculiar to the region.
Names inclosed thus (...) show genera which just enter the region, but are not considered properly to belong to it.
Genera truly belonging to the region are numbered consecutively.

MAMMALIA.

Order, Family, and Genus.	No. of Species.	Range within the Region.	Range beyond the Region.
PRIMATES.			
SIMIIDÆ.			
1. *Simia*	2	Borneo and Sumatra	
2. *Hylobates*... ...	7	Sylhet to Java and S. Ghina	
3. *Siamanga* ...	1	Malacca and Sumatra	
SEMNOPITHECIDÆ.			
4. *Presbytes*	28	Simla to Aracan and E. Thibet, Ceylon, and Java	Moupin, Palæarctic [?]
CYNOPITHECIDÆ.			
5. *Macacus* ·... ...	22	The whole region	S. Palæarctic
6. *Cynopithecus* ..	1	Philippines	Celebes
(*Sub-Order*) *LEMUROIDEA.*			
LEMURIDÆ.			
7. *Nycticebus* ...	3	E. Bengal to Java, and S. China	
8. *Loris*	1	Ceylon and S. India	
TARSIIDÆ.			
9. *Tarsius*	1	Sumatra and Borneo	N. Celebes
CHIROPTERA.			
PTEROPIDÆ.			
10. Pteropus ...	6	The whole region	Tropics of E. Hemisp.
11. Xantharpyia ...	1	The whole region	Austro-Malaya, Ethiop., S. Palæarctic
12. Cynopterus ...	3	The whole region	Tropical Africa
13. *Megœrops*	1	Sumatra	
14. Macroglossus ...	1	Java, Borneo, Philippines	Austro-Malaya
15. Harpyia	1	Philippines	Austro-Malaya
RHINOLOPHIDÆ.			
16. *Aquias*	2	Nepal to Java	

Order, Family, and Genus.	No. of Species.	Range within the Region.	Range beyond the Region.
17. *Phyllotis*	1	Philippines	
18. Rhinolophus ...	10	The whole region	Warmer parts of E. Hem
19. Hipposideros ...	8	The whole region	Austro-Malaya
20. Phyllorhina ...	4	Indo-Malay subregion	Austro-Malaya, Tropica Africa
21. Asellia	1	Java, Sumatra	Amboyna, Egypt
22. *Petalia*	1	Java	
23. *Cœlops*	1	India (Bengal)	
24. *Rhinopoma* ...	1	All India	Egypt, Palestine
25. Megaderma ...	2	The whole region	Ternate, N. Ethiopian
26. Nycteris	1	Java	Ethiopian
VESPERTILIONIDÆ.			
27. Scotophilus ...	10	The whole region	Austral., Nearc., Neotrop
28. Vespertilio ...	12	The whole region	Cosmopolite
29. Keriovula... ...	8	The whole region	S Africa, N. China
30. *Trilatitus* ...	2	Indo-Malaya	?
31. *Noctulina* ...	3	Nepal to Philippines	?
32. Miniopteris ...	3	Java, Philippines, and China	S. Africa, S. Palæarctic Australian
33. *Murina*	2	Himalayas to Java	?
34. Nycticejus ...	8	All India	Trop. Africa, Temp. Amer
35. Harpiocephalus	2	Java and Philippines	
36. Taphozous ...	4	The whole region	Ethiop., Austro-Malayan Neotropical
37. *Myotis*	3	Himalayas	
38. Plecotus	1	Darjeeling	Timor, S. Palæarctic
39. Barbastellus ...	1	Himalayas	Europe
40. Nyctophilus ...	1	Mussoorie	Australian
NOCTILIONIDÆ.			
41. *Chiromeles* ...	1	Indo-Malaya, Siam	
42. Nyctinomus ...		The whole region	Madagascar, America
INSECTIVORA.			
GALEOPITHECIDÆ.			
43. *Galeopithecus* ...	2	Indo-Malay and Philippines, excl. Java	
TUPAIIDÆ.			
44. *Tupaia*	7	S. and E. of India to Borneo	
45. *Hylomys*	2	Tenasserim to Java and Borneo	
46. *Ptilocerus*... ...	1	Borneo	
ERINACEIDÆ.			
47. Erinaceus... ...	2	Hindostan and Formosa	Palæarctic, S. Africa
48. *Gymnura*... ...	1	Malacca, Sumatra, Borneo	
TALPIDÆ.			
49. Talpa	2	Himalayas to Assam, & Formosa	Palæarctic
SORICIDÆ.			
50. Sorex	20	The whole region	All regions but Austral. and S. America

Order, Family, and Genus.	No. of Species	Range within the Region.	Range beyond the Region.
CARNIVORA.			
FELIDÆ.			
51. Felis...	20	The whole region	All regions but Austral.
(Lynx	1	Central India)	Palæarctic, Ethiopian
52. Cynælurus ...	1	S. and W. India	S. Palæarctic, Ethiopian
VIVERRIDÆ.			
53. Viverra	2	The whole region	Ethiopian, Moluccas
54. *Viverricula*	2	India to China and Java	
55. *Prionodon* ...	2	Nepal to Borneo and Java	
56. *Hemigalea* ...	2	Malacca and Borneo	
57. *Arctitis* ...	1	Nepal to Sumatra and Java	
58. *Paradoxurus* ...	8	The whole region	Ke Islands (? introduced)
59. *Paguma*	3	Nepal to Malaya and China	
60. *Arctogale*	1	Tenasserim and Malaya	
61. *Cynogale*... ...	1	Borneo	
62. Herpestes... ...	7	The whole reg., excl. Philippines	S. Palæarctic, Ethiopian
63. *Calogale*	4	India to Cambodjia	Ethiopian
64. *Calictis*	1	Ceylon ?	
65. *Urva*	1	N. India	
66. *Tæniogale* ...	1	Central India	
67. *Onychogale* ...	1	Ceylon	
HYÆNIDÆ.			
68. Hyæna	1	Hindostan, open country	S Palæarctic, Ethiopian
CANIDÆ.			
69. Canis...	2	All India	Almost Cosmopolite
70 *Cuon*	1	India to Java	
71. Vulpes	4	All India	All Continents but S. America and Australia
(Nycterentes ...	1	China)	Japan and Amoorland
MUSTELIDÆ.			
72. Martes	2	India, Ceylon, Java, and China	Palæarctic, Nearctic
73. Mustela	3	Himalayas to Bhotan and China	Palæarc., Ethiop., Nearc.
74. *Gymnopus* ...	2	Nepal to Borneo	
75. *Barangia* ...	1	Sumatra	
76. Lutra	5	The whole region	Palæarctic
77. Aonyx	2	N. India, Malaya	W. and S. Africa
78. *Arctonyx*... ...	1	Nepal to Aracan	
(Meles	1	S. China)	Palæarctic genus
79. *Mydaus*	1	Sumatra, Java	
80. Mellivora... ...	1	Hindostan	Ethiopian
81. *Helictis*	4	Nepal, Formosa, China & Java	
ÆLURIDÆ.			
82. Ælurus	1	E. Himalayas to E. Thibet	Palæarctic !
URSIDÆ.			
83. Ursus	2	Himalayas to China	Palæarctic, Nearctic
84. *Helarctos*.. ...	1	Indo-Malaya	
85. *Melursus*... ...	1	Ganges to Ceylon	

Order, Family, and Genus.	No. of Species	Range within the Region.	Range beyond the Region
CETACEA.			
DELPHINIDÆ.			
86. *Platanista* ...	2	Ganges to India	
SIRENIA.			
MANATIDÆ.			
87. Halicore ...	1	Coasts of W. India, Ceylon, and Indo-Malaya	E. Africa, N. Australia
UNGULATA.			
TAPIRIDÆ.			
88. Tapirus	1	Malay Pen., Sumatra, Borneo	Neotropical
RHINOCEROTIDÆ.			
89. Rhinoceros ...	5	Nepal to Bengal, Siam, & Java	Ethiopian
SUIDÆ.			
90. Sus...	6	The whole region	Palæarc., Austro-Malaya
TRAGULIDÆ.			
91. *Tragulus* ...	5	India and Ceylon to Cambodja and Java	
CERVIDÆ.			
92. Cervus	15	The whole region	Palæarc., Amer., Moluc.
93. *Cervulus*... ...	4	The whole region	
(Moschus... ...	1	Himalayas above 8,000 feet)	Central Asia, Palæarctic
BOVIDÆ.			
94. *Bibos* ...	3	India to Burmah, Formosa, and Java	
95. Bubalus... ..	1	N. and N. Central India	Ethiopian, S. Palæarctic
96. *Portax*	1	Peninsula of India	
97. Gazella	1	Deserts and plains of India	Palæarctic deserts
98. *Antilope*... ...	1	Open country of India	
99. *Tetraceros* ...	2	Hilly districts all over India	
100. Nemorhedus	3	E. Himalayas and Sumatra	N. China and Japan
101. Capra	1	Neilgherries	Palæarctic, Nearctic
PROBOSCIDEA.			
ELEPHANTIDÆ.			
102. Elephas... ...	1	India to Siam, Sumatra & Borneo	Ethiopian
RODENTIA.			
MURIDÆ.			
103. Mus	50	The whole region	The E. Hemisphere
104. Acanthomys ...	1	India	Ethiopian, Australian
105. *Phlœomys* ...	1	Philippines	
106. *Platacanthomys*	1	S. W. India	
107. Meriones ...	2	India and Ceylon	Palæarctic, Ethiopian
108. *Spalacomys* ...	1	India	
109. Arvicola ...	2	Himalayas	Palæarctic, Nearctic

Order, Family, and Genus.	No. of Species.	Range within the Region.	Range beyond the Region.
SPALACIDÆ.			
110. Rhizomys ...	3	Nepal to Canton, Malacca and Sumatra	Abyssinia
SCIURIDÆ.			
111. Sciurus	50	The whole region	Cosmop., excl. Austral. region
112. Sciuropterus...	9	India, and Ceylon to Java, Formosa	N. and E. Palæarctic
113. *Pteromys* ...	9	India & Ceylon to Borneo, Java, Formosa	Japan
(Arctomys ...	2	W. Himalayas above 8,000 ft.)	Palæarctic and Nearctic
HYSTRICIDÆ.			
114. Hystrix	3	India and Ceylon, to Malacca & S. China	S. Palæarctic, Ethiopian
115. Atherura ...	2	India to Malaya	West Africa
116. *Acanthion* ...	2	Nepal to Borneo and Java	
LEPORIDÆ.			
117. Lepus	5	India and Ceylon to S. China and Formosa	All regions but Austral.
ENDENTATA. MANIDIDÆ.			
118. Manis	2	Nepal to Ceylon, S. China and Java	Ethiopian

BIRDS.

PASSERES.			
TURDIDÆ.			
1. *Brachypteryx* ...	8	Himalayas, Ceylon and Java	
2. Oreocincla ...	8	N. W. Himalayas to E. Thibet, Ceylon, Burmah, Malaya, Formosa	Palæarctic, Australian
3. Turdus	26	The whole region	Almost Cosmopolite
4. Geocichla... ..	9	India & Ceylon to Java, Formosa	Celebes, Lombock, to N. Australia
5. Monticola ...	3	The whole region	Palæarctic, Ethiopian, Moluccas
6. Orocœtes	2	N. W. Himalayas, and India	
7. Zoothera	3	W. Himalayas to Aracan, Java	Lombock, Timor ?
SYLVIIDÆ.			
8. *Orthotomus* ...	13	The whole region	
9. *Prinia*	11	The whole reg., excl. Philippines	
10. Drymæca ...	13	The whole reg., excl. Philippines	Ethiopian
11. Cisticola	6	The whole region	Ethiopian Australian
12. *Suya*	5	Nepal to S. China and Formosa	
13. *Megalurus* ...	3	Central India, Java, Philippines	
14. Acrocephalus ...	9	India to Ceylon, S. China, and Philippines	Palæarc.,Ethiop.,Austral.
(Dumeticola ..	2	Nepal and E. Thibet)	A Palæarctic genus

Order, Family, and Genus.	No. of Species.	Range within the Region.	Range beyond the Region
15. { Locustella ...	4	Nepal, Hindostan, S. China	Palæarctic
16. { Horites	2	Himalayas, Formosa	High Himal., E. Thibet
17. (Phylloscopus ...	10	All India and Ceylon, to China Philippines	Palæarctic, Ethiopian
(Gerygone ...	1	Philippine Islands)	Australian genus
(Hypolais ...	1	All India, ? migrant)	Palæarctic genus
18. Abrornis	26	The whole reg., excl. Philippines	Cashmere, E. Thibet
19. Reguloides ...	2	Himalayas and Central India	Palæarctic
(Regulus	1	N. W. Himalayas and E.Thibet)	Palæarctic and Nearctic
(Sylvia	2	India and Ceylon)	Palæarctic genus
(Curruca	2	India)	Palæarctic genus
(Cyanecula ...	1	India)	Palæarctic genus
20. Calliope	2	Himalayas and Central India, Philippine Islands	Palæarctic
21. Ruticilla	8	Himalayas to China and Formosa	Palæarctic, Ethiopian
22. *Chœmarrhornis*	1	Himalayas to Burmah	
23. } *Larvivora* ...	10	W.Himalayas to Ceylon, Malacca and China	
24. *Notodela*	3	Himalayas to Pegu, Java, Formosa	
25. *Tarsiger*	2	Nepal and W. Himalayas	
(Grandala ...	1	Nepal and E. Thibet, high)	Palæarctic genus
26. (Copsychus ...	6	The whole region	Madagascar
27. *Kittacincla* ...	5	The whole region	
28. Thamnobia ...	2	N. W. India, Hindostan, and Ceylon	Ethiopian
{ (Dromolæa ...	1	N. W. India)	Ethiopian genus
(Saxicola ...	2	N. W. India)	Palæarctic and Ethiopian
29. Oreicola ? ...	1	Burmah	Timor
(Cercomela ...	1	N. W. India, a desert genus)	N. E. Africa, S. W. Asia
30. Pratincola ..	5	The whole region	Palæarctic, Ethiopian, Celebes, and Timor
(Accentor ...	2	Himalayas, in winter)	Palæarctic genus
TIMALIIDÆ.			
31. Pomatorhinus...	20	The whole region	Australian
32. Malacocercus ...	14	All India to Burmah, Philippines	Arabia, Nubia
33. Chatarrhæa ...	5	India, Burmah, Philippines	Palestine, Abyssinia
34. *Layardia* ...	3	India and Ceylon	
35. *Acanthoptila* ...	1	Nepal	
36. *Garrulax* ...	22	The whole region	
37. *Janthocincla* ..	8	Himalayas to E. Thibet, Sumatra, Formosa	
38. *Gampsorhynchus*	1	Nepal	
39. *Grammatoptila*	1	N. India	
40. *Trochalopteron*	22	N. W. Himalayas, India, China, Formosa	
41. *Actinodura* ...	3	E. Himalayas, 3,000 to 10,000 ft	
42. *Pellorneum* ...	3	India, Ceylon, Tenasserim	
43. *Dumetia*	2	India and Ceylon	
44. Timalia	10	Malacca to Java	
45. *Stachyris*	6	N. W. Himalayas to China, Formosa, Sumatra	
46. Pyctoris	3	India, Ceylon, and Up. Burmah	
47. Mixornis	8	Himalayas to Borneo and Java	
48. *Malacopteron* ...	3	Malacca to Java	

Order, Family, and Genus.	No. of Species	Range within the Region.	Range beyond the Region.
49. Alcippe	16	The whole region	New Guinea
50. *Macronus* ...	1	Malacca to Java	
51. *Cacopitta*	5	Java, Borneo, Sumatra	
52. Trichastoma ...	9	Nepal, Malacca to Java	Celebes
53. *Napothera* ...	5	Malacca to Java	
54. Drymocataphus	6	Malacca to Java, Ceylon	Timor
55. *Turdinus* ...	4	Tenasserim, Malacca	
56. *Trichixos*	1	Malacca, Borneo	
57. *Sibia*	6	N. W. Himalayas to Tenasserim, Formosa	
PANURIDÆ.			
58. *Paradoxornis* ...	3	Nepal to Aracan and E. Thibet, 3,000–6,000 ft.	
59. Suthora	8	Himalayas to E. Thibet, China, Formosa	N. W. China, E. Thibet
60. *Chlenasicus* ...	1	Sikhim	
CINCLIDÆ.			
61. Cinclus	2	Himalayas, China, and Formosa	Palæarctic and American
62. Eupetes	2	Malacca and Sumatra	New Guinea
63. *Enicurus*	9	N. W. Himalayas (to 11,000 ft.) to Java and West China	
64. *Myiophonus* ...	6	All India (to 9,000 ft. in N. W. Himalayas) S. China, Formosa, Java, Sumatra	Turkestan
TROGLODYTIDÆ.			
65. *Tesia* ...	2	Eastern Himalayas	
66. *Pnocpyga*	6	N. W. Himalayas to E. Thibet, Java	
67. Troglodytes ...	1	Himalayas to E. Thibet	Palæarctic and American
68. *Rimator*	2	Darjeeling	
CERTHIIDÆ.			
69. Certhia	2	Himalayas	Palæarctic and Nearctic
70. *Salpornis* ...	1	Central India	
71. *Rhabdornis* ...	1	Philippine Islands	
(Tichodroma ...	1	Himalayas in winter)	Palæarctic genus
SITTIDÆ.			
72. Sitta...	5	Himalayas to S. India, S. China	Palæarctic and Nearctic
73. *Dendrophila* ...	2	All India and Ceylon to Pegu and Java	
PARIDÆ.			
74. Parus	16	The whole region	Palæarctic and Nearctic
75. *Melanochlora* ...	2	Nepal to Malacca and Sumatra	
76. *Psaltria*	1	West Java	
77. *Ægithaliscus* ..	6	W. Himalayas to China	Afghanistan
78. *Sylviparus* ...	1	W. Himalayas to Central India and E. Thibet	
79. *Cephalopyrus* ...	1	N. W. Himalayas	
LIOTRICHIDÆ.			
80. *Liothrix*	3	Nepal to S. W. China	

Order, Family, and Genus.	No. of Species.	Range within the Region.	Range beyond the Region
81. *Siva*	3	Himalayas :—3,000—7,000 ft.	
82. *Minla*	4	Nepal to E. Thibet; moderate heights	
83. Proparus ...	6	N. W. Himalayas to E. Thibet: high	Perhaps also Palæarctic
84. *Allotrius* ...	7	N. W. Himalayas to Tenasserim E. Thibet and Java	
85. *Cutia*	2	Nepal and Sikhim	
86. *Yuhina*	4	Himalayas to E. Thibet, high	Perhaps Palæarctic
87. *Ixulus*	4	Darjeeling to Tenasserim	
88. *Myzornis* ...	1	Nepal and Sikhim	
PHYLLORNITHIDÆ.			
89. *Phyllornis* ...	10	The whole region ; excluding China and Philippines	
90. *Iora*	5	The whole reg., excl. Philippines	
91. *Erpornis* ...	2	Nepal and Hainan	
PYCNONOTIDÆ.			
92. *Microscelis* ...	5	Burmah, China, Malaya	Japan
93. Pycnònotus ..	40	The whole region	Ethiopian
94. *Hemixus* ...	2	Himalayas and Hainan	
95. Hypsipe tes ...	15	The whole region	Madagascar
96. Criniger	11	India, Ceylon, Malaya, Hainan	Africa, Moluccas
97. *Setornis*	3	Malacca, Sumatra, Borneo	
98. *Iole*	4	Aracan and Malaya	
ORIOLIDÆ.			
99. Oriolus	12	The whole region	Palæarc. Ethiopian, Celebes, Flores
100. *Analcipus* ...	3	Himalayas, Malaya, Formosa, Hainan	
CAMPEPHAGIDÆ.			
101. *Pericrocotus* ...	22	The whole region	Lombock ; the Amoor, migrant
102. Graucalus ...	7	India, Ceylon, Malaya, Philippines, Hainan and Formosa	Australian
103. Campephaga ...	1	Philippine Islands	Celebes to N. Guinea
104. *Volvocivora* ...	7	The whole reg., excl. Philippines	
105. Lalage	2	Malaya and Philippines	Celebes to Pacific Is.
106. *Cochoa*	3	Himalayas and Java	
DICRURIDÆ.			
107. Dicrurus ...	17	The whole region	Ethiop. and Australian
108. *Bhringa* ...	2	Himalayas to Burmah and Java	
109. *Chibia*	1	India to China	Pekin in summer
110. *Chaptia*	3	India to Borneo and Formosa	
111. *Irena*	3	S. India and Ceylon, Assam to Malaya and Philippines	
MUSCICAPIDÆ.			
112. *Muscicapula* ...	6	Cashmere to W. China, S. India	
113. Erythrosterna	7	The whole region, excluding Philippines	Palæarctic and Madagascar

Order, Family, and Genus.	No. of Species.	Range within the Region.	Range beyond the Region.
114. Xanthpygia ...	2	Malacca to China	N. China and Japan
115. *Hemipus* ...	1	India and Ceylon	
116. *Pycnophrys* ...	1	Java	
117. Hemichelidon	3	N. India to Ceylon, and China ; ? Philippines	Eastern Asia
118. *Niltava*	3	Himalayas to W. China	
119. Cyornis	14	The whole region	Celebes and Timor
120. Cyanoptila ...	1	Hainan to Japan	Japan and N. China
121. *Eumyias* ...	8	The whole reg., excl. Philippines	
122. *Siphia*	9	N. W. India, Ceylon, Formosa, E. Thibet	
123. *Anthipes* ...	1	Nepal	
124. *Schwaneria* ...	1	Borneo	
125. *Hypothymis* ...	1	The whole region	Celebes
126. Rhipidura ..	7	All India and Ceylon, Malaya, Philippines	Australian
127. *Chelidorhynx*	1	N. India	
128. *Cryptolopha* ...	1	The whole region	Celebes
129. Tchitrea... ...	6	The whole region	N. China, and Japan, Flores, Ethiopian
130. *Philentoma* ...	4	Malaya and Philippines	
PACHYCEPHALIDÆ.			
131. Hylocharis ...	2	Aracan to Malaya & Philippines	Celebes, Timor
LANIIDÆ.			
132. Lanius	16	The whole region	Nearc., Palæarc., Ethiop.
133. *Laniellus* ..	1	Java	
134. *Tephrodornis*...	5	India, Ceylon, and Malaya ; Hainan	
CORVIDÆ.			
135. *Pityriasis* ...	1	Borneo, Sumatra	
136. *Platylophus* ..	4	Malaya	
137. Garrulus ...	4	Himalayas, S. China, Formosa	Palæarctic
138. *Cissa*	3	Himalayas and Aracan to Java	
139. Urocissa... ...	7	N. W. Himalayas, Ceylon, Burmah, China, Formosa	N. China and Japan
140. *Temnurus* ...	3	Malaya and Cochin China	
141. *Dendrocitta* ...	8	All India to S. China, Formosa, and Sumatra	
142. *Crypsirhina* ...	2	Java and Burmah	
143. Nucifraga ...	2	Himalayas and E. Thibet ;— 8,000—10,000 feet	Palæarctic genus
144. Pica	2	China and Himalayas of Bœtan	Palæarctic and Nearctic
145. Corvus	9	The whole region	Cosmop., excl. S. Am.
(Fregilus ...	2	Himalayas, high)	Palæarctic genus
NECTARINIIDÆ.			
146. *Æthopaga* ...	13	Himalayas to W. China & Java, Central India	Celebes
147. Chalcostetha...	1	Malaya and Siam	Celebes to New Guinea
148. Arachnothera	12	The whole reg., excl. Philippines	Celebes, Lombock, New Guinea
149. Arachnecthera	7	The whole region, excl. China	Celebes to New Ireland

Order, Family, and Genus.	No. of Species.	Range within the Region.	Range beyond the Region.
150. *Nectarophila* ...	4	India, Ceylon, Malaya, Philipp.	Celebes
151. *Anthreptes* ...	1	Malaya and Indo-China	Celebes
DICÆIDÆ.			
152. Dicæum ...	10	The whole region	Australian
153. Pachyglossa ...	1	Nepal	Celebes
154. *Piprisoma* ...	1	India and Ceylon	
155. *Prionochilus* ...	4	Malaya	
156. Zosterops ...	8	The whole region	Ethiopian, Australian
157. *Chalcoparia* ...	1	Aracan to Malaya	
HIRUNDINIDÆ.			
158. Hirundo ...	10	The whole region	Cosmopolite
159. Cotyle	5	India to China	Palæarc., Ethiop., Ame—
160. Chelidon ...	3	India, Borneo	Palæarctic
FRINGILLIDÆ.			
(Fringilla ...	1	Himalayas, in winter)	Palæarctic genus
(Acanthis ...	1	N. W. Himalayas, in winter)	Palæarctic genus
(Procarduelis...	1	High Himalayas)	Palæarctic genus
(Chlorospiza ...	1	China)	Palæarctic and Ethiopia
161. Passer	6	The whole region	Palæarctic and Ethiopia
(Fringillauda	1	High Himalayas)	Palæarctic genus
(Coccothraustes	2	High Himalayas)	Palæarctic and Nearctic
(Mycerobas ...	1	High Himalayas)	Palæarctic genus
162. Eophona ...	1	China	Palæarctic
(Pyrrhula ...	4	Himalayas, winter)	Palæarctic
(Carpodacus ...	4	Himalayas and Central India, in winter)	Palæarctic and Nearctic
(Loxia	1	Snowy Himalayas)	Palæarctic and Nearctic
(Propyrrhula ...	1	Darjeeling, in winter)	[?] Palæarctic
163. *Hæmatospiza*	1	S. E. Himal., 5,000 to 10,000 ft.	
(*S. Fam.* EMBERIZINÆ)			
164. Euspiza	4	N. W. India to Burmah, & China	Palæarctic and Nearctic
165. Emberiza ...	7	All India and China, in winter	Palæarctic genus
PLOCEIDÆ.			
166. Ploceus	4	India & Ceylon, Burmah, Malaya	Ethiopian
167. Munia	20	The whole region	Austro-Malayan
168. Estrilda ...	2	India and Ceylon, Burmah, Java	Ethiopian, Australian
169. Erythrura ...	1	Java, Sumatra	Moluccas to Fiji Islands
STURNIDÆ.			
170. Eulabes	7	The whole reg., excl. Philippines	Flores, Papua
171. *Ampeliceps* ..	1	Tenasserim to Cochin-China	
172. *Gynnops* ...	1	Philippine Islands	
173. Pastor	1	All India to Burmah	S. Palæarctic
174. *Acridotheres* ...	6	The whole region	Celebes
175. *Sturnia*	12	The whole region	N. China&Japan, Celebe—
176. Sturnus	3	India and China	Palæarctic
177. *Sturnopastor* ...	3	Cen. India to Burmah & Malaya	
178. Calornis... ...	2	Malaya and Philippines	[?] Celebes, Moluccas t— Samoan Islands
179. *Saroglossa* ...	1	W. and Central Himalayas	

Order, Family, and Genus.	No. of Species.	Range within the Region.	Range beyond the Region.
ARTAMIDÆ.			
180. Artamus... ...	3	The whole region	Australian
ALAUDIDÆ.			
(Otocorys ...	1	N. India, in winter)	Palæarctic and Nearctic
181. Alauda	7	India and China	Palæarctic and Ethiopian
182. Galerita	2	Central India	Palæarctic
183. Calandrella ...	2	India and Burmah	Palæarctic and Ethiopian
(Melanocorypha	1	N. W. India)	Palæarctic
184. Mirafra	5	India, Ceylon, and Java	Ethiopian
185. Ammomanes...	1	Central India	Palæarctic and Ethiop an
186. Pyrrhulauda...	1	India and Ceylon	Ethiopian
MOTACILLIDÆ.			
187. Motacilla ...	6	India and Ceylon to China and Philippines)	Palæarctic and Ethiopian
188. Budytes	2	China and Philippines	Palæarctic & Ethiopian, Moluccas
189. Calobates ...	1	The whole region	Palæarctic
190. Nemoricola ...	1	India, Ceylon, and Malaya	
191. Authus	3	India and China	Cosmopolite
192. Corydalla ...	8	The whole region	Palæarctic, Australian
193. Heterura ...	1	Himalayas	
EURYLÆMIDÆ.			
194. Eurylœmus ..	2	Malaya	
195. Serilophus ...	1	Himalayas	
196. Psarisomus ..	1	Himalayas	
197. Corydon... ...	1	Malacca, Sumatra, Borneo	
198. Cymbirhynchus	2	Aracan, Siam, and Malaya	
199. Calyptomena...	1	Malacca, Sumatra, Borneo	
PITTIDÆ			
200. Pitta	11	The whole region	Australian, Ethiopian
201. Eucichla ...	3	Malaya	
202. Hydrornis ..	3	Himalayas and Malaya	
PICARIÆ.			
PICIDÆ.			
203. Vivia	1	N. W. Himalayas to E. Thibet, 3,000–6,000 ft.	
204. Sasia	2	Nepal to Malaya and Borneo	
205. Picus	14	The whole region, excl. Philippines	Palæarctic, American
206. Hyopicus ...	1	Himalayas	N. China
207. Yungipicus ...	12	The whole region	N. China, Japan, Celebes
208. Reinwardtipicus	1	Penang to Sumatra and Borneo	
209. Venilia	2	Nepal to Sumatra and Borneo	
210. Chrysocolaptes	8	India, Ceylon, Malaya, Philippines	
211. Hemicercus ...	5	Malabar, Pegu to Malaya	
212. Gecinus	12	All India and Ceylon to Pegu and Malaya	Palæarctic

Order, Family, and Genus.	No. of Species.	Range within the Region.	Range beyond the Region
213. *Mulleripicus*...	5	Malabar, Aracan to Malaya and Philippines	Celebes
214. *Brachypternus*	5	India, Ceylon, and China	
215. *Tiga*	5	India to Malaya	
216. *Gecinulus* ...	2	S. Himalayas to Burmah	
217. *Miglyptes* ...	3	Malaya	
218. *Micropternus*...	8	India and Ceylon, to Borneo and S. China	
YUNGIDÆ.			
219. Yunx	1	Central and S. China	Palæarctic, S Africa
INDICATORIDÆ.			
220. Indicator ...	2	Himalayas and Borneo	Ethiopian
MEGALÆMIDÆ.			
221. *Megalœma* ...	27	The whole region, excl. Philippines	
222. *Xantholœma*...	4	All India and Ceylon to Pegu and Malaya	
223. *Psilopogon* ...	1	Sumatra	
224. *Caloramphus*...	2	Malacca, Sumatra and Borneo	
CUCULIDÆ.			
225. *Phœnicophaës*	1	Ceylon	
226. *Rhinococcyx* ...	1	Java	
227. *Dasylophus* ...	1	Philippine Islands	
228. *Lepidogramnus*	1	Philippine Islands	
229. *Carpococcyx* ...	1	Borneo, Sumatra	
230. *Zanclostomus* ..	1	Malaya	
231. *Rhopodytes* ...	7	Nepal to Ceylon, Hainan and Malaya	
232. *Taccocoua* ...	4	All India, Ceylon, Malacca	
233. *Poliococcyx* ...	1	Malacca, Sumatra, Borneo	
234. *Rhinortha* ...	1	Malacca, Sumatra, Borneo	
235. Centropus ...	14	The whole region	Ethiopian, Australian
236. Cuculus ...	10	The whole region	Palæarc., Ethiop., Aust.
237. Cacomantis ...	9	The whole region	Australian
238. Chrysococcyx	5	The whole region	Ethiopian, Australian
239. *Surniculus* ...	2	India, Ceylon and Malaya	
240. Hierococcyx...	6	The whole region	Celebes, N. China and Amoorland
241. Coccystes ...	2	The whole region, excl. Philippines	Ethiopian
242. Eudynamis ...	2	The whole region	Australian
CORACIIDÆ.			
243. Coracias... ..	2	India, Ceylon and Burmah	Ethiopian, S. Palæarctic Celebes
244. Eurystomus ...	1	The whole region	Ethiopian, Australian
MEROPIDÆ.			
245. *Nyctiornis* ...	3	S. India to Himalayas, Burmah, Sumatra, and Borneo	

Order, Family, and Genus.	No. of Species.	Range within the Region.	Range beyond the Region.
246. Merops	5	The whole region	S. Palæarctic, Ethiopian, Australian
TROGONIDÆ.			
247. *Harpactes* ...	10	The whole region, excl. China	
ALCEDINIDÆ.			
248. Halcyon... ...	10	The whole region	S. Palæarctic, Ethiopian, Australian
249. *Pelargopsis* ...	7	The whole region, excl. China	Celebes and Timor
250. *Carcineutes* ...	2	Burmah, Siam, and Malaya	
251. Ceyx	6	India and Ceylon, Malaya and Philippines	Moluccas & New Guin.
252. Alcedo	5	The whole region	Palæarctic, Ethiopian, Austro-Malayan
253. Alcyone... ...	1	Philippines	Australian genus
254. Ceryle	2	India to S. China	Ethiopian, S. Palæarctic, American
BUCEROTIDÆ.			
255. *Buceros*	4	Nepal to Malaya, S. India, Philippines	
256. *Hydrocissa* ...	7	India, Ceylon and Malaya	
257. Berenicornis ...	1	Sumatra	W. Africa
258. Calao	2	Tenasserim, Malaya	Austro-Malaya
259. *Aceros*	1	S. E. Himalayas	
260. *Cranorrhinus*	2	Malacca to Borneo and Philippines	Celebes
261. *Penelopides* ...	1	Philippines	
262. *Rhinoplax* ...	1	Sumatra, Borneo	
263. *Meniceros* ..	3	India and Ceylon to Tenasserim	
UPUPIDÆ.			
264. Upupa	3	India, Ceylon and Burmah	Ethiopian, S. Palæarctic
PODARGIDÆ.			
265. Batrachostomus	6	India, Ceylon and Malaya	Moluccas
CAPRIMULGIDÆ.			
266. Caprimulgus...	13	The whole region	The Eastern Hemisphere
267. *Lyncornis* ...	4	Burmah, Malaya, & Philippines	Celebes
CYPSELIDÆ.			
268. Cypselus ...	8	The region, excl. Philippines	The Old World & S. Amer.
269. Dendrochelidon	3	Ceylon, India, Malaya, Philipp.	Austro-Malaya
270. Collocalia ...	3	The whole region	Madagascar, Moluccas, Pacific Islands
271. Chætura... ...	3	Ceylon, India, Malaya, Hainan	America, Africa
PSITTACI.			
CACATUIDÆ. (Cacatua ...	1	Philippines)	Australian genus
PALÆORNITHIDÆ.			
272. Palæornis ...	14	N. W. India to Ceylon, Siam & Malaya	Ethiopian

Order, Family, and Genus.	No. of Species.	Range within the Region.	Range beyond the Region
273. Prioniturus ...	1	Philippine Islands	Celebes
274. Cyclopsitta ...	1	Philippine Islands	Papuan Islands
275. *Psittinus* ...	1	Malaya, excl. Java	
276. Tanygnathus...	1	Philippine Islands	Austro-Malaya
277. Loriculus ...	9	Ceylon, India, Malaya, Philippines	Celebes and Molucca Flores
COLUMBÆ.			
COLUMBIDÆ.			
278. Treron	21	The whole region	Ethiopian, Moluccas
279. Ptilopus... ...	3	Malaya and Philippines	Australian
280. Carpophaga ...	10	India and Ceylon to Hainan and Philippines	Australian
281. Columba ...	7	Ceylon and India to Tenasserim	Palæarc., Ethiop., Ame
282. Janthænas ...	3	Philippine, Andaman & Nicobar Islands	Japan, Moluccas to S moan Islands
283. Macropygia ...	6	Nepal, Java, Hainan, Philippines	Austro-Malaya, Austral
284. Turtur	8	The whole region	OldWorld, Austro-Mala
285. Chalcophaps ...	2	India, Ceylon, Malaya, Hainan, Philippines, Formosa	Austro-Malaya, Austral
286. *Phapitreron* ...	2	Philippine Islands	
287. Calœnas... ...	1	Nicobar and Philippine Islands	Austro-Malaya
288. Phlegœnas ...	2	Philippine and Sooloo Islands	Austro-Mal. & Polynesi
289. Geopelia ...	1	Philippine Islands, Java	Austro-Malaya &Austra
GALLINÆ.			
PTEROCLIDÆ.			
290. Pterocles ...	2	Central and S. India	S. Palæarctic, Ethiopia
TETRAONIDÆ.			
291. Francolinus ...	3	Ceylon and India to S. China	S. Palæarctic, Ethiopia
292. *Ortygornis* ...	3	Ceylon to Himalayas, Sumatra & Borneo	
293. Perdix	12	India, Malaya, Philippines, China	Palæarctic
294. Coturnix ...	9	The whole region	The Eastern Hemispher
295. *Rollulus*... ...	2	Malacca, Siam, Borneo, Philipp.	
(Caccabis ...	1	W. Himalayas)	Palæarctic genus
PHASIANIDÆ.			
296. *Pavo*	2	Ceylon to Himalayas,S. W. China and Java	
297. *Argusianus* ...	4	Siam, Malacca, Borneo	
298. *Polyplectron* ...	5	Upper Assam to S. W. China & Sumatra	
(Lophophorus	3	Cashmere and E. Thibet)	Palæarctic genus
(Tetraophasis	1	E. Thibet)	Palæarctic genus
299. Ceriornis ...	5	N. W. Himalayas to W. China	S. E. Palæarctic
(Pucrasia ...	3	N. W. Himalayas to N. China and Mongolia)	Palæarctic genus
300. Phasianus ...	3	W. Himalayas,S. China, Formosa	S. Palæarctic
301. *Euplocamus* ...	13	N. W. Himalayas to China Sumatra and Borneo	
302. *Gallus*	4	The region, excl. China	Celebes and Timor
303. *Galloperdix* ...	3	Central India to Ceylon	

Order, Family, and Genus.	No. of Species.	Range within the Region.	Range beyond the Region.
TURNICIDÆ.			
304. Turnix	9	The whole region	S. Palæarc., Ethiopian, Australian
MEGAPODIIDÆ.			
305. Megapodius ..	2	Nicobar Is., Philippines, N. W. Borneo	Celebes to Samoan Is., N. Australia
ACCIPITRES.			
VULTURIDÆ.			
306. Vultur	1	Himalayas	S. Palæarctic, Ethiopian
307. Gyps	3	India and Siam	3. Palæarctic, Ethiopian
308. *Pseudogyps* ...	1	India and Burmah	N. Ethiopian
309. Neophron ...	1	All India	S. Palæaictic, Ethiopian
FALCONIDÆ.			
310. Circus	4	India and China	Almost Cosmopolite
311. Astur	4	The whole region	Almost Cosmopolite
312. Accipiter ...	2	The whole region	Almost Cosmopolite
313. Buteo	2	India to China	Cosmopolite; excl. Austl.
314. Aquila	4	India to China	Nearc, Palæarc., Ethiop.
315. Nisaëtus ...	2	India and Ceylon	S. Palæar., Ethiop., Aus.
316. Lophotriorchis	1	Indo-Malaya	Neotropical
317. Neopus	1	India to Burmah and Malaya	Celebes and Moluccas
318. Spizaëtus ...	5	India to Malaya and Formosa	Neotropical, Ethiopian, Austro-Malayan
319. Circaëtus ...	1	Indian peninsula	Palæarc., Ethiop., Timor
320. *Spilornis* ...	5	The whole region	Celebes
321. Butastur ...	3	The whole region	N. E. Africa, Celebes, New Guinea
322. Haliæetus ...	2	The whole region	Cosmopolite; excl. Neotropical region
323. Haliastur ...	1	India to Malaya	Austro-Malaya, Austral.
324. Milvus	3	The whole region	The Eastern Hemisphere.
325. Elanus	2	India, Malaya	Africa, Australia
326. Machærhamphus	1	Malacca	S. W. Africa & Madag.
327. Pernis	1	India	Palæarctic and Ethiopian, Celebes
328. Baza	3	India to Malaya	Moluccas and N. Austrl.
329. *Hierax*	4	N. India, Burmah, Malaya	
330. Poliohierax ...	1	Burmah	E. Africa
331. Falco ..	8	The whole region	Almost Cosmopolite
332. Cerchneis ...	3	The whole region	Almost Cosmopolite
PANDIONIDÆ.			
333. Pandion... ...	1	The whole region	Cosmopolite
334. Polioaëtus ...	2	India to Malaya	Indo-Malaya & Polynesia
STRIGIDÆ.			
335. Athene	9	The whole region	The Eastern Hemisphere
336. *Ninox*	7	The whole region	N. China and Japan
337. Bubo	4	India, Ceylon, Malaya and Philip.	Cosmop. exc. Austr. reg.
338. *Ketupa*	3	The whole region	
339. Scops	7	The whole region	Almost Cosmopolite

C C

Order, Family. and Genus.	No. of Species.	Range within the Region.	Range beyond the Region
340. Syrnium ...	6	The whole region	Cosmop. exc. Austr. re
(Asio	2	India)	Palæarc., Ethiop. Amer
341. Strix	4	The whole region	Cosmopolite
342. *Phodilus* ...	2	Nepal, Malaya	

Peculiar or very Characteristic Genera of Wading or Swimming Birds.

GRALLÆ.

RALLIDÆ.

Rallina	10	The whole region	Austro-Malaya

PARRIDÆ.

Hydrophasianus	1	The whole region	

CHARADRIIDÆ.

Æsacus	1	The whole region	Austro-Malayan, Austra

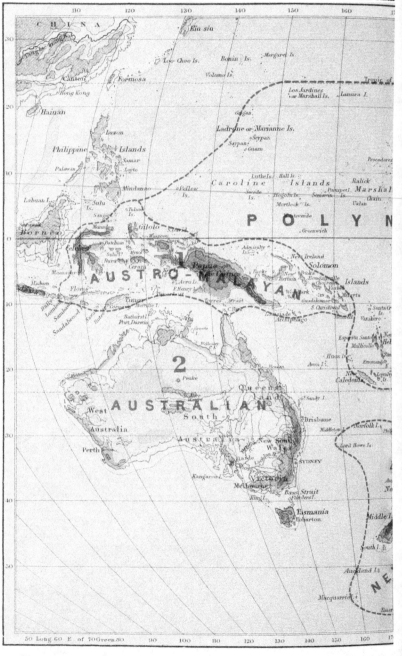

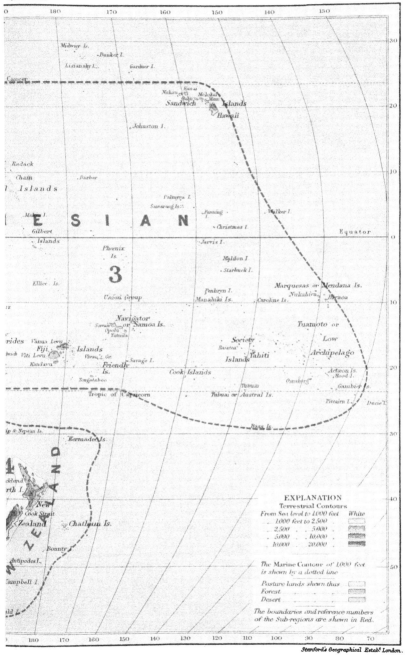

Stanford's Geographical Estab.t London.

CHAPTER XIII.

The Australian is the great insular region of the earth. As a whole it is one of the best marked, and has even been considered to be equal in zoological value to all the rest of the globe; but its separate portions are very heterogeneous, and their limits sometimes ill-defined. Its central and most important masses consist of Australia and New Guinea, in which the main features of the region are fully developed. To the north-west it extends to Celebes, in which a large proportion of the Australian characters have disappeared, while Oriental types are mingled with them to such an extent that it is rather difficult to determine where to locate it. To the south-east it includes New Zealand, which is in some respects so peculiar, that it has even been proposed to constitute it a distinct region. On the east it embraces the whole of Oceania to the Marquesas and Sandwich Islands, whose very scanty and often peculiar fauna, must be affiliated to the general Australian type.

Australia is the largest tract of land in the region, being several times more extensive than all the other islands combined, and it is here that the greatest variety of peculiar types have been developed. This island-continent, being situated in the track of the southern desert zone, and having no central mountains to condense the vapours from the surrounding ocean, has a large portion of its interior so parched up and barren as to be almost destitute of animal life. The most extensive tract of fertile and well-watered country is on the east and south east, where a fine range of mountains reaches, in the Colony of Victoria, the limits of perpetual snow. The west coast also possesses mountains of moderate height, but the climate is very dry and hot. The northern portion is entirely tropical, yet it nowhere presents the luxuriance of vegetation characteristic of the great island of New Guinea immediately to the north of it. Taken as a whole, Australia is characterized by an arid climate and a deficiency of water; conditions which have probably long prevailed, and under which its very peculiar fauna and flora have been developed. This fact will account for some of the marked differences between it and the adjacent sub-regions of New Guinea and the Moluccas, where the climate is moist, and the vegetation luxuriant; and these divergent features must never be lost sight of, in comparing the different portions of the Australian region. In Tasmania alone, which is however, essentially a detached portion of Australia, a more uniform and moister climate prevails; but it is too small a tract of land, and has been too recently severed from its parent mass to have developed a special fauna.

The Austro-Malay sub-region (of which New Guinea is the central and typical mass) is strikingly contrasted with Australia, being subjected to purely equatorial conditions,—a high, but uniform temperature, excessive moisture, and a luxuriant forest vegetation, exactly similar in general features to that which clothes the Indo-Malay Islands, and the other portions of the great equatorial forest zone. Such a climate and vegetation, being the necessary result of its geographical position, must have existed from remote geological epochs with but little change, and must therefore have profoundly affected all the forms of life which have been developed under their influence. Around New Guinea as a centre are grouped a number of important islands, more or less closely agreeing with it in physical features, climate, vegetation, and forms of life. In most immediate connection we place the Aru Islands, Mysol and Waigiou, with Jobie and the other Islands in Geelvinck Bay, all of which are connected with it by shallow seas ; they possess one of its most characteristic groups the Birds of Paradise and have no doubt only recently (in a geological sense) been separated from it. In the next rank come the large islands of the Moluccas on the west, and the range terminating in the Solomon Islands on the east, both of which groups possess a clearly Papuan fauna, although deficient in many of the most remarkable Papuan types.

All these islands agree closely with New Guinea itself in being very mountainous, and covered with a luxuriant forest vegetation; but to the south-west we find a set of islands extending from Timor to Lombock, which agree more nearly with Australia, both in climate and vegetation; being arid and abounding in eucalypti, acacias, and thickets of thorny shrubs. These, like the Moluccas, are surrounded by deep sea, and it is doubtful whether they have either of them been actually connected with New Guinea or Australia in recent geological times; but the general features of their zoology oblige us to unite all these islands with New Guinea as forming the Austro-Malay sub-division of the Australian region. Still further west however, we have the large island of Celebes, whose position is very difficult to determine. It is mountainous, but has also extensive plains and low lands. Its climate is somewhat arid in the south, where the woods are often scattered and thorny, while in the north it is moister, and the forests are luxuriant. It is surrounded by deep seas, but also by coralline and volcanic islets, indicating former elevations and subsidences. Its fauna presents the most puzzling relations, showing affinities to Java, to the Philippines, to the Moluccas, to New Guinea, to continental India, and even to Africa; so that it is almost impossible to decide whether to place it in the Oriental or the Australian region. On the whole the preponderance of its relations appears to be with the latter, though it is undoubtedly very anomalous, and may, with almost as much propriety, be classed with the former. This will be better understood when we come to discuss its zoological peculiarities.

The next sub-region consists of the extensive series of islands scattered over the Pacific, the principal groups being the Sandwich Islands, the Marquesas and Society Islands, the Navigators', Friendly, and Fiji Islands. New Caledonia and the New

Hebrides have rather an uncertain position, and it. is difficult to decide whether to class them with the Austro-Malay Islands, the Pacific Islands, or Australia. The islands of the west Pacific, north of the equator, also probably come into this region, although the Ladrone Islands may belong to the Philippines; but as the fauna of all these small islets is very scanty, and very little known, they are not at present of much importance.

There remains the islands of New Zealand, with the surrounding small islands, as far as the Auckland, Chatham, and Norfolk Islands. These are situated in the south temperate forest-zone. They are mountainous, and have a moist, equable, and temperate climate. They are true oceanic islands, and the total absence of mammalia intimates that they have not been connected with Australia or any other continent in recent geological times. The general character of their zoology, no less than their botany, affiliates them however, to Australia as portions of the same zoological region.

General Zoological Characteristics of the Australian Region.— For the purpose of giving an idea of the very peculiar and striking features which characterise the Australian region, it will be as well at first to confine ourselves to the great central land masses of Australia and New Guinea, where those features are manifested in their greatest force and purity, leaving the various peculiarities and anomalies of the outlying islands to be dealt with subsequently.

Mammalia.—The Australian region is broadly distinguished from all the rest of the globe by the entire absence of all the orders of non-aquatic mammalia that abound in the Old World, except two—the winged bats (Chiroptera), and the equally cosmopolite rodents (Rodentia). Of these latter however, only one family is represented—the Muridæ—(comprising the rats and mice), and the Australian representatives of these are all of small or moderate size—a suggestive fact in appreciating the true character of the Australian fauna. In place of the Quadrumana, Carnivora, and Ungulates, which abound in endless variety in all the other regions under equally favourable conditions, Australia possesses two new orders (or perhaps sub-classes)—

Marsupialia and Monotremata, found nowhere else on the globe except a single family of the former in America. The Marsupials are wonderfully developed in Australia, where they exist in the most diversified forms, adapted to different modes of life. Some are carnivorous, some herbivorous; some arboreal, others terrestrial. There are insect-eaters, root-gnawers, fruit-eaters, honey-eaters, leaf or grass-feeders. Some resemble wolves, others marmots, weasels, squirrels, flying squirrels, dormice or jerboas. They are classed in six distinct families, comprising about thirty genera, and subserve most of the purposes in the economy of nature, fulfilled in other parts of the world by very different groups; yet they all possess common peculiarities of structure and habits which show that they are members of one stock, and have no real affinity with the Old-World forms which they often outwardly resemble.

The other order, Monotremata, is only represented by two rare and very remarkable forms, *Ornithorhynchus* and *Echidna*, probably the descendants of some of those earlier developments of mammalian life which in every other part of the globe have long been extinct.

The bats of Australia all belong to Old-World genera and possess no features of special interest, a result of the wandering habits of these aerial mammals. The Rodents are more interesting. They are all more or less modified forms of mice or rats. Some belong to the widely distributed genus *Mus*, others to four allied genera, which may be all modifications of some common Old-World form. They spread all over Australia, and allied species occur in Celebes, so that although not yet known from New Guinea or the Moluccas, there can be little doubt that some of them exist there.

Birds.—The typical Australian region, as above defined, is almost as well characterized by its birds, as by its mammalia; but in this case the deficiencies are less conspicuous, while the peculiar and characteristic families are numerous and important. The most marked deficiency as regards wide-spread families, is the total absence of Fringillidæ (true finches), Picidæ (woodpeckers), Vulturidæ (vultures), and Phasianidæ (pheasants).

and among prevalent Oriental groups, Pycnonotidæ (bulbuls), Phyllornithidæ (green bulbuls), and Megalæmidæ (barbets) are families whose absence is significant. Nine families are peculiar to the region, or only just pass its limits in the case of single species. These are Paridiseidæ (paradise-birds), Meliphagidæ (honey-suckers), Menuridæ (lyre-birds), Atrichidæ (scrub-birds), Cacatuidæ (cockatoos), Platycercidæ (broad-tailed and grass-paroquets), Trichoglossidæ (brush-tongued paroquets, Megapodiidæ (mound-makers), and Casuariidæ (cassowaries). There are also eight very characteristic families, of which four,—Pachycephalidæ (thick-headed shrikes), Campephagidæ (caterpillar shrikes), Dicæidæ (flower-peckers), and Artamidæ (swallow-shrikes)—are feebly represented elsewhere, while the other four —Ploceidæ (weaver-finches), Alcædinidæ (kingfishers), Podargidæ (frog-mouths), and Columbidæ (pigeons)—although widely distributed, are here unusually abundant and varied, and (except in the case of the Ploceidæ) better represented in the Australian than in any other region. Of all these the Meliphagidæ (honey-suckers) are the most peculiarly and characteristically Australian. This family abounds in genera and species; it extends into every part of the region from Celebes and Lombock on the west, to the Sandwich Islands, Marquesas, and New Zealand on the east, while not a single species overpasses its limits, with the exception of one (*Ptilotis limbata*) which abounds in all the islands of the Timorese group, and has crossed the narrow strait from Lombock to Baly; but this can hardly be considered to impugn the otherwise striking fact of wide diffusion combined with strict limitation, which characterizes it. This family is the more important, because, like the Trichoglossidæ or brush-tongued paroquets, it seems to have been developed in co-ordination with that wealth of nectariferous flowering shrubs and trees which is one of the marked features of Australian vegetation. It probably originated in the extensive land-area of Australia itself, and thence spread into all the tributary islands, where it has become variously modified, yet always in such close adaptation to the other great features of the Australian fauna, that it seems unable to maintain itself when subject to the competition of the more varied forms of life in the Oriental region ; to which, possessing great powers of flight, some species must occasionally have emigrated. Its presence or absence serves therefore to define and limit the Australian region with a precision hardly to be equalled in the case of any other region or any other family of birds.

The Trichoglossidæ, as already intimated, are another of these peculiarly organized Australian families,—parrots with an extensile brush-tipped tongue, adapted to extract the nectar and pollen from flowers. These are also rigidly confined to this region, but they do not range so completely over the whole of it, being absent from New Zealand (where however they are represented by a closely allied form *Nestor*), and from the Sandwich Islands. The Paradiseidæ (birds of paradise and allies) are another remarkable family, confined to the Papuan group of Islands, and the tropical parts of Australia. The Megapodiidæ (or mound-builders) are another most remarkable and anomalous group of birds, no doubt specially adapted to Australian conditions of existence. Their peculiarity consists in their laying enormous eggs (at considerable intervals of time) and burying them either in the loose hot sand of the beach above high-water mark, or in enormous mounds of leaves, sticks, earth, and refuse of all kinds, gathered together by the birds, whose feet and claws are enlarged and strengthened for the work. The warmth of this slightly fermenting mass hatches the eggs ; when the young birds work their way out, and thenceforth take care of themselves, as they are able to run quickly, and even to fly short distances, as soon as they are hatched. This may perhaps be an adaptation to the peculiar condition of so large a portion of Australia, in respect to prolonged droughts and scanty water-supply, entailing a periodical scarcity of all kinds of food. In such a country the confinement of the parents to one spot during the long period of incubation would often lead to starvation, and the consequent death of the offspring. But the same birds with free power to roam about, might readily maintain themselves. This peculiar constitution and habit, which enabled the Megapodii to maintain an existence under the unfavourable conditions of their original habitat gives them a great advantage in the luxuriant islands of the Moluccas, to which they have spread. There they abound to a remarkable extent, and their eggs furnish a luxurious repast to the natives. They have also reached many of the smallest islets, and have spread beyond the limits of the region to the Philippines, and North-Western Borneo, as well as to the remote Nicobar Islands.

The Platycercidæ, or broad-tailed paroquets, are another wide-spread Australian group, of weak structure but gorgeously coloured, ranging from the Moluccas to New Zealand and the Society Islands, and very characteristic of the region, to which they are strictly confined. The Cockatoos have not quite so wide a range, being confined to the Austro-Malayan and Australian sub-regions, while one species extends into the Philippine Islands. The other two peculiar families are more restricted in their range, and will be noticed under the sub-regions to which they respectively belong.

Of the characteristic families, the Pachycephalidæ, or thick-headed shrikes, are especially Australian, ranging over all the region, except New Zealand; while only a single species has spread into the Oriental, and one of doubtful affinity to the Ethiopian region. The Artamidæ, or swallow-shrikes, are also almost wholly confined to the region, one species only extending to India. They range to the Fiji Islands on the east, but only to Tasmania on the south. These two families must be considered as really peculiar to Australia. The Podargidæ, or frog-mouths—large, thick-billed goat-suckers—are strange birds very characteristic of the Australian region, although they have representatives in the Oriental and Neotropical regions. Campephagidæ (caterpillar-shrikes) also abound, but they are fairly represented both in India and Africa. The Ploceidæ, or weaver-birds, are the finches of Australia, and present a variety of interesting and beautiful forms.

We now come to the kingfishers, a cosmopolitan family of birds, yet so largely developed in the Australian region as to deserve special notice. Two-thirds of all the genera are found here, and no less than 10 out of the 19 genera in the family are peculiar to the Australian region. Another of the universally distributed families which have their metropolis here, is that of the Columbidæ or pigeons. Three-fourths of the genera have representatives in the Australian region, while two-fifths of the whole are confined to it; and it possesses as many species of pigeons as any other two regions combined. It also possesses the most remarkable forms, as exemplified in the great crowned pigeons (*Goura*) and the hook-billed *Didunculus*, while the green fruit-pigeons (*Ptilopus*) are sometimes adorned with colours vying with those of the gayest parrots or chatterers. This enormous development of a family of birds so defenceless as the pigeons, whose rude nests expose their eggs and helpless young to continual danger, may perhaps be correlated, as I have suggested elsewhere (Ibis, 1865, p. 366), with the entire absence of monkeys, cats, lemurs, weasels, civets and other arboreal mammals, which prey on eggs and young birds. The very prevalent green colour of the upper part of their plumage, may be due to the need of concealment from their only enemies,—birds of prey; and this is rendered more probable by the fact that it is among the pigeons of the small islands of the Pacific (where hawks and their allies are exceedingly scarce) that we alone meet with species whose entire plumage is a rich and conspicuous yellow. Where the need of concealment is least, the brilliancy of colour has attained its maximum. We may therefore look upon the genus *Ptilopus*, with its fifty species whose typical coloration is green, with patches of bright blue, red, or yellow on the head and breast, as a special development suited to the tropical portion of the Australian region, to which it is almost wholly confined.

It will be seen from the sketch just given, that the ornithological features of the Australian region are almost as remarkable as those presented by its Mammalian fauna; and from the fuller development attained by the aërial class of birds, much more varied and interesting. None of the other regions of the earth can offer us so many families with special points of interest in structure, or habits, or general relations. The paradise-birds, the honeysuckers, the brush-tongued paroquets, the mound-builders, and the cassowaries—all strictly peculiar to the region—with such remarkable developments as we have indicated in the kingfishers and pigeons, place the Australian region in the first rank for the variety, singularity, and interest of its birds, and only second to South America as regards numbers and beauty.

Reptiles.—In Reptiles the peculiarity of the main Australian region is less marked, although the fauna is sufficiently distinct. There is no family of snakes confined to the region, but many peculiar genera of the families Pythonidæ and Elapidæ. About two-thirds of the Australian snakes belong to the latter family, and are poisonous; so that although the Crotalidæ and Viperidæ are absent, there are perhaps a larger proportion of poisonous to harmless snakes than in any other part of the world. According to Mr. Gerard Krefft the proportion varies considerably in the different colonies. In Victoria, New South Wales, and Queensland the proportion is about two to one; in West Australia three to one; and in South Australia six to one. In Tasmania there are only 3 species and all are poisonous. The number of species, as in other parts of the world, seems to increase with temperature. The 3 in Tasmania have increased to 12 in Victoria, 15 in South Australia and the same in West Australia; 31 in New South Wales, and 42 in sub-tropical Queensland.

The lizards of Australia have lately been catalogued by Dr. Günther in the concluding part of the "Voyage of the Erebus and Terror," issued in 1875. They belong to 8 families, 3 of which are peculiar; 57 genera of which 36 are peculiar; and about 140 species, all but 2 or 3 of which are peculiar. The scinks and geckoes form the great bulk of the Australian lizards, with a few Agamidæ, Gymnopthalmidæ, and Varanidæ. The three peculiar families are the Pygopodidæ, Aprasiidæ and Lialidæ; comprising only 4 genera and 7 species. The above all belong to Australia proper. Those of the other sub-regions are few in number and will be noticed under their respective localities. They will perhaps bring up the number of genera to 70. West and South Australia seem to offer much peculiarity in their lizards; these districts possessing 12 peculiar genera, while a much smaller number are confined to the East and South-East, or to the North.

Among the fresh-water turtles of the family Chelydidæ there are three peculiar genera—*Chelodina, Chelemys,* and *Elseya,* all from Australia.

Amphibia.—No tailed amphibians are known from the whole region, but no less than eleven of the families of tail-less Batrachians (toads and frogs) are known to inhabit some part or other of it. A peculiar family (Xenorhinidæ), consisting of a single species, is found in New Guinea; the true toads (Bufonidæ) are only represented by a single species of a peculiar genus in Australia, and by a *Bufo* in Celebes. Nine of the families are represented in Australia itself, and the following genera are peculiar to it:—*Pseudophryne* (Phryniscidæ), *Pachybatrachus,* and *Chelydobatrachus* (Engystomydæ); *Helioporus* (Alytidæ); *Pelodyras* and *Chirodyras* (Pelodryadæ); *Notaden* (Bufonidæ).

Fresh-water Fish.—There is only one peculiar family of freshwater fishes in this region—the Gadopsidæ—represented by a single genus and species. The other species of Australia belong to the families Trachinidæ, Atherinidæ, Mugillidæ, Siluridæ, Homalopteræ, Haplochitonidæ, Galaxidæ, Osteoglossidæ, Symbranchidæ, and Sirenoidei; most of the genera being peculiar. The large and widely-distributed families, Cyprinodontidæ and Cyprinidæ, are absent. The most remarkable fish is the recently discovered *Ceratodus,* allied to the *Lepidosiren* of Tropical America, and *Protopterus* of Tropical Africa, the three species constituting the Sub-class Dipnoi, remains of which have been found fossil in the Triassic formation.

Summary of Australian Vertebrata.—In order to complete our general sketch of Australian zoology, and to afford materials for comparison with other regions, we will here summarize the distribution of Vertebrata in the entire Australian region, as given in detail in the tables at the end of this chapter. When an undoubted Oriental family or genus extends to Celebes only we do not count it as belonging to the Australian region, that island being so very anomalous and intermediate in character.

The Australian region, then, possesses examples of 18 families of Mammalia, 8 of which are peculiar ; 71 of Birds, 16 being peculiar; 31 of Reptiles, 4 being peculiar; 11 of Amphibia, with 1 peculiar ; and 11 of Fresh-water fish, with 1 peculiar. In all, 142 families of Vertebrates, 30 of which are almost or quite confined to it, or between one-fourth and one-fifth of the whole number.

The genera of Mammalia occurring within the limits of this region are 70, of which 45 are almost, or quite, confined to it.

Of Land-Birds there are 296 genera, 196 of which are equally limited. The proportion is in both cases very nearly five-eighths.

This shows a considerable deficiency both in families of Vertebrates and genera of Mammalia, as compared with the Oriental and Ethiopian regions ; while in genera of Birds it is a little superior to the latter in total numbers, and considerably so in the proportion of peculiar types.

Supposed Land Connection between Australia and South America.

We may now consider how far the different classes and orders of vertebrates afford indications that during past ages there has been some closer connection between Australia and South America than that which now exists.

Among Mammalia we have the remarkable fact of a group of marsupials inhabiting South America, and extending even into the temperate regions of North America, while they are found in no other part of the globe beyond the limits of the Australian region ; and this has often been held to be evidence of a former connection between the two countries. A preliminary objection to this view is, that the opossums seem to be rather a tropical group, only one species reaching as far as 42° south latitude on the west coast of South America ; but whatever evidence we have which seems to require a former union of these countries shows that it took place, if at all, towards their cold southern limits, the tropical faunas on the whole showing no similarity. This is not a very strong objection, since climates may have changed in the south to as great an extent as we know they have in the north. Perhaps a more important consideration is, that *Didelphys* is a family type unknown in Australia; and this implies that the point of common origin is very remote in geological time. But the most conclusive fact is that in the Eocene and Miocene periods this very family, Didelphyidæ, existed in Europe, while it only appeared in America in the Post-pliocene or perhaps the Pliocene period; so that it is really an Old-World group, which, though long since extinct in its birthplace, has survived in America, to which country it is a comparatively recent emigrant. Primeval forms of marsupials we know abounded in Europe during much of the Secondary epoch, and no doubt supplied Australia with the ancestors of the present fauna. It is clear, therefore, that in this case there is not a particle of evidence for any former union between Australia and South America; while it is almost demonstrated that both derived their marsupials from a common source in the northern hemisphere.

Birds offer us more numerous but less clearly defined cases of this kind. Among Passeres, the wonderful lyre bird (*Menura*) is believed by some ornithologists to be decidedly allied to the South American Pteroptochidæ, while others maintain that it is altogether peculiar, and has no such affinity. The Australian Pachycephalidæ have also been supposed to find their nearest allies in the American Vireonidæ, but this is, perhaps, equally problematical. That the mound-makers (Megapodiidæ) of the Australian region are more nearly allied to the South American curassows (Cracidæ) than to any other family, is perhaps better established; but if proved, it is probably due, as in the case of the marsupials, to the survival of an ancient and once wide-spread type, and thus lends no support to the theory of a land connection between the two regions. A recent author, Professor Garrod, classes *Phaps* and other Australian genera of pigeons along with *Zenaida* and allied South American forms; but here again the affinity, if it exists, is so remote that the explanation already given will suffice to account for it. There remain only the penguins of the genus *Eudyptes;* and these have almost certainly passed from one region to the other, but no actual land connection is required for birds which can cross considerable arms of the sea.

Reptiles again seem to offer no more support to the view than do mammalia or birds. Among snakes there are no families in common that have not a very wide distribution. Among lizards the Gymnopthalmidæ are the only family that favour the notion, since they are found in Australia and South America, but not in the Oriental region. Yet they occur in both the Palæarctic and Ethiopian regions, and their distribution is altogether too erratic to be of any value in a case of this kind; and the same remarks apply to the tortoises of the family Chelydidæ.

The Amphibia, however, furnish us with some more decided facts. We have first the family of tree-frogs, Pelodryade, confined to the two regions; *Litoria*, a genus of the family Hylidæ peculiar to Australia, but with one species in Paraguay ; and in the family Discoglossidæ, the Australian genus *Chiroleptes* has its nearest ally in the Chilian genus *Calyptocephalus*.

Fresh-water fishes give yet clearer evidence. Three groups are exclusively found in these two regions; *Aphritis*, a fresh-water genus of Trachinidæ, has one species in Tasmania and two others in Patagonia; the Haplochitonidæ inhabit only Terra del Fuego, the Falkland Islands and South Australia; while the genus *Galaxias* (forming the family Galaxidæ) is confined to South Temperate America, Australia, and New Zealand. We have also the genus *Osteoglossum* confined to the tropical rivers of Eastern South America, the Indo-Malay Islands and Australia.

It is important here to notice that the heat-loving Reptilia afford hardly any indications of close affinity between the two regions, while the cold-enduring amphibia and fresh-water fish, offer them in abundance. Taking this fact in connection with the absence of all indications of close affinity among the mammalia and terrestrial birds, the conclusion seems inevitable that there has been no land-connection between the two regions within the period of existing species, genera, or families. Yet some interchange of amphibia and fresh-water fishes, as of plants and insects, has undoubtedly occurred, but this has been effected by other means. If we look at a globe we see at once how this interchange may have taken place. Immediately south of Cape Horn we have the South Shetland Islands and Graham's land, which is not improbably continuous, or nearly so, with South Victoria land immediately to the south of New Zealand. The intervening space is partly occupied by the Auckland, Campbell, and Macquaries' Islands, which, there is reason to believe are the relics of a great southern extension of New Zealand. At all events they form points which would aid the transmission of many organisms; and the farthest of the Macquaries' group, Emerald Island, is only 600 miles from the outlying islets of Victoria land. The ova of fish will survive a considerable time in the air, and the successful transmission of salmon ova to New Zealand packed in ice, shows how far they might travel on icebergs. Now there is evidently some means by which ova or young fishes are carried moderate distances, from the fact that remote alpine lakes and distinct river systems often have the same species. Glaciers and icebergs generally have pools of fresh water on their surfaces; and whatever cause transmits fish to an isolated pond might occasionally stock these pools, and by this means introduce the fishes of one southern island into another. Batrachians, which are equally patient of cold, might be transported by similar means; while, as Mr. Darwin has so well shown, (*Origin of Species*, 6th Ed. p. 345) there are various known modes by which plants might be transmitted, and we need not therefore be surprised that botanists find a much greater similarity between the production of the several Southern lands and islands, than do zoologists. It is important to notice that, however this intercommunication was effected, it has continued down to the epoch of existing species; for Dr. Günther finds the same species of fresh-water fish (*Galaxias attenuatus*) inhabiting Tasmania, New Zealand, the Falkland Islands, and Temperate South America; while another species is common to New Zealand and the Auckland Islands. We cannot believe that a land connection has existed between all these remote lands within the period of existence of this one species of fish,

D D

not only on account of what we know of the permanence of
continents and deep oceans, but because such a connection must
have led to much more numerous and important cases of similarity of natural productions than we actually find. And if
within the life of *species* such interchange may have taken
place across seas of greater or less extent, still more easy is it
to understand, how, within the life of *genera* and *families*, a number of such interchanges may have occurred ; yet always limited
to those groups whose conditions of life render transmission
possible. Had an actual land connection existed within the
temperate zone, or during a period of warmth in the Antarctic
regions, there would have been no such strict limitations to the
inter-migration of animals. It may be held to support the view
that floating ice has had *some* share in the transmission of fish
and amphibia, when we find that in the case of the narrow
tropical sea dividing Borneo from Celebes and the Moluccas, no
proportionate amount of transmission has taken place, but
numerous species, genera, and whole families, terminate abruptly
at what we have other reasons for believing to be the furthest
limits of an ancient continent. We can hardly suppose, however, that this mode of transmission would have sufficed for
such groups as tree-frogs, which are inhabitants of the more
temperate or even warm portions of the two southern lands.
Some of these cases may perhaps be explained by the supposition of a considerable extent of land in the South-Temperate and
Antarctic regions now submerged, and by a warm or temperate
climate analogous to that which prevailed in the Arctic regions
during some part of the Miocene epoch; while others may be
due to cases of survival in the two areas of once wide-spread
groups, a view supported in the case of the Amphibia by the
erratic manner in which many of the groups are spread over
the globe.
   From an examination of the facts presented by the various classes of vertebrates, we are, then, led to the conclusion,
that there is no evidence of a former land-connection between the Australian and Neotropical regions; but that the
various scattered resemblances in their natural productions

that undoubtedly occur, are probably due to three distinct causes.

First, we have the American Didelphyidæ, among Mammals, and the Cracidæ, among birds, allied respectively to the Marsupials and the Megapodiidæ of Australia. This is probably more a coincidence than an affinity, due to the preservation of ancient wide-spread types in two remote areas, each cut off from the great northern continental masses, in which higher forms were evolved leading to the extinction of the lower types. In each of these southern isolated lands the original type would undergo a special development; in the one case suited to an arboreal existence, in the other to a life among arid plains.

The second case is that of the tree-frogs, and the genus *Ostroglossum* among fishes; and is most likely due to the extension and approximation of the two southern continents, and the existence of some intermediate lands, during a warm period when facilities would be afforded for the transmission of a few organisms by the causes which have led to the exceptional diffusion of fresh-water productions in all parts of the world. As however *Osteoglossum* occurs also in the Sunda Islands, this may be a case of survival of a once wide-spread group.

The third case is that of the same genera and even species of fish, and perhaps of frogs, in the two countries; which may be due to transmission from island to island by the aid of floating ice, with or without the assistance of more intervening lands than now exist.

Having arrived at these conclusions from a consideration of the vertebrata, we shall be in a position to examine how far the same causes will explain, or agree with, the distribution of the invertebrate groups, or elucidate any special difficulties we may meet with in the relations of the sub-regions.

Insects.

The insects of the Australian region are as varied, and in some respects as peculiar as its higher forms of life. As we have already indicated in our sketch of the Oriental region, a vast number of forms inhabit the Austro-Malay sub-region which are absent from Australia proper. Such of these as are common to the Malay archipelago as a whole, have been already noted; we shall here confine ourselves more especially to the groups peculiar to the region, which are almost all either Australian or Austro-Malayan, the Pacific Islands and New Zealand being very poor in insect life.

Lepidoptera.—Australia itself is poor in butterflies, except in its northern and more tropical parts, where green *Ornithopterœ* and several other Malayan forms occur. In South Australia there are less than thirty-five species, whereas in Queensland there are probably over a hundred. The peculiar Australian forms are few. In the family Satyridæ, *Xenica* and *Heteronympha*, with *Hypocista* extending to New Guinea; among the Lycæ-nidæ, *Ogyris* and *Utica* are confined to Australia proper, and *Hypochrysops* to the region; and in Papilionidæ, the remark-able *Eurycus* is confined to Australia, but is allied to *Euryades*, a genus found in Temperate South America (La Plata), and to the *Parnassius* of the North-Temperate zone.

The Austro-Malay sub-region has more peculiar forms. *Hama-dryas*, a genus of Danaidæ, approximates to some South American forms; *Hyades* and *Hyantis* are remarkable groups of Morphidæ; *Mynes* and *Prothoë* are fine Nymphalidæ, the former extending to Queensland; *Dicallaneura*, a genus of Erycinidæ, and *Elodina*, of Pieridæ, are also peculiar forms. The fine *Ægeus* group of *Papilio*, and *Priamus* group of *Ornithoptera*, also belong exclu-sively to this region.

Xois is confined to the Fiji Islands, *Bletogona* to Celebes, and *Acropthalmia* to New Zealand, all genera of Satyridæ. Seven-teen genera in all are confined to the Australian region.

Among the Sphingina, *Pollanisus*, a genus of Zygænidæ, is Australian; also four genera of Castniidæ—*Synemon*, *Euschemon*, *Damias*, and *Cocytia*, the latter being confined to the Papuan islands. The occurrence of this otherwise purely South American family in the Australian region, as well as the affinity of *Eurycus* and *Euryades* noticed above, is interesting; but as we have seen that the genera and families of insects are more permanent than those of the higher animals, and as the groups in question are confined to the warmer parts of both countries, they may be best explained as cases of survival of a once wide-spread type, and may probably date back to the period when the ancestors of the Marsupials and Megapodii were cut off from the rest of the world.

Coleoptera.—The same remark applies here as in the Lepidoptera, respecting the affinity of the Austro-Malay fauna to that of Indo-Malay Islands; but Australia proper is much richer in beetles than in butterflies, and exhibits much more speciality. Although the other two parts of the Australian region (Polynesia and New Zealand) are very poor in beetles, it will, nevertheless, on the whole compare favourably with any of the regions except the very richest.

Cicindelidæ are not very abundant. *Therates* and *Tricondyla* are the characteristic genera in Austro-Malaya, but are absent from Australia, where we have *Tetracha* as the most characteristic genus, with one species of *Megacephala* and two of *Distypsidera*, a genus which is found also in New Zealand and some of the Pacific Islands. The occurrence of the South American genus, *Tetracha*, may perhaps be due to a direct transfer by means of intervening lands during the warm southern period; but considering the permanence of coleopterous forms (as shown by the Miocene species belonging almost wholly to existing genera), it seems more probable that it is a case of the survival of a once wide-spread group.

Carabidæ are well represented, there being no less than 94 peculiar genera, of which 19 are confined to New Zealand. The Australian genera of most importance are *Carenum* (68 species), *Promecoderus* (27 species), *Silphomorpha* (32 species), *Adelotopus* (27 species), *Scaraphites* (25 species), *Notonomus* (18 species), *Gnathoxys* (12 species), *Eutoma* (9 species), *Ænigma* (15 species), *Lacordairea* (8 species), *Pamborus* (8 species), *Catadromus* (4 species),—the latter found in Australia and Celebes. Common to Australia and New Zealand are *Mecodema* (14 species), *Homalosoma* (32 species), *Dicrochile* (12 species), and *Scopodes* (5 species). The larger genera, confined to New Zealand only, are *Metaglymma* (8 species), and *Demetrida* (3 species). The curious genus *Pseudomorpha* (10 species), is divided between California, Brazil, and Australia; and the Australian genera, *Adelotopus*, *Silpho-morpha*, and *Sphallomorpha*, form with it a distinct tribe of Cole-optera. These being all confined to the warmer regions, and having so scattered a distribution, are no doubt the relics of a wide-spread group. The Australian genus, *Promecoderus*, has, how-ever, closely allied genera (*Cascelius* and its allies), in Chili and Patagonia; while two small genera confined to the Auckland Islands (*Heterodactylus* and *Pristancyclus*) are allied to a group found only in Terra-del-Fuego and the Falkland Islands, (*Migadops*); and in these cases we may well believe that a direct transmission has taken place by some of the various means already indicated.

In Lucanidæ, Australia is only moderately rich, having 7 peculiar genera. The most important are *Ceratognathus* and *Rhys-sonotus*, confined to Australia; *Lissotes* to Australia and New Zealand; *Lamprima* to Australia and Papua. *Mitophyllus* and *Dendroblax* inhabit New Zealand only; while *Syndesus* is found in Australia, New Caledonia, and tropical South America.

The beautiful Cetoniidæ are poorly represented, there being only 3 peculiar genera;—*Schizorhina*, mainly Australian, but extending to Papua and the Moluccas; *Anacamptorhina*, con-fined to New Guinea, and *Sternoplus* to Celebes. *Lomaptera* is very characteristic of the Austro-Malay Islands. This almost tropical family shows no approximations between the Australian and Neotropical faunas.

In Buprestidæ, the Australian region is the richest, possessing no less than 47 genera, of which 20 are peculiar to it. Of these, 15 are peculiar to Australia itself, the most important being *Stig-modera* (212 species), *Ethon* (13 species), and *Nascio* (3 species); *Cisseis* (17 species), and the magnificent *Calodema* (3 species), are common to Australia and Austro-Malaya; while *Sambus* (10 species) and *Anthaxomorpha* (4 species), with some smaller groups, are peculiarly Austro-Malayan. In this family occur several points of contact with the Neotropical region. *Stigmo-dera* is said to have a species in Chili, while there are undoubt-edly several allied genera in Chili and South Temperate America. The genus *Curis* has 5 Australian and 3 Chilian species, and

Acherusia has 2 species in Brazil, 1 in Australia. These resemblances may probably have arisen from intercommunication during the warm southern period, when floating timber would occasionally transmit a few larvæ of this family from island to island across the antarctic seas. When the cold period returned, they would spread northward, and become more or less modified under the new physical conditions and organic competition, to which they were subjected.

We now come to the very important group of Longicorns, in which the Australian region as a whole, is very rich, possessing 360 genera, of which 263 are peculiar to it. Of these about 50 are confined to the Austro-Malay Islands, 12 to New Zealand, and the remainder to Australia proper with Tasmania. Of the genera confined to, or highly characteristic of Australia, the following are the most important :— *Cnemoplites*, belonging to the Prionidæ; *Phoracantha*, to the Cerambycidæ; *Zygocera, Hebecerus, Symphyletes,* and *Rhytidophora,* to the Lamiidæ. Confined to the Austro-Malay Islands are *Tethionea* (Cerambycidæ) : *Tmesisternus, Arrhenotus, Micracantha,* and *Sybra* (Lamiidæ); but there are also such Malayan genera as *Batocera Gnoma, Praonetha,* and *Sphenura,* which are very abundant in the Austro-Malay sub-region. A species of each of the Australian genera, *Zygocera, Syllitus,* and *Pseudocephalus,* is said to occur in Chili, and one of the tropical American genus, *Hammatochœrus,* in tropical Australia; an amount of resemblance which, as in the case of the Buprestidæ, may be imputed to trans-oceanic migration during the Southern warm period. This concludes our illustrations of the distribution of some of the more important groups of Australian insects ; and it will be admitted that we have not met with any such an amount of identity with the fauna of Temperate South America, as to require us to modify the conclusions we arrived at from a consideration of the vertebrate groups.

Land-Shells.—The distribution of many of the larger genera of land-shells is very erratic, while others are exceedingly restricted, so that it requires an experienced conchologist to investigate the affinities of the several groups, and thus work out the important facts of distribution. All that can be done here is to note the characteristic and peculiar genera, and any others presenting features of special interest.

In the great family of the snails (Helicidæ), the only genera strictly confined to the region are, *Partula*, now containing above 100 species, and ranging over the Pacific from the Solomon Isles on the west, to the Sandwich Islands and Tahiti on the east; and *Achatinella*, now containing nearly 300 species, and wholly confined to the Sandwich Islands. *Pfeifferia* is confined to the Philippine Islands and Moluccas; *Cochlostyla* to the Indo-Malay Islands and Australia; *Bulimus* occurs in most of the insular groups, including New Zealand, but is absent from Australia.

Among the Aciculidæ, the widely-scattered *Truncatella* is the only genus represented. Among Diplommatinidæ, *Diplommatina* is the characteristic genus, ranging over the whole region, and found elsewhere as far as India, with one species in Trinidad. The extensive family Cyclostomidæ, is not well represented. Seven genera reach the Austro-Malay Islands, one of which, *Registoma*, is confined to the Philippines, Moluccas, New Caledonia, and the Marshall Islands. *Omphalotropis* is the most characteristic genus, ranging over the whole region; *Callia* is confined to the Philippines, Ceram, and Australia; *Realia* to New Zealand and the Marquesas. The genus *Helicina* alone represents the Helicinidæ, and is found in the whole region except New Zealand. The number of species known from Australia is perhaps about 300; while the Polynesian sub-region, according to Mr. Harper Pease, contains over 600; the Austro-Malay Islands will furnish probably 200; and New Zealand about 100; making a total of about 1,200 species for the whole region.

AUSTRALIAN SUB-REGIONS.

Few of the great zoological regions comprise four divisions so strongly contrasted as these, or which present so many interesting problems. We have first the Austro-Malay Islands, an equatorial forest-region teeming with varied and beautiful forms of life; next we have Australia itself, an island-continent with its satellite

Tasmania, both tropical and temperate, but for the most part arid, yet abounding in peculiar forms in all the classes of animals; then come the Polynesian Islands, another luxuriant region of tropical vegetation, yet excessively poor in most of the higher groups of animals as well as in some of the lower; and lastly, we have New Zealand, a pair of temperate forest-clad islands far in the southern ocean, with a very limited yet strange and almost wholly peculiar fauna. We have now to consider the general features and internal relations of the faunas of each of these sub-regions, together with any external relations which have not been discussed while treating the region as a whole.

I. Austro-Malayan Sub-region.

The central mass on which almost every part of this sub-region is clearly dependent, is the great island of New Guinea, inhabited by the Papuan race of mankind; and this, with the surrounding islands, which are separated from it by shallow seas and possess its most marked zoological features, are termed Papua. A little further away lie the important groups of the Moluccas on one side and the Eastern Papuan Islands on the other, which possess a fauna mainly derivative from New Guinea, yet wanting many of its distinctive types; and, in the case of the Moluccas possessing many groups which are not Australian, but derived from the adjacent Oriental region. To the south of these we have the Timor group, whose fauna is clearly derivative, from Australia, from Java, and from the Moluccas. Lastly comes Celebes, whose fauna is most complex and puzzling, and, so far as we can judge, not fundamentally derivative from any of the surrounding islands.

Papua, or the New Guinea Group.—New Guinea is very deficient in Mammalia as compared with Australia, though this apparent poverty may, in part, depend on our very scanty knowledge. As yet only four of the Australian families of Marsupials are known to inhabit it, with nine genera, several of which are peculiar. It also possesses a peculiar form of wild pig; but as yet no other non-marsupial terrestrial mammal has been discovered, except a rat, described by Dr. Gray as *Uromys aruensis*, but about the locality of which there seems some doubt.[1] Omitting bats, of which our knowledge is very imperfect, the Papuan Mammals are as follows :—

Family.	Genus.	Species.	
Suidæ	*Sus*	1	Eastern limit of the genus.
Muridæ	*Uromys*	1	Aru Islands (?)
Dasyuridæ ...	*Phascogale*	1	Australian genus.
„ ...	*Antechinus*	1	„ „
„ ...	*Dactylopsila*	1	To North Australia only.
„ ...	*Myoictis*	1	Aru islands only.
Peramelidæ ...	*Perameles*	1	New Guinea only.
Macropodidæ ...	*Dendrolagus*	2	New Guinea only.
„ ...	*Dorcopsis*	2	Papua only.
Phalangistidæ...	*Cuscus*	7	Celebes to New Guinea.
„ ...	*Belideus*	1	Australia and Moluccas.

We have here no sign of any approach to the Mammalian fauna of the Oriental region, for though *Sus* has appeared, the Muridæ (rats and mice) seem to be wanting.

In Birds the case is very different, since we at once meet with important groups, either wholly, or almost peculiar to the Papuan fauna. According to a careful estimate, embodying the recent discoveries of Meyer and D'Albertis, there are 350 species of Papuan land-birds comprised in 136 genera. About 300 of the species are absolutely peculiar to the district, while 39 of the genera are exclusively Papuan or just extend into the Moluccas, or into North Australia where it closely approaches New Guinea. In analysing the genera we may set aside 31 as having a wide range, and being of no significance in distribution; such are most of the birds of prey, with the genera *Hirundo, Caprimulgus, Zosterops;* and others widely spread in both the Oriental and Australian regions, as *Dicæum, Munia, Eudynamis,* &c. Of the remainder, as above stated, about 39 are peculiar to the Papuan fauna, 50 are characteristic Australian genera; 9 are more especially Malayan, and as much Australian as Oriental; while 7 only, appear to be typically Oriental with a discontinuous distribution, none of them occurring in the Moluccas.

[1] See *Ann. Nat. Hist.*, 1873, p. 418, where the species is said to inhabit the Aru Islands and Celebes, which renders it not improbable that it may have been carried to the former islands from the latter.

This Papuan fauna is so interesting and remarkable, that it seems advisable to give lists of these several classes of generic types.

I. Genera occurring in the Papuan Islands which are characteristic of the Australian region (89). Those marked with an asterisk are exclusively Papuan.

Sylviidæ...	...	*Malurus, Gerygone, Petroica, Orthonyx.*
Certhiidæ	...	*Climacteris.*
Sittidæ ...	...	*Sittella.*
Oriolidæ...	...	*Mimeta.*
Campephagidæ	...	*Graucalus, Lalage.*
Dicruridæ	...	**Chætorhynchus.*
Muscicapidæ	...	**Peltops, Monarcha, *Leucophantes, Micræca, Sisura, Myiagra, *Machærirhynchus, Rhipidura, *Todopsis.*
Pachycephalidæ...		*Pachycephala.*
Laniidæ ...	...	**Rectes.*
Corvidæ ...	...	*Cracticus, *Gymnocorvus.*
Paradiseidæ	...	**Paradisea, *Manucodia, *Astrapia, *Parotia, *Lophorina, *Diphyllodes, *Xanthomelus, *Cicinnurus, *Paradigalla, *Epimachus, *Drepanornis,*Seleucides, Ptilorhis,Æluræ-dus, *Amblyornis.*
Meliphagidæ	...	*Myzomela, Entomophila, Glicyphila, Ptilotis, *Melidectes, *Melipotes, *Melirrhophetes, Anthochæra, Philemon, *Euthyrhynchus, Melithreptes.*
Nectariniidæ	...	*Chalcostetha, *Cosmetira.*
Artamidæ	...	*Artamus.*
Pittidæ ...	...	**Melampitta.*
Cuculidæ	...	**Caliechthrus.*
Alcedinidæ	...	*Alcyone, *Syma, Dacelo, *Tanysiptera, *Melidora.*
Podargidæ	...	*Podargus, Ægotheles.*
Caprimulgidæ	...	*Eurostopodus.*
Cacatuidæ	...	*Cacatua, *Microglossus, Licmetis, *Nasiterna.*
Platycercidæ	...	*Aprosmictus*
Palæornithidæ	...	*Tanygnathus, Eclectus, Geoffroyus, *Cyclopsitta.*
Trichoglossidæ	...	*Trichoglossus, *Charmosyna, Eos, Lorius.*
Nestoridæ	...	**Dasyptilus.*
Columbidæ	...	*Ptilopus, Carpophaga, Ianthænas, Reinward-tænas, *Trugon, *Henicophaps, Phlogænas, *Otidiphaps, *Goura.*
Megapodiidæ	...	*Talegallus, Megapodius.*
Falconidæ	...	**Henicopernis.*
Casuariidæ	...	*Casuarius.*

The chief points of interest here are the richness and specialization of the parrots, pigeons, and kingfishers; the wonderful paradise-birds; the honeysuckers; and some remarkable flycatchers.

The most prominent deficiencies, as compared with Australia, are in Sylviidæ, Timaliidæ, Ploceidæ, Platycercidæ, and Falconidæ.

II. The genera which are characteristic of the whole Malay Archipelago are the following (10) :—

1. *Erythrura*	...	(Ploccidæ)	6. *Loriculus*	...	(Psittacidæ)
2. *Pitta*...	...	(Pittidæ)	7. *Macropygia*	...	(Columbidæ)
3. *Ceyx* ...	...	(Alcedinidæ)	8. *Chalcophaps*	...	,,
4. *Calao*	...	(Bucerotidæ)	9. *Calœnas*	...	,,
5. *Dendrochelidon*		(Cypselidæ)	10. *Baza* ...	...	(Falconidæ)

III. The curious set of genera apparently of Indo-Malayan origin, but unknown in the Moluccas, are as follows :—

1. *Eupetes*	...	(Cinclidæ)	4. *Arachnothera*	(Nectariniidæ)
2. *Alcippe*	...	(Timaliidæ)	5. *Prionochilus*...	(Dicæidæ)
3. *Pomatorhinus*		,,	6. *Eulabes* ...	(Sturnidæ)

The above six birds are very important as indicating past changes in the Austro-Malay Islands, and we must say a few words about each.   (1) *Eupetes* is very remarkable, since the New Guinea birds resemble in all important characters that which is confined to Malacca and Sumatra. They are probably the survivors of a once wide-spread Malayan group. (2) *Alcippe* or *Drymocataphus* (for in which genus the birds should be placed is doubtful) seems another clear case of a typical Indo-Malayan form occurring in New Guinea and Java, but in no intervening island.   (3) *Pomatorhinus* is a most characteristic Himalayan and Indo-Malayan genus, occurring again in New Guinea and also in Australia, but in no intermediate island.   The New Guinea bird seems as nearly related to Oriental as Australian species.   (4) *Arachnothera* is exactly parallel to *Alcippe*, occurring nowhere east of Borneo except in New Guinea.   (5) *Prionochilus*, a small black bird, sometimes classed as a distinct genus, but evidently allied to the *Prionochili* of the Indo-Malay Islands.   (6) *Eulabes*, the genus which contains the well known Mynahs of India, extends east of Java as far as Flores, but is not found in Celebes or the Moluccas. The two New Guinea species are sometimes classed in different genera, but they are undoubtedly allied to the Mynahs of India and Malaya.

We find then, that while the ornithology of New Guinea is

preeminently Australian in character and possesses many peculiar developments of Australian types, it has also—as might be expected from its geographical position, its climate, and its vegetation—received an infusion of Malayan forms. But while one group of these is spread over the whole Archipelago, and occasionally beyond it, there is another group which presents the unusual and interesting feature of discontinuous distribution, jumping over a thousand miles of island-studded sea from Java and Borneo to New Guinea itself. It is a parallel case to that of Java in the Oriental region, which we have already discussed, but the suggested explanation in that case is more difficult to apply here. The recent soundings by the *Challenger* show us, that although the several islands of the Moluccas are surrounded by water from 1,200 to 2,800 fathoms deep, yet these seas form inclosed basins· with rims not more than from 400 to 900 fathoms deep, suggesting the idea of great lakes or inland seas which have sunk down bodily with the surrounding land, or that enormous local and restricted elevations and subsidences have here occurred. We have also the numerous small islands and coral banks south of Celebes and eastward towards Timor-Laut and the Aru Islands, indicating great subsidence; and it is possible that there was an extension of Papua to the west, approaching sufficiently near to Java to receive occasional straggling birds of Indo-Malay type, altogether independent of the Moluccas to the north.

Bright Colours and Ornamental Plumage of New Guinea Birds. —One of the most striking features of Papuan ornithology is the large proportion which the handsome and bright-coloured birds bear to the more obscure species. That this is really the case has been ascertained by going over my own collections, made at Aru and New Guinea, and comparing them with my collection made at Malacca—a district remarkable for the number of handsome birds it produces. Using, as nearly as possible, the same standard of beauty, about one-third of the Malacca birds may be classed as handsome,[1] while in Papua the proportion comes out exactly one-half. This is due, in part to the great abundance of

---

[1] I also find about this proportion in my Amazonian collections, even counting all the humming-birds, parrots, and toucans as handsome birds.

parrots, cockatoos, and lories, almost all of which are beautiful; and of pigeons, more than half of which are very beautiful; as well as to the numerous kingfishers, most of which are excessively brilliant. Then we have the absence of thrushes, and the very small numbers of the warblers, shrikes, and Timaliidæ, which are dull-coloured groups; and, lastly, the presence of numerous gay pittas, flycatchers, and the unequalled family of paradise-birds. A large number of birds adorned with metallic plumage is also a marked feature of this fauna, more than a dozen genera being so distinguished Among the remarkable forms are *Peltops*, a fly-catcher, long classed as one of the Indo-Malayan Eurylæmidæ, which it resembles both in bill and coloration; *Machærirhynchus*, curious little boat-billed flycatchers; and *Todopsis*, a group of ter-restrial flycatchers with the brilliant colours of *Pitta* or *Malurus*. The paradise-birds present the most wonderful developments of plumage and the most gorgeous varieties of colour, to be found among passerine birds. The great whiskered-swift, the handsomest bird in the entire family, has its head-quarters here. Among king-fishers the elegant long-tailed *Tanysipteræ* are preeminent, whether for singularity or beauty. Among parrots, New Guinea possesses the great black cockatoo, one of the largest and most singular birds in the order; *Nasiterna*, the smallest of known parrots; and *Charmosyna*, perhaps the most elegant. Lastly, among the pigeons we have the fine crowned-pigeons, the largest and most remarkable group of the order.

Plate X. Illustrating the Ornithology of New Guinea.—The wonderful ornithological fauna we have just sketched, could only be properly represented in a series of elaborate coloured plates. We are obliged here to confine ourselves to representing a few of the more remarkable types of form, as samples of the great number that adorn this teeming bird-land. The large central figure is the fine twelve-wired paradise-bird (*Epimachus albus*), one of the most beautiful and remarkable of the family. Its general plumage appears, at first sight, to be velvety black; but on closer examination, and by holding the bird in various lights, it is found that every part of it glows with the most ex-quisite metallic tints—rich bronze, intense violet, and, on the

PLATE X.

SCENE IN NEW GUINEA, WITH CHARACTERISTIC ANIMALS.

edges of the breast-feathers, brilliant green. An immense tuft of dense plumes of a fine orange-buff colour, springs from each side of the body, and six of these on each side terminate in a black curled rachis or shaft, which form a perfectly unique adornment to this lovely bird. To appreciate this wonderful family (of which no good mounted collection exists) the reader should examine the series of plates in Mr. Elliot's great work on the Paradiseidæ, where every species is figured of the size of life, and with a perfection of colouring that leaves little to be desired.

Below the *Epimachus* is one of the elegant racquet-tailed king-hunters (*Tanysiptera galatea*) whose plumage of vivid blue and white, and coral-red bill, combined with the long spatulate tail, renders this bird one of the most attractive of the interesting family of kingfishers. On a high branch is seated the little Papuan parroquet (*Charmosyna papuensis*), one of the Trichoglossidæ, or brush-tongued parrots,—richly adorned in red and yellow plumage, and with an unusually long and slender tail. On the ground is the well-known crowned pigeon (*Goura coronata*,) a genus which is wholly confined to New Guinea and a few of the adjacent islands. One of the very few Papuan mammals, a tree-kangaroo (*Dendrolagus inustus*), is seated on a high branch. It is interesting, as an arboreal modification of a family which in Australia is purely terrestrial; and as showing how very little alteration of form or structure is needed to adapt an animal to such a different mode of life.

Reptiles and Amphibia.—Of these classes comparatively little is at present known, but there is evidence that the same intermixture of Oriental and Australian forms that occurs in birds and insects, is also found here. Dr. A. B. Meyer, the translator of this work into German, and well known for his valuable discoveries in New Guinea, has kindly furnished me with a manuscript list of Papuan reptiles, from which most of the information I am able to give is derived.

Of Snakes, 24 genera are known, belonging to 11 families. Six of the genera are Oriental,—*Calamaria, Cerberus, Chrysopelea, Lycodon, Chersydrus*, and *Ophiophagus*. Four are Australian,

—*Morelia, Liasis, Diemenia,* and *Acanthophis;* while four others are more especially Papuan,—*Dibamus* (Typhlopidæ), *Brachyorros*—a sub-genus of the wide-spread *Rhabdosoma* (Calamariidæ), found also in Timor; *Nardoa* and *Enygrus* (Pythonidæ), ranging from the Moluccas to the Fiji Islands. The rest are either common to the Oriental and Australian regions or of wide range.

Of Lizards also, 24 genera are recorded, belonging to 5 families. Three only are peculiarly Oriental,—*Eumeces, Tiaris,* and *Nycteridium;* but another, *Gonyocephalus,* is Malayan, ranging from Java and Borneo to the Pelew Islands. Three are Australian,—*Cyclodus, Heteropus,* and *Gehyra;* while six are especially Papuan, —*Kencuxia* (extending to the Philippines), *Elania, Carlia* (to North Australia), *Lipinia* (to the Philippine Islands), and *Tribolonotus,*—all belonging to the Scincidæ; and *Arua* belonging to the Agamidæ. We must add *Cryptoblepharus,* which is confined to the Australian region, except a species in Mauritius. The other genera have a wider distribution.

The preponderant Oriental element in the snakes as compared with the lizards, is suggestive of the dispersal of the former being dependent on floating trees, or even on native canoes, which for an unknown period have traversed these seas, and in which various species of snakes often secrete themselves. This seems the more probable, as snakes are usually more restricted in their range than lizards, and exhibit less numerous examples of widespread genera and species. The other orders of reptiles present no features of interest.

Of Amphibia only 8 genera are known, belonging to 6 families. *Rana, Hylarana,* and *Hyla* are wide-spread genera, the former being, however, absent from Australia. *Hyperolius, Pelodryas, Litoria,* and *Asterophrys* are Australian; while *Platymantis* is Polynesian, with a species in the Philippine Islands. Hence it appears that the amphibia, so far as yet known, exhibit no Oriental affinity; and this is a very suggestive fact. We have seen (p. 29) that salt water is almost a complete barrier to the dispersal of these creatures; so that the wholly Australian character of the Papuan batrachia is what we might expect, if, as here advocated, no actual land connection between the Oriental and Australian regions, has probably occurred during the entire Tertiary and Post-tertiary periods.

Insects.—The general character of the Papuan insects has been sufficiently indicated in our sketch of the Entomology of the region. We will here only add, that the metallic lustre so prevalent among the birds, is also apparent in such insects as *Sphingnotus mirabilis,* a most brilliant metallic Longicorn; *Lomaptera wallacei* and *Anacamptorhina fulgida,* Cetonii of intense lustre; *Calodema wallacei* among the Buprestidæ; and the elegant blue *Eupholi* among the weevils. Even among moths we have *Cocytia durvillii,* remarkable for its brilliant metallic colours.

The Moluccas.—The islands of Gilolo, Bouru, and Ceram, with several smaller islands adjacent, together with Sanguir, and perhaps Tulour or Salibaboo to the north-west, and the islands from Ke to Timor-Laut to the south-east, form the group of the Moluccas or Spice-Islands, remarkable for the luxuriance of their vegetation and the extreme beauty of their birds and insects. Their Mammalia are of Papuan character, with some foreign intermixture. Two genera of the New Guinea marsupials, *Belideus* and *Cuscus,* abound; and we have also the widespread *Sus.* But besides these, we find no less than five genera of placental Mammals quite foreign to the Papuan or Australian faunas. These are 1. *Cynopithecus nigrescens,* found only in the small island of Batchian, and probably introduced from Celebes, where the same ape occurs. 2. *Viverra tangalunga,* a common Indo-Malayan species of civet, probably introduced. 3. *Cervus hippelaphus,* var. *Moluccensis,* a deer abundant in all the islands, very close to a Javan species and almost certainly introduced by man, perhaps very long ago. 4. *Babirusa alfurus,* the babirusa, found only in the island of Bouru, and perhaps originally introduced from Celebes. 5. *Sorex* sp., small shrews. With the exception of the last, *all* these species are animals habitually domesticated and kept in confinement by the Malays; and when we consider that none of the smaller Mammalia of Java and Borneo, numbering at least fifty different species, are found

E E

in any of the Moluccas, we can hardly suppose that such large animals as the deer and ape, could have reached them by natural means. There is every reason to believe, therefore, that the indigenous Mammalia of the Moluccas are wholly of Papuan stock, and very limited in number.

The birds are much more varied and interesting. About 200 species of land-birds are now known, belonging to 85 genera. Of the species about 15 are Indo-Malayan, 32 Papuan, and about 140 peculiar. Of the genera only two are peculiar,—*Semioptera*, a paradise bird, and *Lycocorax*, a singular form of Corvidæ; but there is also a peculiar rail-like wader, *Habroptila*. One genus, *Basilornis*, is found only in Ceram and Celebes; another, *Scythrops*, is Australian, and perhaps a migrant. About 30 genera are characteristic Papuan types, and 37 others, of more or less wide range, are found in New Guinea and were therefore probably derived thence. There remains a group of birds which are not found in New Guinea, and are either Palæarctic or Oriental. These are 13 in number as follows :—

1. Monticola.	8. Corydalla.
2. Acrocephalus.	9. Hydrornis.
3. Cisticola.	10. Batrachostomus.
4. Hypolais.	11. Loriculus.
5. Criniger.	12. Treron.
6. Butalis.	13. Neopus.
7. Budytes.	

Of these the *Monticola*, found only in Gilolo, appears to be a straggler or migrant from the Philippine islands. *Acrocephalus*, of which four species occur, is a wide-spread group; one of the Moluccan birds is an Australian and another a North-Asian species, which perhaps indicates that there has long been some migration southward from island to island, across the Moluccas. *Cisticola* is a genus of very wide range, extending to Australia. *Hypolais* is probably a modified form of a Chinese or Java-nese species. *Criniger* is a pure Indo-Malay form, represented here by three fine species. *Butalis* is a Chinese species, no doubt straggling southward. *Budytes* and *Corydalla* are wide-spread Oriental and Palæarctic species or slight modifications of them. *Hydrornis* is a Malayan form of Pittidæ. *Batrachostomus* is a distinct representative of a purely Indo-Malay genus. *Lori-*

culus is Malayan, and especially Philippine, but it reaches as far as Mysol. *Treron* is here at its eastern limit, and is represented in Bouru and Ceram by one of the most beautiful species. *Neopus*, a Malayan eagle, is said to occur in the Moluccas. We find then only three characteristic Indo-Malay types in the Moluccas,—*Criniger, Batrachostomus*, and *Treron*. All are represented by distinct and well marked species, indicating a somewhat remote period since their ancestors entered the district but all are birds of considerable powers of flight, so that a very little extension of the islands in a south-westerly direction would afford the means of transmission, but this could not well have been by way of Celebes, because the two former genera are unknown in that island.

It is evident, therefore, that the Moluccas are wholly Papuan in their zoology ; yet they are no less clearly derivative, and must have obtained their original immigrants under conditions that rendered a full representation of the fauna impossible. Such remarkable and dominant types as the eleven genera of Paradiseidæ, with *Cracticus, Rectes, Todopsis, Machœrirhynchus, Gerygone, Dacelo, Podargus, Cyclopsitta, Microglossum, Nasiterna, Chalcopsitta*, and *Goura*,—all characteristic Papuan groups, found in almost all the islands and most of them very abundant, are yet totally absent from the Moluccas. Taking this, in conjunction with the absence of the two genera of Papuan kangaroos and the other smaller groups of marsupials, and we must be convinced that the Moluccas cannot be mere fragments of the old Papuan land, or they would certainly, in some one or other of their large and fertile islands, have preserved a more complete representation of the parent fauna. Most of the Moluccan birds are very distinct from the allied species of New Guinea ; and this would imply that the entrance of the original forms took place at a remote period. The two peculiar genera with clearly Papuan affinities, show the same thing. The cassowary, found only in the large island of Ceram and distinct from any Papuan species, would however seem to have required a land connection for its introduction, almost as much as any of the larger mammalia.

E E 2

Taking all the facts into consideration, I would suggest as the most probable explanation, that if the Moluccas ever formed part of the main Papuan land, they were separated at an early date, and subsequently so greatly submerged as to destroy a large proportion of their fauna. They have since risen, and have probably been larger than at present, and rather more closely approximated to the parent land, whence they received a considerable immigration of such animals as were adapted to cross narrow seas. This gave them several Papuan forms, but still left them without a number of the types more especially confined to the forest depths, or powerful enough to combat the gales which often blow weaker flyers out to sea. Most of the birds whose absence from the Moluccas is so conspicuous belong to one or other of these classes.

Among the most characteristic birds of the Moluccas are the handsome crimson lories of the genera *Lorius* and *Eos*. These are found in every island (but not in Celebes or the Timor group); and a fine species of *Eos*, peculiar to the small islands of Siau and Sanguir, just north of Celebes, obliges us to place these with the Moluccas instead of with the former island, to which they seem most naturally to belong. The crimson parrots of the genus *Eclectus* are almost equally characteristic of the Moluccas, and add greatly to the brilliancy of the ornithology of these favoured islands.

Reptiles.—The Reptiles, so far as known, appear to agree in their distribution with the other vertebrates. In some small collections from Ceram there were no less than six of the genera peculiar to the Australian region, and which were before only known from Australia itself. These are, of snakes, *Liasis* and *Enygrus*, genera of Pythonidæ; with *Diemenia* and *Acanthophis* (Elapidæ); of lizards, *Cyclodus*, a genus of Scincidæ; and of Amphibia, a tree-frog of the genus *Pelodryas.*

Insects—Peculiarities of the Moluccan Fauna.—In insects the Moluccas are hardly, if at all, inferior to New Guinea itself. The islands abound in grand *Papilios* of the largest size and extreme beauty; and it is a very remarkable fact, that when the closely-allied species of the Moluccas and New Guinea are compared, the former are almost always the largest. As examples may be mentioned, *Ornithoptera priamus* and *O. helena* of the Moluccas, both larger than the varieties (or species) of Papua; *Papilio ulysses* and *deiphobus* of Amboyna, usually larger than their allies in New Guinea; *Hestia idea*, the largest species of the genus; *Diadema pandarus* and *Charaxes euryalus*, both larger than any other species of the same genera in the whole archipelago. It is to be noted also, that in the Moluccas, the very largest specimens or races seem always to come from the small island of Amboyna; even those of Ceram, the much larger island to which it is a satellite, being almost always of less dimensions. Among Coleoptera, the Moluccas produce *Euchirus longimanus*, one of the largest and most remarkable of the Lamellicornes; *Sphingnotus dunningi*, the largest of the Austro-Malayan Tmesisterninæ; a *Sphenura*, the largest and handsomest of an extensive genus; an unusually large *Schizorhina* (Cetoniidæ); and some of the most remarkable and longest-horned Anthotribidæ. Even in birds the same law may be seen at work,—in the *Tanysiptera nais* of Ceram, which has a larger tail than any other in the genus; in *Centropus goliath* of Gilolo, being the largest and longest-tailed species; in *Hydrornis maximus* of Gilolo, the largest and perhaps the most elegantly and conspicuously coloured of all the Pittidæ; in *Platycercus-amboinensis*, being pre-eminent in its ample blue tail; in the two Moluccan lories and *Eos rubra*, being more conspicuously red than the allied New Guinea species; and in *Megapodius wallacei* of Bouru, being the only species of the genus conspicuously marked and banded.

All these examples, of larger size, of longer tails or other appendages, and of more conspicuous colouring, are probably indications of a less severe struggle for existence in these islands than in the larger tract of New Guinea, with a more abundant and more varied fauna; and this may apply even to the smaller islands, as compared with the larger in the immediate vicinity. The limited number of forms in the small islands compared with a similar area in the parent land, implies, perhaps, less competition and less danger; and thus allows, where all other conditions are favourable, an unchecked and continuous development in size, form, and colour, until they become positively injurious. This law may not improbably apply to the New Guinea fauna itself, as compared with that of Borneo or any other similar country; and some of its peculiarities (such as its wonderful paradise-birds) may be due to long isolation, and consequent freedom from the influence of any competing forms. The difference between the very sober colours of the Coleoptera, and in a less degree of the birds, of Borneo, as compared with their brilliancy in New Guinea, always struck me most forcibly, and was long without any, even conjectural, explanation. It is not the place here to go further into this most curious and interesting subject. The reader who wishes for additional facts to aid him in forming an opinion, should consult Mr. Darwin's *Descent of Man*, chapters x. to xv.; and my own *Contributions to the Theory of Natural Selection*, chapters iii. and iv.

Timor Group.—Mammalia.—In the group of islands between Java and Australia, from Lombok to Timor inclusive, we find a set of mammals similar to those of the Moluccas, but some of them different species. A wide-spread species of *Cuscus* represents the Papuan element. A *Sorex* and a peculiar species of wild pig, we may also accept as indigenous. Three others have almost certainly been introduced. These are, (1.) *Macacus cynomolgus*, the very commonest Malay monkey, which may have crossed the narrow straits from island to island between Java and Timor, though it seems much more probable that it was introduced by Malays, who constantly capture and rear the young of this species. (2.) *Cervus timoriensis*, a deer, said to be a distinct species, inhabits Timor, but it is probably only a variety of the *Cervus hippelaphus* of Java. This animal is, however, much more likely to have crossed the sea than the monkey. (3.) *Paradoxurus fasciatus*, takes the place of *Viverra tangalunga* in the Moluccas, both common and wide-spread civets which are often kept in confinement by the Malays. The *Felis megalotis*, long supposed to be a native of Timor, has been ascertained by Mr. Elliot to belong to a different country altogether.

Birds.—The birds are much more interesting, since they are sufficiently numerous to allow us to determine their relations, and trace their origin, with unusual precision. There are 96 genera and 160 species of land-birds known to inhabit this group of islands ; and on a careful analysis, they are found to be almost equally related to the Australian and Oriental regions, 30 genera being distinctly traceable to the former, and the same number to the latter. Their connection with the Moluccas is shown by the presence of the genera *Mimeta*, *Geoffroyus*, *Cacatua*, *Ptilopus*, and *Ianthœnas*, together with *Megapodius* and *Cerchneis* represented by Moluccan species. *Turacœna* shows a connection with Celebes, and *Scops* is represented by a Celebesian species. The connection with Australia is shown by the genera *Sphœcothera*, *Gerygone*, *Myiagra*, *Pardalotus*, *Gliciphila*, *Amadina*, and *Aprosmictus* ; while *Milvus*, *Hypotriorchis*, *Eudynamis*, and *Eurystomus*, are represented by Australian species. Other genera confined to or characteristic of the Australian region, are *Rhipidura*, *Monarcha*, *Artamus*, *Campephaga*, *Pachycephala*, *Philemon*, *Ptilotis*, and *Myzomela*.

We now come to the Indo-Malay or Javan element represented by the following genera :

1. Turdus (T.)	11. Oriolus.	21. Yungipicus.
2. Geocichla (T.)	12. Pericrocotus.	22. Merops.
3. Zoothera.	13. Cyornis (T.)	23. Pelargopsis.
4. Megalurus (T.)	14. Hypothymis.	24. Ceyx.
5. Orthotomus.	15. Tchitrea.	25. Loriculus.
6. Pratincola (T.)	16. Lanius (T.)	26. Treron (T.)
7. Oreicola (T.)	17. Anthreptes.	27. Iotreron (s.g. of *Ptilopus*).
8. Drymocataphus (T.)	18. Eulabes.	28. Chalcophaps (T.)
9. Parus.	19. Estrilda (T.)	29. Gallus (T.)
10. Pycnonotus.	20. Erythrura (T.)	30. Strix.

Such genera as *Merops* and *Strix*, which are as much Australian as Oriental, are inserted here because they are represented by Javan species. The list is considerably swelled by genera which have reached Lombok across the narrow strait from Baly, but have passed no further. Such are *Zoothera*, *Orthotomus*, *Pycnonotus*, *Pericrocotus* and *Strix*. A much larger number (12) stop short at Flores, leaving only 13, indicated in the list by (T) after their names, which reach Timor. It is evident, therefore, that these islands have been stocked from three chief sources,—the

Moluccas (with New Guinea and Celebes,) Australia, and Java. The Moluccan forms may well have arrived as stragglers from island to island, aided by whatever facilities have been afforded by lands now submerged. Most of the remainder have been derived either from Australia or from Java; and as their relations to these islands are very interesting, they must be discussed with some detail.

Origin of the Timorese Fauna.—We must first note, that 80 species, or exactly one-half of the land-birds of the islands, are peculiar and mostly very distinct, intimating that the immigration commenced long enough back to allow of much specific modification. There is also one peculiar genus of kingfishers, *Caridonax*, found only in Lombok and Flores, and more allied to Australian than to Oriental types. The fine white-banded pigeons (s. g. *Leucotreron*) are also almost peculiar; one other less typical species only being known, a native of N. Celebes. In order to compare the species with regard to their origin, we must first take away those of wide distribution from which no special indications can be obtained. In this case 49 of the land-birds must be deducted, leaving 111 species which afford good materials for comparison. These, when traced to their origin, show that 62 came from some part of the Australian region, 49 from Java or the Oriental region. But if we divide them into two groups, the one containing the species *identical* with those of the Australian or Oriental regions, the other containing *allied* or *representative* species peculiar to the islands, we have the following result:

Species common to the Timorese Islands and the Oriental Region   30
Peculiar Timorese species allied to those of the Oriental Region   19

      Total   49

Species common to the Timorese Islands and the Australian
    Region   18
Peculiar Timorese species allied to those of the Australian Region   44

      Total   62

This table is very important, as indicating that the connection with Australia was probably earlier than that with Java; since the majority of the Australian species have become modified, while the majority of the Oriental species have remained unchanged. This is due, no doubt, in part to the continued immigration of fresh individuals from Java, after that from Australia, the Moluccas and New Guinea had almost wholly ceased. We must also notice the very small proportion of the genera, either of Australia or Java, that have found their way into these islands, many of the largest and most wide-spread groups in both countries being altogether absent. Taking these facts into consideration, it is pretty clear that there has been no close and long-continued approximation of these islands to any part of the Australian region; and it is also probable that they were fairly stocked with such Australian groups as they possess before the immigration from Java commenced, or a larger number of characteristic Oriental forms would have been able to have established themselves.

On looking at our map, we find that a shallow submerged bank extends from Australia to within about twenty miles of the coast of Timor; and this is probably an indication that the two countries were once only so far apart. This would have allowed the purely Australian types to enter, as they are not numerous; there being about 6 Australian species, and 10 or 12 representatives of Australian species, in Timor. All the rest may have been derived from the Moluccas or New Guinea, being mostly wide-spread genera of the Australian region; and the extension of Papua in a south-west direction towards Java (which was suggested as a means of providing New Guinea with peculiar Indo-Malay types not found in any other part of the region) may have probably served to supply Timor and Flores with the mass of their Austro-Malayan genera across a narrow strait or arm of the sea. Lombok, Baly, and Sumbawa were probably not then in existence, or nothing more than small volcanic cones rising out of the sea, thus leaving a distance of 300 miles between Flores and Java. Subsequently they grew into islands, which offered an easy passage for a number of Indo-Malay genera into such scantily stocked territories as Flores and Timor. The

north coast of Australia then sank, cutting off the supply from
that country; and this left the Timorese group in the position it
now occupies.

The reptiles and fishes of this group are too little known to
enable us to make any useful comparison.

Insects.—The insects, though not numerous, present many fine
species, some quite unlike any others in the Archipelago. Such
are—*Papilio liris, Pieris læta, Cirrochroa lamarckii* and *C. lesche-
naultii* among butterflies. The Coleoptera are comparatively little
known, but in the insects generally the Indo-Malay element pre-
dominates. This may have arisen from the peculiar vegetation
and arid climate not being suitable to the Papuan insects. Why
Australian forms did not establish themselves we cannot conjecture; but the field appears to have been open to immigrants from
Java, the climate and vegetation of which island at its eastern extremity approximates to that of the Timorese group. The insects
are, however, so peculiarly modified as to imply a very great antiquity, and this is also indicated by a group of Sylviine birds here
classed under *Oreicola,* but some of which probably form distinct
genera. There may, perhaps, have been an earlier and a later
approximation to Java, which, with the other changes indicated,
would account for most of the facts presented by the fauna of
these islands. One deduction is, at all events, clear: the extreme paucity of indigenous mammals along with the absence of
so many groups of birds, renders it certain that the Timorese
islands did not derive their animal life by means of an actual
union with any of the large islands either of the Australian or
the Oriental regions.

Celebes Group.—We now come to the Island of Celebes, in
many respects the most remarkable and interesting in the whole
region, or perhaps on the globe, since no other island seems to
present so many curious problems for solution. We shall therefore give a somewhat full account of its peculiar fauna, and
endeavour to elucidate some of the causes to which its zoological
isolation may be attributed.

Mammalia.—The following is the list of the mammalia of

Celebes as far as at present known, though many small species may yet be discovered.

1. Cynopithecus nigrescens.	7. Barbirusa alfurus.
2. Tarsius spectrum.	8. Sciurus (5 peculiar sp.)
3. Viverra tangalunga.	9. Mus (2 peculiar sp.)
4. Cervus hippelaphus.	10. Cuscus (2 peculiar sp.)
5. Anoa depressicornis	Also 7 species of bats, of
6. Sus celebensis.	which 5 are peculiar.

The first—a large black ape—is itself an anomaly, since it is not closely allied to any other form of quadrumana. Its flat projecting muzzle, large superciliary crests and maxillary ridges, with the form and appearance of its teeth, separate it altogether from the genus *Macacus*, as represented in the Indo-Malay islands, and ally it closely to the baboons of Africa.[1] We have already seen reason to suppose that it has been carried to Batchian, and there is some doubt about the allied species or variety (*C. niger*) of the Philippines being really indigenous there ; in which case this interesting form will remain absolutely confined to Celebes. (2.) The tarsier is a truly Malayan species, but it is said to occur in a small island at the northern extremity of Celebes. It might possibly have been introduced there. (3) and (4)—a civet and a deer—are, almost certainly, as in the Moluccas, introduced species. (5.) *Anoa depressicornis*. This is one of the peculiar Celebesian types; a small straight-horned wild-bull, anatomically allied to the buffaloes, and somewhat resembling the bovine antelopes of Africa, but having no near allies in the Oriental region. (6.) *Sus Celebensis*; a peculiar species of wild-pig. (7.) *Babirusa alfurus ;* another remarkable type, having no near allies. It differs in its dentition from the typical Suidæ, and seems to approach the African Phacochœridæ. The manner in which the canines of the upper tusks are reversed, and grow directly upwards in a spiral curve over the eyes, is unique among mammalia. (8.) Five squirrels inhabit Celebes, and all are peculiar species. (9.) These are forest rats of the sub-genus *Gymnomys*, allied to Australian species. 10. *Cuscus.* This typical

---

[1] The general form of the skull agrees best with that of *Cynocephalus mormon*, the largest and most typical of the African baboons ; while the position of the nostrils brings it nearer the macaques.

Australian form is represented in Celebes by two peculiar species.

Leaving out the Indo-Malay *species*, which may probably have been introduced by man, and are at all events comparatively recent immigrants, and the wild pig, a genus which ranges over the whole archipelago and which has therefore little significance, we find two genera which have come from the Australian side, —*Cuscus* and *Mus ;* and four from the Oriental side,—*Cynopithecus, Anoa, Babirusa,* and *Sciurus.* But *Sciurus* alone corresponds to *Cuscus,* as a genus still inhabiting the adjacent islands; the other three being not only peculiar to Celebes, but incapable of being affiliated to any specially Oriental group. We seem, then, to have indications of two distinct periods ; one very ancient, when the ancestors of the three peculiar genera roamed over some unknown continent of which Celebes formed, perhaps, an outlying portion ;—another more recent, when from one side there entered *Sciurus,* and from the other *Cuscus.* But we must remember that the Moluccas to the east, possess scarcely any indigenous mammals except *Cuscus ;* whereas Borneo and Java on the west, have nearly 50 distinct genera. It is evident then, that the facilities for immigration must have been much less with the Oriental than with the Australian region, and we may be pretty certain that at this later period there was no land connection with the Indo-Malay islands, or some other animals than squirrels would certainly have entered. Let us now see what light is thrown upon the subject by the birds.

Birds.—The total number of birds known to inhabit Celebes is 205, belonging to about 150 genera. We may leave out of consideration the wading and aquatic birds, most of which are wide-ranging species. There remain 123 genera and 152 species of land-birds, of which 9 genera and 66 species are absolutely confined to the island, while 20 more are found also in the Sula or Sanguir Islands, so that we may take 86 to be the number of peculiar Celebes species. Lord Walden, from whose excellent paper on the birds of Celebes (*Trans. Zool. Soc.* vol. viii. p. 23) most of these figures are obtained, estimates, that of the species which are not peculiar to Celebes, 55 are of Oriental and 22 of

Australian origin, the remainder being common to both regions. This shows a preponderant recent immigration from the West and North, which is not to be wondered at when we look at the long coast line of Java, Borneo, and the Philippine islands, with an abundant and varied bird population, on the one side, and the small scattered islands of the Moluccas, with a comparatively scanty bird-fauna, on the other.

But, adopting the method here usually followed, let us look at the relations of the *genera* found in Celebes, omitting for the present those which are peculiar to it. I divide these genera into two series:—those which are found in Borneo or Java but not in the Moluccas, and those which inhabit the Moluccas and not Borneo or Java; these being the respective sources from which, *primâ facie*, the species of these genera must have been derived. Genera which range widely into both these districts are rejected, as teaching us nothing of the origin of the Celebesian fauna. In a few cases, sub-genera which show a decided eastern or western origin, are given.

GENERA DERIVED FROM BORNEO AND JAVA.

1. Geocichla.	9. Nectarophila.	17. Hydrocissa.
2. Pratincola (sp.)	10. Anthreptes (sp.)	18. Cranorrhinus.
3. Trichastoma.	11. Munia (sp.)	19. Lyncornis.
4. Oriolus (sp.)	12. Acridotheres.	20. Treron (sp.)
5. Cyornis.	13. Yungipicus.	21. Gallus (sp.)
6. Hypothymis.	14. Mulleripicus.	22. Spilornis.
7. Hylocharis.	15. Rhamphococcyx.	23. Butastur.
8. Æthopyga.	16. Hierococcyx.	24. Pernis.

GENERA DERIVED FROM THE MOLUCCAS OR TIMOR.

1. Graucalus (sp.)	6. Tanygnathus.	11. Myristicivora (s. g.)
2. Chalcostetha.	7. Trichoglossus.	12. Ducula (s. g.)
3. Myzomela.	8. Scythrops (sp.)	13. Zonœnas (s. g.)
4. Munia (sp.)	9. Turacœna.	14. Lamproteron (s. g.)
5. Cacatua (sp.)	10. Reinwardtœnas (sp.)	15. Megapodius.

These tables show a decided preponderance of Oriental over Australian forms. But we must remember that the immediately adjacent lands from whence the supply was derived, is very much richer in the one case than in the other. The 24 genera derived from Borneo and Java are only about *one fourth* of the characteristic genera of those islands; while the 15 Moluccan and Timorese genera are fully *one third* of their characteristic types. The *proportion* derived from the Australian, is greater than that derived from the Oriental side.

We shall exhibit this perhaps more clearly, by giving a list of the important groups of each set of islands which are absent from Celebes.

Important Families of Java and Borneo absent from Celebes.	Important Families of the Moluccas absent from Celebes.
1. Eurylæmidæ. 5. Laniidæ. 2. Timaliidæ. 6. Megalæmidæ. 3. Phyllornithidæ. 7. Trogonidæ. 4. Pycnonotidæ 8. Phasianidæ.	1. Meliphagidæ.

Additional important genera of Java or Borneo absent from Celebes.	Important genera of the Moluccas absent from Celebes.
1. Orthotomus.	1. Mimeta.
2. Copsychus.	2. Monarcha.
3. Enicurus.	3. Rhipidura.
4. Tchitrea.	4. Pachycephala.
5. Pericrocotus.	5. Lycocorax.
6. Irena.	6. Alcyone.
7. Platylophus.	7. Tanysiptera.
8. Dendrocitta.	8. Geoffroyus.
9. Eulabes.	9. Eclectus.
10. Hemicercus.	10. Platycercus.
11. Chrysocolaptes.	11. Eos.
12. Tiga.	12. Lorius.
13. Micropternus.	
14. Batrachostomus.	
15. Palæornis.	
16. Rollulus.	

If we reckon the absent families to be each represented by only two important genera, we shall find the deficiency on the Oriental side much the greatest; yet those on the side of the Moluccas are sufficiently remarkable. The Meliphagidæ are not indeed absolutely wanting, since a *Myzomela* has now been found in Celebes; but all its larger and more powerful forms which range over almost the entire region, are absent. This may be balanced by the absence of the excessively abundant Timaliidæ of the Indo-Malay islands, which are represented by

only a single species; and by the powerful Phasianidæ, represented only by the common Malay jungle fowl, perhaps introduced. The entire absence of Pycnonotidæ is a very anomalous
fact, since one of the largest genera, *Criniger*, is well represented
in several islands of the Moluccas, and one has even been found
in the Togian islands in the great northern inlet of Celebes;
but yet it passes over Celebes itself. *Ceyx*, a genus of small
kingfishers, is a parallel case, since it is found everywhere from
India to New Guinea, leaving out only Celebes; but this comes
among those curiosities of the Celebesian fauna which we shall
notice further on. In the list of genera derived from Borneo or
Java, no less than 6 are represented by identical species (indicated by sp. after the name); while in the Moluccan list 5
are thus identical. These must be taken to indicate, either that
the genus is a recent introduction, or that stragglers still occasionally enter, crossing the breed, and thus preventing specific
modification. In either case they depend on the existing state
of things, and throw no light on the different distribution of
land and sea which aided or checked migration in former times;
and they therefore to some extent diminish the weight of the
Indo-Malay affinity, as measured by the relations of the peculiar
species of Celebes.

From our examination of the evidence thus far,—that is, taking
account firstly, of the *species*, and, secondly, of the *genera*, which
are common to Celebes and the groups of islands between which
it is situated, we must admit that the connexion seems rather
with the Oriental than with the Australian region; but when we
take into account the *proportion* of the genera and species present, to those which are absent, and giving some weight to the
greater extent of coast line on the Indo-Malay side, we seem
justified in stating that the Austro-Malay element is rather the
most fully represented. This result applies both to birds and
mammals; and it leads us to the belief, that during the epoch of
existing species and genera, Celebes has never been united with
any extensive tract of land either on the Indo-Malay or Austro-
Malay side, but has received immigrants from both during a very
long period, the facilities for immigration having been rather the

greatest on the Austro-Malay or Australian side. We have now to consider what further light can be thrown on the subject by the consideration of the *peculiar genera* of Celebes, and of those curiosities or anomalies of distribution to which we have referred.

Nine genera of birds are altogether peculiar to Celebes; three more are found only in one other island, and seem to be typically Celebesian; while one is found in the Sula islands (which belongs to the Celebes group) and probably exists in Celebes also. The following is a list of these 13 genera:

1. *Artamides*...	(Campephagidæ)	8. *Monachalcyon* (Alcedinidæ)
2. *Streptocitta*..	(Corvidæ)	9. *Cittura* ... ,,
3. *Charitornis*..	,,	10. *Ceycopsis* ... ,,
4. *Gazzola*, (s. g.)	,,	11. *Meropogon* .. (Meropidæ)
5. *Basilornis* ..	(Sturnidæ)	12. *Prioniturus*. (Psittacidæ)
6. *Enodes* ...	,,	13. *Megacephalon* (Megapodiidæ)
7. *Scissirostrum*	,,	

Of the above, *Artamides, Monachalcyon, Cittura,* and *Megacephalon,* are modifications of types characteristic of the Australian region. All are peculiar to Celebes except *Cittura,* found also in the Sanguir islands to the northward, but which seems to belong to the Moluccan group. *Streptocitta, Charitornis,* and *Gazzola,* are peculiar types of Corvidæ; the two former allied to the magpies, the latter to the jackdaws. *Charitornis* is known only from the Sula islands east of Celebes, and is closely related to *Streptocitta.* There is nothing comparable to these three groups in any of the Malay islands, and they seem to have relations rather with the Corvidæ of the old-world northern continent. *Basilornis, Enodes,* and *Scissirostrum,* are remarkable forms of Sturnidæ. *Basilornis* has a beautiful compressed crest, which in the allied species found in Ceram is elongated behind. *Enodes* has remarkable red superciliary streaks, but seems allied to *Calornis. Scissirostrum* seems also allied to *Calornis* in general structure, but has a very peculiarly formed bill and nostrils. We can hardly say whether these three forms show more affinity to Oriental or to Australian types, but they add to the weight of evidence as to the great antiquity and isolation of the Celebesian fauna. *Scissirostrum* has been classed with *Euryceros,* a Mada-

gascar bird, and with *Buphaga*, an African genus ; but the peculiar beak and nostrils approximate more to *Cracticus* and its
allies, of the Australian region, which should probably form a
distinct family. *Ceycopsis* is undoubtedly intermediate between
the Malayan *Ceyx* and the African *Ispidina*, and is therefore especially interesting. *Meropogon* is a remarkable form of bee-
eater, allied to the Indo-Malayan *Nyctiornis.* *Prioniturus* (the
raquet-tailed parrots) of which two species inhabit Celebes, and
one the Philippines, appears to be allied to the Austro-Malayan
Geoffroyus.

We must finally notice a few genera found in Celebes, whose
nearest allies are not in the surrounding islands, and which thus
afford illustrations of discontinuous distribution. The most remarkable, perhaps, is *Coracias*, of which a fine species inhabits
Celebes ; while the genus is quite unknown in the Indo-Malay
sub-region, and does not appear again till we reach Burmah and
India ; and the species has no closer affinity for Indian than for
African forms. *Myialestes*, a small yellow flycatcher, is another
exmple; its nearest ally (*M. cinereocapilla*) being a common Indian
bird, but unknown in the Malay islands. The Celebesian bird
described by me as *Prionochilus aureolimbatus*, is probably a
third case of discontinuous distribution, if (as a more careful
examination seems to show) it is not a *Prionochilus*, but con-
generic with *Pachyglossa*, a bird only found in the Himalayas.
The fine pigeon, *Carpophaga forsteni*, belongs to a group found in
the Philippines, Australia, and New Zealand ; but the Celebes
species is very distinct from all the others, and seems, if anything, more allied to that of New Zealand.

The Sula islands (Sula-mangola, Sula-taliabo, and Sula-besi)
lie midway between Celebes and the Moluccas, being 80 miles
from the nearest part of Celebes, with several intervening
islands, and 40 miles from Bouru, all open sea. Their birds
show, as might be expected, a blending of the two faunas, but
with a decided preponderance of that of Celebes. Out of 43
land birds which have been collected in these islands, we may
deduct 6 as of wide range and no significance. Of the 37 remaining, 21 are Celebesian species, and 4 are new species but

allied to those of Celebes; while there are 10 Moluccan species and 2 new species allied to those of the Moluccas. It is curious that no less than 3 Moluccan genera, quite unknown in Celebes itself, occur here,—*Monarcha, Pachycephala*, and *Criniger;* but all these, as well as several other of the Moluccan birds, are rather weak flyers, and such as are likely to have been carried across by strong winds. Of the *genera*, 23 are from Celebes, 10 from the Moluccas. These facts show, that the Sula islands form part of the Celebes group, although they have received an infusion of Moluccan forms, which will perhaps in time spread to the main island, and diminish the remarkable individuality that now characterises its fauna.

Insects.—Of the reptiles and fishes of Celebes we have not sufficient information to draw any satisfactory conclusions. I therefore pass to the insects of which something more is known.

The Butterflies of Celebes are not very numerous, less than 200 species in all having been collected; but a very large proportion of them, probably three-fourths of the whole, are peculiar. There is only one peculiar genus, *Amechania*, allied to *Zethera* (a group confined to the Philippine Islands), with which it should perhaps be united. Most of the genera are of wide distribution in the archipelago, or are especially Malayan, only two truly Australian genera, *Elodina* and *Acropthalmia*, reaching Celebes. On the other hand, 7 peculiar Oriental genera are found in Celebes, but not further east, viz., *Clerome, Adolias, Euripus, Apatura, Limenitis, Iolaus*, and *Leptocircus*. There are also several indications of a direct affinity with the continent rather than with Malaya, as in the cases already enumerated among birds. A fine butterfly, yet unnamed, almost exactly resembles *Dichorragia nesimachus*, a Himalayan species. *Euripus robustus* is closely allied to *E. halitherses* of N. India; there are no less than 5 species of *Limenitis*, all quite unlike those found in other parts of the archipelago. The butterflies of Celebes are remarkably distinguished from all others in the East, by peculiarities of form, size, and colour, which run through groups of species belonging to different genera. Many Papilionidæ and Pieridæ, and some

Nymphalidæ, have the anterior wings elongated, with the apex often acute, and, what is especially remarkable, an abrupt bend or shoulder near the base of the wing. (See *Malay Archipelago*, 3rd Ed. p. 281, woodcut.) No less than 13 species of *Papilio*, 10 Pieridæ, and 4 or 5 Nymphalidæ, are thus distinguished from their nearest allies in the surrounding islands or in India.   In size again, a large number of Celebesian butterflies stand pre-eminent over their allies. The fine Papilios—*adamantius, blumei*, and *gigon*—are perfect giants by the side of the closely-allied forms of Java ; while *P. androcles* is the largest and longest-tailed, of all the true swallow-tailed group of the Old World.   Among Nymphalidæ, the species of *Rhinopalpa* and *Euripus*, peculiar to Celebes, are immensely larger than their nearest allies ; and several of the Pieridæ are also decidedly larger, though in a less marked degree. In colour, many of the Celebesian butterflies differ from the nearest allied species ; so that they acquire a singularity of aspect which marks them off from the rest of the group. The most curious case is that of three butterflies, belonging to three distinct genera (*Cethosia myrina*, *Messaras mæonides*, and *Atella celebensis*) all having a delicate violet or lilac gloss in lines or patches, which is wholly wanting in every allied species of the surrounding islands. These numerous peculiarities of Celebesian butterflies are very extraordinary ; and imply isolation from surrounding lands, almost as much as do the strange forms of mammals and birds, which more prominently characterise this interesting island.

Of the Coleoptera we know much less, but a few interesting facts may be noted. There are a number of fine species of *Cicindela*, some of peculiar forms ; and one *Odontochila*, a South American genus ; while *Collyris* reaches Celebes from the Oriental region.   In Carabidæ it has one peculiar genus, *Dicraspeda ;* and a species of the fine Australian genus *Catadromus*. In Lucanidæ it has the Oriental genus, *Odontolabris*.   In Cetoniidæ it has a peculiar genus, *Sternoplus*, and several fine *Cetoniæ;* but the characteristic Malayan genus, *Lomaptera*, found in every other island of the archipelago from Sumatra to New Guinea, is absent—an analogous fact to the case of *Ceyx* among birds.

In Buprestidæ, the principal Austro-Malay genus, *Sambus*, is found here; while *Sponsor*, a genus 8 species of which inhabit Mauritius, has one species here and one in New Guinea. In Longicorns there are four peculiar genera, *Comusia*, *Pytholia*, *Bityle*, and *Ombrosaga;* but the most important features are the occurrence of the otherwise purely Indo-Malayan genera *Agelasta*, *Nyctimene*, and *As:athes;* and of the purely Austro-Malayan *Arrhenotus*, *Trysimia*, *Xenolea*, *Amblymora*, *Diallus*, and *Ægocidnus*. The remaining genera range over both portions of the archipelago. In the extensive family of Curculionidæ we can only notice the elegant genus, *Celebia*, allied to *Eupholus*, which, owing to its abundance and beauty, is a conspicuous feature in the entomology of the island.

Origin of the fauna of Celebes.—We have now to consider, briefly, what past changes of physical geography are indicated by the curious assemblage of facts here adduced. We have evidently, in Celebes, a remnant of an exceedingly ancient land, which has undergone many and varied revolutions; and the stock of ancient forms which it contains must be taken account of, when we speculate on the causes that have so curiously limited more recent immigrations. Going back to the arrival of those genera which are represented in Celebes by peculiar species, and taking first the Austro-Malay genera, we find among them such groups as *Zonœnas* (s.g.), *Phlogœnas*, *Leucotreron* (s.g.), and *Turacœna*, which are not found in the Moluccas at all; and *Myzomela*, found in Timor and Banda, but not in Ceram or Bouru, which are nearest to Celebes. This, combined with the curious absence of so many of the commonest Moluccan genera, leads to the conclusion that the Austro-Malay immigration took place by way of Timor and the southern part of New Guinea. It will be remembered, that to account for the Indo-Malayan forms in New Guinea, we suggested an extension of that country in a westerly direction just north of Timor. Now this is exactly what we require, to account for the stocking of Celebes with the Australian forms it possesses. At this time Borneo did not approach so near, and it was at a somewhat later period that the last great Indo-Malay migration set in; but finding the country already fairly stocked, comparatively few groups were able to establish themselves.

Going back a little farther, we come to the entrance of those few birds and insects which belong to India or Indo-China; and this probably occurred at the same time as that continental extension southward, which we found was required to account for a similar phenomenon in Java. Celebes, being more remote, received only a few stragglers. We have now to go much farther back, to the time when the ancestors of the peculiar Celebesian genera entered the country, and here our conjectures must necessarily be less defined.

On the Australian side we have to account for *Megacephalon*, and the other genera of purely Papuan type It may perhaps be sufficient to say, that we do not yet know that these genera, or some very close allies, do not still exist in New Guinea; in which case they may well have entered at the same time with the *species*, already referred to. If, on the other hand, they are really as isolated as they appear to be, they represent an earlier communication, either by an approximation of the two islands over the space now occupied by the Moluccas; or, what is perhaps more probable, through a former extension of the Moluccas, which have since undergone so much subsidence, as to lead to the extinction of a large proportion of their ancient fauna. The wide-spread volcanic action, and especially the prevalence of raised coral-reefs in almost all the islands, render this last supposition very probable.

On the Oriental side the difficulty is greater; for here we find, what seem to be clear indications of a connection with Africa, as well as with Continental Asia, at some immensely remote epoch. *Cynopithecus, Babirusa,* and *Anoa; Ceycopsis, Streptocitta,* and *Gazzola* (s. g.), and perhaps *Scissirostrum,* may be well explained as descendants of ancestral types in their respective groups, which also gave rise to the special forms of Africa on the one hand, and of Asia on the other. For this immigration we must suppose, that at a period before the formation of the present Indo-Malay Islands, a great tract of land extended in a north-westerly direction, till it met the old Asiatic continent. This may have been before the Himalayas had risen to any great height, and when a large part of what are now the cold plateaus of Central Asia may have teemed with life, some forms of which are preserved in Africa, some in Malaya, and a few in Celebes.   Here may have lived the common ancestor of *Sus, Babirusa,* and *Phacochœrus;* as well as of *Cynopithecus, Cynocephalus,* and *Macacus;* of *Anoa* and *Bubalus;* of *Scissirostrum* and *Euryceros;* of *Ceyx, Ceycopsis,* and *Ispidina.*   Such an origin accounts, too, for the presence of the North-Indian forms in Celebes ; and it offers less difficulties than a direct connection with continental Africa, which once appeared to be the only solution of the problem.   If this south-eastward extension of Asia occurred at the same time as the north-eastward extension of South Africa and Madagascar, the two early continents may have approached each other sufficiently to have allowed of some interchange of forms : *Tarsius* may be the descendant of some Lemurine animal that then entered the Malayan area, while the progenitors of *Cryptoprocta* may then have passed from Asia to Madagascar.

It is true that we here reach the extremest limits of speculation ; but when we have before us such singular phenomena as are presented by the fauna of the island of Celebes, we can hardly help endeavouring to picture to our imaginations by what past changes of land and sea (in themselves not improbable) the actual condition of things may have been brought about.

II. Australia and Tasmania, or the Australian Sub-region.

A general sketch of Australian zoology having been given in the earlier part of this chapter, it will not be necessary to occupy much time on this sub-region, which is as remarkably homogeneous as the one we have just left is heterogeneous.   Although much of the northern part of Australia is within the tropics, while Victoria and Tasmania are situated from 36° to 43° south latitude, there is no striking change in the character of the fauna throughout the continent; a number of important genera extending over the whole country, and giving a very uniform character to its zoology.   The eastern parts, including the colonies of New South Wales and Queensland, are undoubtedly the richest, several

PLATE XI.

A SCENE IN TASMANIA, WITH CHARACTERISTIC MAMMALS.

peculiar types being found only here. The southern portion is somewhat poorer, and has very few peculiar forms; and Tasmania being isolated is poorer still, yet its zoology has much resemblance to that of Victoria, from which country it has evidently not been very long separated. The north, as far as yet known, is characterised by hardly any peculiar forms, but by the occurrence of a number of Papuan types, which have evidently been derived from New Guinea.

Mammalia.—The Australian sub-region contains about 160 species of Mammalia, of which 3 are Monotremata, 102 Marsupials, 23 Chiroptera, 1 Carnivora (the native dog, probably not indigenous), and 31 Muridæ. The north is characterised by a species of the Austro-Malayan genus *Cuscus*. *Phascolarctos* (the koala, or native bear) is found only in the eastern districts; *Phascolomys* (the wombat) in the south-east and Tasmania; *Petaurista* (a peculiar form of flying opossum) in the east. *Thylacinus* (the zebra-wolf), and *Sarcophilus* (the "native devil"), two carnivorous marsupials, are confined to Tasmania. West Australia, the most isolated and peculiar region botanically, alone possesses the curious little honey-eating *Tarsipes*, and the *Peragalea*, or native rabbit. The remarkable *Myrmecobius*, a small ant-eating marsupial, is found in the west and south; and *Onychogalea*, a genus of kangaroos, in West and Central Australia. All the other genera have a wider distribution, as will be seen by a reference to the list at the end of this chapter.

Plate XI. A Scene in Tasmania, with Characteristic Mammalia. —As some of the most remarkable Mammalia of the Australian region are now found only in Tasmania, we have chosen this island for the scene of our first illustration of the fauna of the Australian sub-region. The pair of large striped animals are zebra-wolves (*Thylacinus cynocephalus*), the largest and most destructive of the carnivorous marsupials. These creatures used to be tolerably plentiful in Tasmania, where they are alone found. They are also called "native tigers," or "native hyænas;" and being destructive to sheep, they have been destroyed by the farmers and will doubtless soon be exterminated. In the foreground on the left is a bandicoot (*Perameles gunnii*). These are delicate little animals allied to the kangaroos; and they are found in all parts of Australia, and Tasmania, to which latter country this species is confined. On the right is the wombat (*Phascolomys wombat*), a root-eating marsupial, with large incisor teeth like those of our rodents. They inhabit south-east Australia and Tasmania. In the foreground is the porcupine ant-eater (*Echidna setosa*), belonging to a distinct order of mammalia, Monotremata, of which the only other member is the duck-billed *Ornithorhynchus*. These animals are, however, more nearly allied to the marsupials, than to the insectivora or edentata of the rest of the world, which in some respects they resemble. An allied species (*Echidna hystrix*) inhabits south-east Australia.

Birds.—Australia (with Tasmania) possesses about 630 species of birds, of which 485 are land-birds. Not more than about one-twentieth of these are found elsewhere, so that it has a larger proportion of endemic species than any other sub-region on the globe. These birds are divided among the several orders as follows:

Passeres	...	...	306	Accipitres ...	...	36
Picariæ	...	..	41	Grallæ	...	...	77
Psittaci	...	...	60	Anseres	...	...	65
Columbæ	...	...	24	Struthiones	...	3
Gallinæ	...	...	15			

The Psittaci, we see, are very richly represented, while the Picariæ are comparatively few; and the Columbæ are scarce as compared with their abundance in the Austro-Malay sub-region.

Birds seem to be very evenly distributed over all Australia; comparatively few genera of importance being locally restricted. In the eastern districts alone, we find *Origma*, and *Orthonyx* (Sylviidæ); *Sericulus* and *Ptilorhynchus* (Paradiseidæ); *Leucosarcia* (Columbidæ); and *Talegalla* (Megapodiidæ). *Nectarinia, Pitta, Ptilorhis, Chlamydodera*, and *Sphecotheres*, range from the north down the east coasts. *Nanodes* (Psittacidæ), and *Lipoa* (Megapodiidæ), are southern forms, the first extending to Tasmania; which island appears to possess no peculiar genus of birds except *Eudyptes*, one of the penguins. West Australia has no wholly peculiar genus except *Geopsittacus*, a curious form of ground parroquet; the singular *Atrichia*, first found here, having been discovered in the east. In North Australia, *Emblema* (Ploceidæ) is the only peculiar Australian genus, but several Austro-Malayan and Papuan genera enter,— as, *Syma* and *Tanysiptera* (Alcedinidæ) ; *Machærihynchus* (Muscicapidæ); *Calornis* (Sturnidæ) ; *Manucodia, Ptilorhis*, and *Æluroedus* (Paradiseidæ) ; *Megapodius;* and *Casuarius.* The presence of a species of bustard (*Eupodotis*) in Australia. is very curious, its nearest allies being in the plains of India and Africa. Among waders the genus *Tribonyx*, a thick-legged bird somewhat resembling the *Notornis* of New Zealand, though not closely allied to it, is the most remarkable. The district where the typical Australian forms most abound is undoubtedly the eastern side of the island. The north and south are both somewhat poorer, the west much poorer, although it possesses a few very peculiar forms, especially among Mammalia. Tasmania is the poorest of all, a considerable number of genera being here wanting ; but, except the two peculiar carnivorous marsupials, it possesses nothing to mark it off zoologically from the adjacent parts of the main land. It is probable that its insular climate, more moist and less variable than that of Australia, may not be suitable to some of the absent forms ; while others may require more space and more varied conditions, than are offered by a comparatively small island.

The remaining classes of animals have been already discussed in our sketch of the region as a whole (p. 396).

Plate XII. Illustrating the Fauna of Australia.—In this plate we take New South Wales as our locality, and represent chiefly, the more remarkable Australian types of birds. The most conspicuous figure is the wonderful lyre-bird (*Menura superba*), the elegant plumage of whose tail is altogether unique in the whole class of birds. The unadorned bird is the female. In the centre is the emu (*Dromæus novæ-hollandiæ*), the representative in Australia, of the ostrich in Africa and America, but be-

longing to a different family, the Casurariidæ.   To the right are
a pair of crested pigeons (*Ocyphaps lophotes*), one of the many singular forms of the pigeon family to which the Australian region gives birth.   In every other part of the globe pigeons are
smooth-headed birds, but here they have developed three distinct forms of crest, as seen in this bird, the crowned pigeon
figured in Plate X., and the double-crested pigeon (*Lopholœmus
antarcticus*).   The large bird on the tree is one of the Australian
frog-mouthed goat-suckers (*Podargus strigoides*), which are
called in the colony "More-pork," from their peculiar cry.   They
do not capture their prey on the wing like true goat-suckers, but
hunt about the branches of trees at dusk, for large insects, and
also for unfledged birds.   A large kangaroo (*Macropus giganteus*)
is seen in the distance ; and passing through the air, a flying
opossum (*Petaurus sciureus*), a beautiful modification of a marsupial, so as to resemble in form and habits the flying squirrels
of the northern hemisphere.

III. The Pacific Islands, or Polynesian Sub-region.

Although the area of this sub-region is so vast, and the
number of islands it contains almost innumerable, there is a
considerable amount of uniformity in its forms of animal life.
From the Ladrone islands on the west, to the Marquesas on the
east, a distance of more than 5,000 miles, the same characteristic
genera of birds prevail ; and this is the only class of animals on
which we can depend, mammalia being quite absent, and reptiles
very scarce.   The Sandwich Islands, however, form an exception
to this uniformity ; and, as far as we yet know, they are so
peculiar that they ought, perhaps, to form a separate sub-region.
They are, however, geographically a part of Polynesia ; and a
more careful investigation of their natural history may show
more points of agreement with the other islands.   It is therefore
a matter of convenience, at present, to keep them in the Polynesian sub-region, which may be divided into Polynesia proper
and the Sandwich Islands.

Polynesia proper consists of a number of groups of islands of
some importance, and a host of smaller intermediate islets.

PLATE XII.

THE PLAINS OF NEW SOUTH WALES, WITH CHARACTERISTIC ANIMALS.

For the purpose of zoological comparison, we may class them in four main divisions. 1. The Ladrone and Caroline Islands; 2. New Caledonia and the New Hebrides; 3. The Fiji, Tonga, and Samoa Islands; 4. The Society, and Marquesas Islands. The typical Polynesian fauna is most developed in the third division; and it will be well to describe this first, and then show how the other islands diverge from it, and approximate other sub-regions.

Fiji, Tonga, and Samoa Islands.—The land-birds inhabiting these islands belong to 41 genera, of which 17 are characteristic of the Australian region, and 9 more peculiarly Polynesian. The characteristic Australian genera are the following: *Petroica* (Sylviidæ); *Lalage* (Campephagidæ); *Monarcha, Myiagra, Rhipidura* (Muscicapidæ); *Pachycephala* (Pachycephalidæ); *Rectes* (Laniidæ); *Myzomela, Ptilotis, Anthochœra* (Meliphagidæ); *Amadina, Eythrura,* (Ploceidæ); *Artamus* (Artamidæ); *Lorius* (Trichoglossidæ); *Ptilopus, Phlogœnas* (Columbidæ); *Megapodius* (Megapodiidæ).

The peculiar Polynesian genera are :—*Tatare, Lamprolia* (Sylviidæ) ; *Aplonis, Sturnodes* (Sturnidæ) ; *Todiramphus* (Alcedinidæ) ; *Pyrhulopsis, Cyanoramphus,* (Platycercidæ); *Coriphilus* (Trichoglossidæ) ; *Didunculus* (Didunculidæ).

The wide-spread genera are *Turdus, Zosterops, Hirundo, Halcyon, Collocalia, Eudynamis Cuculus, Ianthœnas, Carpophaga, Turtur, Haliœetus, Astur, Circus, Strix, Asio.* The aquatic birds are fifteen in number, all wide-spread species except one—a form of moor-hen (Gallinulidæ), which has been constituted a new genus *Pareudiastes.*

Society, and Marquesas Islands.—Here, the number of genera of land-birds has considerably diminished, amounting only to 16 in all. The characteristic Australian genera are 5 ;—*Monarcha, Anthochœra, Trichoglossus, Ptilopus,* and *Phlogœnas.* The Polynesian genera are 4 ;—*Tatare, Todiramphus, Cyanoramphus, Coriphilus,* and one recently described genus, *Serresius,* an extraordinary form of large fruit pigeon, here classed under *Carpophaga.* These remote groups have thus all the character of Oceanic islands, even as regards the rest of Polynesia, since they possess hardly anything, but what they might have received by immigration over a wide extent of ocean.

Ladrone, and Caroline Islands.—These extensive groups of small islands are very imperfectly known, yet a considerable number of birds have been obtained. They possess two peculiar Polynesian genera, *Tatare* and *Sturnodes;* one peculiar sub-genus, *Psammathia* (here included under *Acrocephalus*); and ten of the typical Australian genera found in Polynesia,— *Lalage, Monarcha, Myiagra, Rhipidura, Myzomela, Erythrura, Artamus, Phlogœnas, Ptilopus,* and *Megapodius,* as well as the Papuan genus *Rectes,* and the Malayan *Calornis;*—so that they can be certainly placed in the sub-region. Genera which do not occur in the other Polynesian islands are, *Acrocephalus,* (s.g. *Psammathia*) originally derived perhaps from the Philippines; and *Caprimulgus,* a peculiar species, allied to one from Japan.

New Caledonia, and the New Hebrides.—Although these islands seem best placed with Polynesia, yet they form a transition to Australia proper, and to the Papuan group. They possess 30 genera of land-birds, 18 of which are typical of the Australian region; but while 13 are also Polynesian, there are 5 which do not pass further east. These are *Acanthiza, Eopsaltria, Glici-phila, Philemon,* and *Ianthœnas.* The peculiar Polynesian genus, *Aplonis,* of which three species inhabit New Caledonia, link it to the other portions of the sub-region. The following are the genera at present known from New Caledonia :—*Turdus, Acan-thiza, Campephaga, Lalage, Myiagra, Rhipidura, Pachycephala, Eopsaltria, Corvus, Physocorax* (s.g. of *Corvus,* allied to the jack-daws), *Glicphila, Anthochœra, Philemon, Zosterops, Erythrura, Aplonis, Artamus, Cuculus, Halcyon, Collocalia, Cyanoramphus, Trichoglossus, Ptilopus, Carpophaga, Macropygia, Ianthœnas, Chalcophaps, Haliastur, Accipiter.* The curious *Rhinochetus jubatus,* forming the type of a distinct family of birds (Rhino-chetidæ), allied to the herons, is only known from New Cale-donia.

It thus appears, that not more than about 50 genera and 150 species of land-birds, are known from the vast number of islands that are scattered over the Central Pacific, and it is not probable that the number will be very largely increased. Some of the species, as the *Eudynamis taitensis* and *Tatare longirostris*, range over 40° of longitude, from the Fiji Islands to the Marquesas. In other genera, as *Cyanoramphus* and *Ptilopus*, each important island or group of islands, has its peculiar species. The connection of all these islands with each other, on the one hand, and their close relation to the Australian region, on the other, are equally apparent; but we have no sufficient materials for speculating with any success, on the long series of changes that have brought about their existing condition, as regards their peculiar forms of animal life.

Sandwich Islands.—This somewhat extensive group of large islands, is only known to contain 11 genera and 18 species of indigenous land-birds; and even of this small number, two birds of prey are wide ranging species, which may well have reached the islands during their present isolated condition. These latter are, *Strix delicatula*, an owl spread over Australia and the Pacific; and *Asio accipitrinus*, a species which has reached the Galapagos from S. America, and thence perhaps the Sandwich Islands. Of the remaining 8 genera, one is a crow (*Corvus hawaiensis*), and another a fishing eagle (*Pandion solitarius*), of peculiar species; leaving 7 genera, which are all (according to Mr. Sclater) peculiar. First we have *Chasiempis*, a genus of Muscicapidæ, containing two species (which may however belong to distinct genera); and as the entire family is unknown on the American continent these birds must almost certainly be allied to some of the numerous Muscicapine forms of the Australian region. Next we have the purely Australian family Meliphagidæ, represented by two genera,—*Moho*, an isolated form, and *Chætoptila*, a genus established by Mr. Sclater for a bird before classed in *Entomyza*, an Australian group. The four remaining genera are believed by Mr. Sclater to belong to one group, the Drepanididæ, altogether confined to the Sandwich Islands. Two of them, *Drepanis* and *Hemignathus*, with three species each, are undoubtedly allied; the other two, *Loxops* and *Psittirostra*, have usually been classed as finches. The former seem to approach the Dicæidæ; and all resemble this group in their coloration.

The aquatic birds and waders all belong to wide-spread genera, and only one or two are peculiar species.

The Sandwich Islands thus possess a larger proportion of peculiar genera and species of land-birds than any other group of islands, and they are even more strikingly characterised by what seems to be a peculiar family. The only other class of terrestrial animals at all adequately represented on these islands, are the land shells; and here too we find a peculiar family, subfamily, or genus (Achatinella or Achatinellidæ) consisting of a number of genera, or sub-genera,—according to the divergent views of modern conchologists,—and nearly 300 species. The Rev. J. T. Gulick, who has made a special study of these shells on the spot, considers that there are 10 genera, some of which are confined to single islands. The species are so restricted that their average range is not more than five or six square miles, while some are confined to a tract of only two square miles in extent, and very few range over an entire island. Some species are confined to the mountain ridges, others to the valleys; and each ridge or valley possesses its peculiar species. Considerably more than half the species occur in the island of Oahu, where there is a good deal of forest. Very few shells belonging to other groups occur, and they are all small and obscure; the Achatinellæ almost monopolising the entire archipelago.

Remarks on the probable past history of the Sandwich Islands. —The existence of these peculiar groups of birds and landshells in so remote a group of volcanic islands, clearly indicates that they are but the relics of a more extensive land; and the reefs and islets that stretch for more than 1,000 miles in a west-north-west direction, may be the remains of a country once sufficiently extensive to develope these and many other, now extinct, forms of life.[1]

Some light may perhaps be thrown on the past history of the

---

[1] A new genus of Beetles (*Apterocyclus*) of the family Lucanidæ, has recently been described from the Sandwich Islands, and it is said to be most nearly related to a group inhabiting Chili,—an indication either of the great antiquity of the fauna, or of the varied accidental migrations from which it has had its origin.

Sandwich Islands, by the peculiar plants which are found on their mountains. The peak of Teneriffe produces no Alpine plants of European type, and this has been considered to prove that it has been always isolated; whereas the occurrence of North Temperate forms on the mountains of Java, accords with other evidence of this island having once formed part of the Asiatic continent. Now on the higher summits of the Sandwich Islands, nearly 30 genera of Arctic and North Temperate flowering plants have been found. Many of these occur also in the South Temperate zone, in Australia or New Zealand; but there are others which seem plainly to point to a former connection with some North Temperate land, probably California, as a number of islets are scattered in the ocean between the two countries. The most interesting genera are the following :—*Silene*, which is wholly North Temperate, except that it occurs in S. Africa; *Vicia*, also North Temperate, and in South Temperate America; *Fragaria*, with a·similar distribution; *Aster*, widely spread in America, otherwise North Temperate only; *Vaccinium*, wholly confined to the northern hemisphere, in cold and temperate climates. None of these are found in Australia or New Zealand; and their presence in the Sandwich Islands seems clearly to indicate a former approximation to North Temperate America, although the absence of any American forms of vertebrata renders it certain that no actual land connection ever took place.

Recent soundings have shown, that the Sandwich Islands rise from a sea which is 3,000 fathoms or 18,000 feet deep; while there is a depth of at least 2,000 fathoms all across to California on one side, and to Japan on the other. Between the Fiji Islands, New Caledonia, the Solomon Islands, and Australia, the depth is about 1,300 fathoms, and between Sydney and New Zealand 2,600 fathoms; showing, in every case, a general accordance between the depth of sea and the approximation of the several faunas. In a few more years, when it is to be hoped we shall know the contour of the sea-bottom better than that of the continents, we shall be able to arrive at more definite and trustworthy conclusions as to the probable changes of land and sea by which the phenomena of animal distribution in the Pacific have been brought about.

Reptiles of the Polynesian Sub-region.—The researches of Mr. Darwin on Coral Islands, proved, that large areas in the Pacific Ocean have been recently subsiding; but the peculiar forms of life which they present, no less clearly indicate the former existence of some extensive lands. The total absence of Mammalia, however, shows either that these lands never formed part of the Australian or Papuan continents, or if they did, that they have been since subjected to such an amount of subsidence as to exterminate most of their higher terrestrial forms of life. It is a remarkable circumstance, that although Mammalia (except bats) are wanting, there are a considerable number of reptiles ranging over the whole sub-region. Lizards are the most numerous, five families and fourteen genera being represented, as follows:—

1. Cryptoblepharus (Gymnopthalmidæ) Fiji Islands.
2. Ablepharus   ...   „   All the islands.
3. Lygosoma   ... (Scincidæ)   ... Pelew Islands, New Caledonia.
4. Mabouya   ...   „   ... Samoa Islands.
5. Euprepes   ...   „   ... Pacific Islands.
6. *Dactyloperus* ... (Geckotidæ)   ... Sandwich Islands,
7. *Doryura*   ... (Geckotidæ)   ... Pacific Islands.
8. Gehyra   ...   „   ... Fiji Islands.
9. *Amydosaurus* ...   „   ... Tahiti.
10. Heteronota   ...   .,   ... Fiji Islands.
11. *Correlophus* ...   „   ... New Caledonia.
12. *Brachylophus* ... (Iguanidæ)   ... Fiji Islands.
13. Lophura   ... (Agamidæ)   ... Pelew Islands.
14. *Chloroscartes* ...   „   ... Fiji Islands.

The first five are wide-spread genera, represented mostly by peculiar species; but sometimes the species themselves have a wide range, as in the case of *Ablepharus pœcilopleurus*, which (according to Dr. Günther) is found in Timor, Australia, New Caledonia, Savage Island (one of the Samoa group), and the Sandwich Islands! *Gehyra* and *Heteronota* are Australian genera; while *Lophura* has reached the Pelew Islands from the Moluccas. The remainder (printed in italics), are peculiar genera; *Brachylophus* being especially interesting as an example of an otherwise peculiar American family, occurring so far across the Pacific.

Snakes are much less abundant, only four genera being represented, one of them marine. They are, *Anoplodipsas*, a peculiar genus of Amblycephalidæ from New Caledonia; *Enygrus*, a genus of Pythonidæ from the Fiji Islands; *Ogmodon*, a peculiar genus of Elapidæ, also from the Fiji Islands, but ranging to Papua and the Moluccas; and *Platurus*, a wide-spread genus of sea-snakes (Hydrophidæ). In the more remote Sandwich and Society Islands there appear to be no snakes. This accords with our conclusion that lizards have some special means of dispersal over the ocean which detracts from their value as indicating zoo-geographical affinities; which is further proved by the marvellous range of a single species (referred to above) from Australia to the Sandwich Islands.

A species of *Hyla* is said to inhabit the New Hebrides, and several species of *Platymantis* (tree-frogs) are found in the Fiji Islands; but otherwise the Amphibians appear to be unrepresented in the sub-region, though they will most likely be found in so large an island as New Caledonia.

From the foregoing sketch, it appears, that although the reptiles present some special features, they agree on the whole with the birds, in showing, that the islands of Polynesia all belong to the Australian region, and that in the Fiji Islands is to be found the fullest development of their peculiar fauna.

IV. New Zealand Sub-region.

The islands of New Zealand are more completely oceanic than any other extensive tract of land, being about 1,200 miles from Australia and nearly the same distance from New Caledonia and the Friendly Isles. There are, however, several islets scattered around, whose productions show that they belong to the same sub-region;—the principal being, Norfolk Island, Lord Howe's Island, and the Kermadec Isles, on the north; Chatham Island on the east; the Auckland and Macquarie Isles on the south;—and if these were once joined to

New Zealand, there would have been formed an island-continent not much inferior in extent to Australia itself.

New Zealand is wholly situated in the warmer portion of the Temperate zone, and enjoys an exceptionally mild and equable climate. It has abundant moisture, and thus comes within the limits of the South-Temperate forest zone; and this leads to its productions often resembling those of the tropical, but moist and wooded, islands of the Pacific, rather than those of the temperate, but arid and scantily wooded plains of Australia. The two islands of New Zealand are about the same extent (approximately) as the British Isles, but the difference in the general features of their natural history is very great. There are, in the former, no mammalia, less than half as many birds, very few reptiles and fresh-water fishes, and an excessive and most unintelligible poverty of insects; yet, considering the situation of the islands and their evidently long-continued isolation, the wonder rather is that their fauna is so varied and interesting as it is found to be. Our knowledge of this fauna, though no doubt far from complete, is sufficiently ample; and it will be well to give a pretty full account of it, in order to see what conclusions may be drawn as to its origin.

Mammalia.—The only mammals positively known as indigenous to New Zealand are two bats, both peculiar to it,—*Scotophilus tuberculatus* and *Mystacina tuberculata*. The former is allied to Australian forms; the latter is more interesting, as being a peculiar genus of the family Noctilionidæ, which does not exist in Australia; and in having decided resemblances to the Phyllostomidæ of South America, so that it may almost be considered to be a connecting link between the two families. A forest rat is said to have once abounded on the islands, and to have been used for food by the natives; but there is much doubt as to what it really was, and whether it was not an introduced species. The seals are wide-spread antarctic forms which have no geographical significance.

Birds.—About 145 species of birds are natives of New Zealand, of which 88 are waders or aquatics, leaving 57 land-birds belonging to 34 genera. Of this latter number, 16, or nearly half, are peculiar; and there are also 5 peculiar genera of waders and aquatic birds, making 21 in all. Of the remaining genera of land-birds, four are cosmopolite or of very wide range, while the remainder are characteristic of the Australian region. The following is a list of the Australian genera found in New Zealand: *Sphenœacus, Gerygone, Orthonyx* (Sylviidæ); *Graucalus* (Campephagidæ); *Rhipidura* (Muscicapidæ); *Anthochœra* (Meliphagidæ); *Zosterops* (Dicæidæ); *Cyanoramphus* (Platycercidæ); *Carpophaga* (Columbidæ); *Hieracidea* (Falconidæ); *Tribonyx* (Rallidæ). Besides these there are several genera of wide range, as follows:—*Anthus* (Motacillidæ); *Hirundo* (Hirundinidæ); *Chrysococcyx, Eudynamis* (Cuculidæ); *Halcyon* (Alcedinidæ); *Coturnix* (Tetraonidæ); *Circus* (Falconidæ); *Athene* (Strigidæ).

Most of the above genera are represented by peculiar New Zealand species, but in several cases the species are identical with those of Australia, as in the following: *Anthochœra carunculata, Zosterops lateralis, Hirundo nigricans,* and *Chrysococcyx lucidus;* also one—*Eudynamis taitensis*—which is Polynesian.

We now come to the genera peculiar to New Zealand, which are of especial interest:

LIST OF GENERA OF BIRDS PECULIAR TO NEW ZEALAND.

Family and Genus.		No. of Species.	Remarks.
SYLVIIDÆ.			
1. Myiomoira	... ...	3	Allied to Petroica, an Australian genus
2. Miro ...	... ...	2	„ „ „ „
TIMALIIDÆ (?)			
3. Turnagra	... ...	2	Of doubtful affinities.
SITTIDÆ.			
4. Xenicus	... ...	3	Of doubtful affinities.
5. Acanthisitta	...	1	Of doubtful affinities.
PARIDÆ.			
6. Certhiparus	... ...	2	Of doubtful affinities.
MELIPHAGIDÆ.			
7. Prosthemadera	...	1	Peculiar genera of honeysuckers, a
8. Pogonornis	... ...	1	family which is confined to the
9. Anthornis	... ...	3	Australian Region.

G G 2

Family and Genus.			No. of Species.	Remarks.
STURNIDÆ.				
10. Creadion	...	...	2	These three genera are probably
11. Heterolocha	...	...	1	allied, and perhaps form a dis-
12. Callæas	...	...	2	tinct family.
NESTORIDÆ.				
13. Nestor...	...	...	3	A peculiar family of Parrots.
STRINGOPIDÆ.				
14. Stringops	...	...	1	A peculiar family of Parrots.
STRIGIDÆ.				
15. (Sceloglaux)	...	...	1	s.g. of Athene.
RALLIDÆ.				
16. Ocydromus	...	...	6	Allied to *Eulabeornis*, an Australian genus.
17. Notornis	...	...	1	Allied to *Porphyrio*, a genus of wide range.
CHARADRIIDÆ.				
18. Thinornis	...	...	1	
19. Anarhynchus	...	...	1	
ANATIDÆ.				
20. Hymenolæmus	...	...	1	Allied to *Malacorhynchus*, an Australian genus.
APTERYGIDÆ.				
21. Apteryx	...	...	4	Forming a peculiar family.

We have thus a wonderful amount of speciality; yet the affinities of the fauna, whenever they can be traced, are with Australia or Polynesia. Nine genera of New Zealand birds are characteristically Australian, and the eight genera of wide range are Australian also. Of the peculiar genera, 7 or 8 are undoubtedly allied to Australian groups. There are also four Australian and one Polynesian *species*. Even the peculiar *family*, Nestoridæ, is allied to the Australian Trichoglossidæ. We have therefore every gradation of similarity to the Australian fauna, from identical species, through identical genera, and allied genera, to distinct but allied families; clearly indicating very long continued yet rare immigrations from Australia or Polynesia; immigrations which are continued down to our day. For resident ornithologists believe, that the *Zosterops lateralis* has found its way to New Zealand within the last few years, and that the two cuckoos now migrate annually, the one from Australia, the other from some part of Polynesia, distances of more than 1,000 miles ! These facts seem, however, to have been accepted on insufficient evidence and to be in themselves extremely improbable. It is observed that the cuckoos appear annually in certain districts and again disappear ; but their course does not seem to have been traced, still less have they ever been actually seen arriving or departing across the ocean. In a country which has still such wide tracts of unsettled land, it is very possible that the birds in question may only move from one part of the islands to another.

Islets of the New Zealand Sub-region.

We will here notice the smaller islands belonging to the sub-region, as it is chiefly their birds that possess any interest.

Norfolk Island.—The land-birds recorded from this island amount to 15 species, of which 8 are Australian, viz.: *Climacteris scandens, Symmorphus leucopygius, Zosterops tenuirostris* and *Z. albogularis, Halcyon sanctus, Platycercus pennanti, Carpophaga spadicea, Phaps picata* and *P. chalcoptera.* Of the peculiar species three belong to Australian genera ; *Petroica, Gerygone,* and *Rhipidura ;* one to a cosmopolitan genus, *Turdus.* So far the affinity seems to be all Australian, and there remain only three birds which ally this island to New Zealand,—*Nestor productus, Cyanoramphus rayneri,* and *Notornis alba.* The former inhabited the small Phillip Island (close to Norfolk Island) but is now extinct. Being a typical New Zealand genus, quite incapable of flying across the sea, its presence necessitates some former connexion between the two islands, and it is therefore perhaps of more weight than all the Australian genera and species, which are birds capable of long flights. The *Cyanoramphus* is allied to a New Zealand broad-tailed parroquet. The *Notornis alba* is extinct, but two specimens exist in museums, and it is even a stronger case than the *Nestor,* as showing a former approximation or union of this island with New Zealand. A beautiful figure of this bird is given in the *Ibis* for 1873.

Lord Howe's Island.—This small island, situated half-way between Australia and Norfolk Island, is interesting, as containing a peculiar species of the New Zealand genus *Ocydromus,* or wood-hen (*O. sylvestris*). There is also a peculiar thrush, *Turdus vinitinctus*. Its other birds are wholly of Australian types, and most of them probably Australian species. The following have been observed, and no doubt constitute nearly its whole indigenous bird fauna. *Acanthiza* sp., *Rhipidura* sp., *Pachycephala gutturalis*, *Zosterops strennuus* and *Z. tephropleurus*, *Strepera* sp., *Halcyon* sp., and *Chalcophaga chrysochlora*. The two species of *Zosterops* are peculiar. The *Ocydromus* is important enough to ally this island to New Zealand rather than to Australia; and if the white bird seen there is, as supposed, the *Notornis alba* which is extinct in Norfolk Island, the connection will be rendered still more clear.

Chatham Islands.—These small islands, 450 miles east of New Zealand, possess about 40 species of birds, of which 13 are landbirds. All but one belong to New Zealand genera, and all but five are New Zealand species. The following are the genera of the land-birds : *Sphenœacus*, *Gerygone*, *Myiomoira*, *Rhipidura*, *Zosterops*, *Anthus*, *Prosthemadera*, *Anthornis*, *Chrysococcyx*, *Cyanoramphus*, *Carpophaga*, *Circus*. The peculiar species are *Anthornis melanocephala*, *Myiomoira*, *diffenbachi* and *M. traversi*, *Rhipidura flabellifera*, and a peculiar rail incapable of flight, named by Captain Hutton *Cabalus modestus*. It is stated that the *Zosterops* differs from that of New Zealand, and is also a migrant; and it is therefore believed to come every year from Australia, passing over New Zealand, a distance of nearly 1,700 miles! Further investigation will perhaps discover some other explanation of the facts. It is also stated, that the pigeon and one of the small birds (? *Gerygone* or *Zosterops*) have arrived at the islands within the last eight years. The natives further declare, that both the *Stringops* and *Apteryx* once inhabited the islands, but were exterminated about the year 1835.

The Auckland Islands.—These are situated nearly 300 miles south of New Zealand, and possess six land-birds, of which three are peculiar,—*Anthus aucklandicus*, *Cyanoramphus aucklandicus*, and *C. malherbii*, the others being New Zealand species of *Myiomoira*, *Prosthemadera*, and *Anthornis*. It is remarkable that two peculiar parrots of the same genus should inhabit these

PLATE XIII.

small islands; but such localities seem favourable to the Platy-cercidæ, for another peculiar species is found in the remote Macquarie Islands, more than 400 miles farther south. A peculiar species and genus of ducks, *Nesonetta aucklandica,* is also found here, and as far as yet known, nowhere else. A species of the northern genus *Mergus* is also found on these islands, and has been recently obtained by Baron von Hügel.

Plate XIII. Illustrating the peculiar Ornithology of New Zealand.—Our artist has here depicted a group of the most remarkable and characteristic of the New Zealand birds. In the middle foreground is the Owl-parrot or Kakapoe (*Stringops habroptilus*), a nocturnal burrowing parrot, that feeds on fern-shoots, roots, berries, and occasionally lizards; that climbs but does not fly; and that has an owl-like mottled plumage and facial disc. The wings however are not rudimentary, but fully developed; and it seems to be only the muscles that have become useless for want of exercise. This would imply, that these birds have not long been inhabitants of New Zealand only, but were developed in other countries (perhaps Australia) where their wings were of use to them.

Beyond the Kakapoe are a pair of the large rails, *Notornis mantelli;* heavy birds with short wings quite useless for flight, and with massive feet and bill of a red colour. On the right is a pair of Kiwis (*Apteryx australis*), one of the queerest and most unbird-like of living birds. It has very small and rudimentary wings, entirely concealed by the hair-like plumage, and no tail. It is nocturnal, feeding chiefly on worms, which it extracts from soft earth by means of its long bill. The genus *Apteryx* forms a distinct family of birds, of which four species are now known, besides some which are extinct. They are allied to the Cassowary and to the gigantic extinct *Dinornis.* On the wing are a pair of Crook-billed Plovers (*Anarhynchus frontalis*), remarkable for being the only birds known which have the bill bent sideways. This was at first thought to be a malformation; but it is now proved to be a constant character of the species, as it exists even in the young chicks; yet the purpose served by such an anomalous structure is not yet discovered.

No country on the globe can offer such an extraordinary set of birds as are here depicted.

Reptiles.—These consist almost wholly of lizards, there being no land-snakes and only one frog. Twelve species of lizards are known, belonging to three genera, one of which is peculiar, as are all the species. *Hinulia*, with two species, and *Mocoa*, with four species (one of which extends to the Chatham Islands), belong to the Scincidæ; both are very wide-spread genera and occur in Australia. The peculiar genus *Naultinus*, with six species, belongs to the Geckotidæ, a family spread over the whole world.

The most extraordinary and interesting reptile of New Zealand is, however, the *Hatteria punctata*, a lizard-like animal living in holes, and found in small islands on the north-east coast, and more rarely on the main land. It is somewhat intermediate in structure between lizards and crocodiles, and also has bird-like characters in the form of its ribs. It constitutes, not only a distinct family, Rhyncocephalidæ, but a separate order of reptiles, RHYNCOCEPHALINA. It is quite isolated from all other members of the class; and is probably a slightly modified representative of an ancient and generalised form, which has been superseded in larger areas by the more specialized lizards and saurians.

The only representatives of the Ophidia are two sea-snakes of Australian and Polynesian species, and of no geographical interest.

Amphibia.—The solitary frog indigenous to New Zealand, belongs to a peculiar genus, *Liopelma*, and to the family Bomburatoridæ, otherwise confined to Europe and temperate South America.

Fresh-water Fishes.—There are, according to Captain Hutton, 15 species of fresh-water fish in New Zealand, belonging to 7 genera; six species, and one genus (*Retropinna*), being peculiar. *Retropinna richardsoni* belongs to the Salmonidæ, and is the only example of that family occurring in the Southern hemisphere, where it is confined to New Zealand and the Chatham Islands. The wide distribution of *Galaxias attenuatus*—from the

Chatham Islands to South America—has already been noticed ; while another species, *G. fasciatus*, is found in the Chatham and Auckland Isles as well as New Zealand. A second genus peculiar to New Zealand, *Neochanna*, allied to *Galaxias*, has recently been described. *Prototroctes oxyrhynchus* is allied to an Australian species, but belongs to a family (Haplochitonidæ) which is otherwise South American. An eel, *Anguilla latirostris*, is found in Europe, China, and the West Indies, as well as in New Zealand! while the genus *Agonostoma* ranges to Australia, Celebes, Mauritius, and Central America.

Insects.—The great poverty of this class is well shown by the fact, that only eleven species of butterflies are known to inhabit New Zealand. Of these, six are peculiar, and one, *Argyrophenga* (Satyridæ), is a peculiar genus allied to the Northern genus *Erebia*. The rest are either of wide range, as *Pyrameis cardui* and *Diadema bolina ;* or Australian, as *Hamdyaas zoilus* ; while one, *Danais erippus*, is American, but has also occurred in Australia, and is no doubt a recent introduction into both countries. Only one *Sphinx* is recorded, and no other species of the Sphingina except the British currant-moth, *Ægeria tipuliformis*, doubtless imported. Coleoptera are better represented, nearly 300 species having been described, all or nearly all being peculiar. These belong to about 150 genera, of which more than 50 are peculiar. No less than 14 peculiar genera belong to the Carabidæ, mostly consisting of one or two species, but *Demetrida* has 3, and *Metaglymma* 8 species. Other important genera are *Dicrochile, Homalosoma, Mecodema*, and *Scopodes*, all in common with Australia. *Mecodema* and *Metaglymma* are the largest genera. Even the Auckland Islands have two small genera of Carabidæ found nowhere else.

Cicindelidæ are represented in New Zealand by 6 species of *Cicindela*, and 1 of *Dystipsidera*, a genus peculiar to the Australian region.

The Lucanidæ are represented by two peculiar genera, *Dendroblax* and *Oxyomus ;* two Australian genera, *Lissotes* and *Ceratognathus ;* and by the almost cosmopolite *Dorcus*.

The Scarabeidæ consist of ten species only, belonging to four genera, two of which are peculiar (*Odontria* and *Stethaspis*) ; and two Australian (*Pericoptus* and *Calonota*).   There are no Cetoniidæ.

There is only one Buprestid, belonging to the Australian genus *Cisseis*. The Elateridæ, (about a dozen species,) belong mostly to Australian genera, but two, *Metablax* and *Ochosternus*, are peculiar.

There are 30 species of Curculionidæ, belonging to 22 genera. Of the genera, 12 are peculiar ; 1 is common to New Zealand and New Caledonia ; 5 belong to the Australian region, and the rest are widely distributed.

Longicorns are, next to Carabidæ, the most numerous family, there being, according to Mr. Bates (*Ann. Nat. Hist.*, 1874), about 35 genera, of which 26 are peculiar or highly characteristic, and 7 of the others Australian. The largest and most character- istic genera are *Æmona* and *Xyloteles*, both being peculiar to New Zealand ; few of the remainder having more than one or two species. *Demonax* extends to the Moluccas and S. E. Asia. A dozen of the genera have no near relations with those of any other country.

Phytophaga are remarkably scarce, only two species of *Colaspis* being recorded ; and there is only a single species of *Coccinella*.

The other orders of Insects appear to be equally deficient. Hymenoptera are very poorly represented, only a score of species being yet known ; but two of the genera are peculiar, as are all the species. The Neuroptera and Heteroptera are also very scarce, and several of the species are wide-spread forms of the Australian region. The few species of Homoptera are all peculiar. The Myriapoda afford some interesting facts. There are nine or ten species, all peculiar. One genus, *Lithobius*, ranges over the northern hemisphere as far south as Singapore, and probably through the Malay Archipelago, but is not found in Australia. *Henicops* occurs elsewhere only in Tasmania and Chili. *Cryptops*, only in the north temperate zone ; while two others, *Cermatia* and *Cormocephalus*, both occur in Australia.

Land-Shells.—Of these, 114 species are known, 97 being peculiar. Three species of *Helix* are also found in Australia, and five more in various tropical islands of the Pacific. *Nanina, Lymnæa,* and *Assiminea,* are found in Polynesia or Malaya, but not in Australia. *Amphibola* is an Australian genus, as is *Janella. Testacella* and *Limax* belong to the Palæarctic region.

From the Chatham Islands, 82 species of shells are known, all being New Zealand species, except nine, which are peculiar.

The Ancient Fauna of New Zealand.—One of the most remarkable features of the New Zealand fauna, is the existence, till quite recent times, of an extensive group of wingless birds, —called Moas by the natives—many of them of gigantic size, and which evidently occupied the place which, in other countries, is filled by the mammalia. The most recent account of these singular remains, is that by Dr. Haast, who, from a study of the extensive series of specimens in the Canterbury museum, believes, that they belong to two families, distinguished by important differences of structure, and constitute four genera,— *Dinornis* and *Miornis,* forming the family Dinornithidæ ; *Palapteryx* and *Euryapteryx,* forming the family Palapterygidæ. These were mostly larger birds than the living *Apteryx,* and some of them much larger even than the African ostrich, and were more allied to the Casuariidæ and Struthionidæ than to the Apterygidæ. No less than eleven species of these birds have been discovered; all are of recent geological date, and there are indications that some of them may have been in existence less than a century ago, and were really exterminated by man. Remains have been found (of apparently the same recent date) of species of *Apteryx, Stringops, Ocydromus,* and many other living forms, as well as of *Harpagornis,* a large bird of prey, and *Cnemiornis,* a gigantic goose. Bodies of the *Hatteria punctata* have also been found along with those of the Moa, showing that this remarkable reptile was once more abundant on the main islands than it is now.

The Origin of the New Zealand Fauna.—Having now given

an outline sketch of the main features of the New Zealand fauna and of its relations with other regions, we may consider what conclusions are fairly deducible from the facts. As the outlying Norfolk, Chatham, and Lord Howe's Islands, are all inhabited (or have recently been so) by birds of New Zealand type or even identical species, almost incapable of flight, we may infer that these islands show us the former minimum extent of the land-area in which the peculiar forms which characterise the sub-region were developed. If we include the Auckland and Macquarie Islands to the south, we shall have a territory of not much less extent than Australia, and separated from it by perhaps several hundred miles of ocean. Some such ancient land must have existed to allow of the development and specialization of so many peculiar forms of birds, and it probably remained with but slight modifications for a considerable geological period. During all this time it would interchange many of its forms of life with Australia, and there would arise that amount of identity of genera between the two countries which we find to exist. Its extension southwards, perhaps considerably beyond the Macquaries, would bring it within the range of floating ice during colder epochs, and within easy reach of the antarctic continent during the warm periods ; and thus would arise that interchange of genera and species with South America, which forms one of the characteristic features of the natural history of New Zealand.

Captain F. W. Hutton (to whose interesting paper on the Geographical relations of the New Zealand Fauna we are indebted for some of our facts) insists upon the necessity of former land-connections in various directions, and especially of an early southern continental period, when New Zealand, Australia, Southern Africa, and South America, were united. Thus he would account for the existence of Struthious birds in all these countries, and for the various other groups of birds, reptiles, fishes, or insects which have no obvious means of traversing the ocean,—and this union must have occurred before mammalia existed in any of these countries. But such a supposition is quite unnecessary, if we consider that all wingless land-birds and some water-birds (as the Gare-fowl

and Steamer Duck) are probably cases of abortion of useless organs, and that the common ancestors of the various
forms of Struthiones may have been capable of a moderate
degree of flight; or they may have originated in the northern
hemisphere, as already explained in Chap. XI. p. 287. The existence of two, if not three, distinct families of these birds in New
Zealand, proves that the original type was here isolated at a
very early date, and being wholly free from the competition of
mammalia, became more differentiated than elsewhere. The
Hatteria is probably coeval with these early forms, and is the
only relic of a whole order of reptiles, which once perhaps
ranged far over the globe.

Still less does any other form of animal inhabiting New Zealand, require a land connection with distant countries to account
for its presence. With the example before us of the Bermudas
and Azores, to which a great variety of birds fly annually over vast
distances, and even of the recent arrival of new birds in New
Zealand and Chatham Island, we may be sure that the ancestors
of every New Zealand bird could easily have reached its shores
during the countless ages which elapsed while the *Dinornis* and
Apteryx were developing. The wonderful range of some of the
existing species of lizards and fresh-water fish, as already given,
proves that they too possess means of dispersal which have
sufficed to spread them, within a comparatively recent period,
over countries separated by thousands of miles of ocean; and the
fact that a group like the snakes, so widely distributed and for
which the climate of New Zealand is so well adapted, does not
exist there, is an additional proof that land connection had nothing
to do with the introduction of the existing fauna. We have
already (p. 398), discussed in some detail the various modes in
which the dispersal of animals in the southern hemisphere has
been effected; and in accordance with the principles there established, we conclude, that the New Zealand fauna, living and
extinct, demonstrates the existence of an extensive tract of land
in the vicinity of Australia, Polynesia, and the Antarctic continent, without having been once actually connected with either
of these countries, since the period when mammalia had peopled

all the great continents. That event certainly dates back to Secondary, if not to Palæozoic, times, because so dominant a group must soon have spread over the whole continuous land-area of the globe. We have no reason for believing that birds were an earlier development; and certainly cannot, with any probability, place the origin of the Struthiones before that of Mammals.

Causes of the Poverty of Insect-life in New Zealand : its Influence on the Character of the Flora.—The extreme paucity of insects in New Zealand, to which we have already alluded, seems to call for some attempt at explanation. No other country in the world, in which the conditions are equally favourable for insect-life, and which has either been connected with, or is in proximity to, any of the large masses of land, presents a similar phenomenon. The only approach to it is in the Galapagos, and in some of the islands of the Pacific; and in each of these cases the absence of mammals leads us to infer, that no connection with a continent has ever taken place. Yet the fauna of New Zealand evidently dates back to a remote geological epoch, and it seems strange that an abundance of indigenous insects have not been developed, especially when we consider the vast antiquity that most of the orders and families, and many of the genera, of insects possess (see p. 156), and that they must always have reached the country in greater numbers and variety than any of the higher animals. The undoubted fact that such an indigenous insect-fauna has not arisen, would therefore lead us to conclude, that insects find the conditions requisite for their development only in the great continental masses of land, in strict adaptation to, and dependance on, a varied fauna and flora of ever-increasing richness and complexity. A small number of widely-separated forms, introduced into a country where the fauna and flora are alike scanty and unrelated to them, seem to have little tendency to vary and branch out into that vast network of insect-life which enriches all the great continents and their once connected islands.

It is a striking confirmation on a large scale, of Mr. Darwin's beautiful theory—that the gay colours of flowers have mostly, or perhaps, wholly been produced, in order to attract insects which aid in their fertilization—that in New Zealand, where insects are so strikingly deficient in variety, the flora should be almost as strikingly deficient in gaily-coloured blossoms. Of course there are some exceptions, but as a whole, green, inconspicuous, and imperfect flowers prevail, to an extent not to be equalled in any other part of the globe ; and affording a marvellous contrast to the general brilliancy of Australian flowers, combined with the abundance and variety of its insect-life. We must remember, too, that the few gay or conspicuous flowering-plants possessed by New Zealand, are almost all of Australian, South American, or European *genera ;* the peculiar New Zealand or Antarctic genera being almost wholly without conspicuous flowers. In the tropical Galapagos the same thing occurs. Mr. Darwin notices the wretched weedy appearance of the vegetation ; and states that it was some time before he discovered that most of the plants were in flower at the time of his visit ! And the insect-life was correspondingly deficient, consisting mainly of a few terrestrial beetles.

The poverty of insect-life in New Zealand must, therefore, be a very ancient feature of the country ; and it furnishes an additional argument against the theory of land-connection with, or even any near approach to, either Australia, South Africa, or South America. For in that case numbers of winged insects would certainly have entered, and the flowers would then, as in every other part of the world, have been rendered attractive to them by the development of coloured petals ; and this character once acquired would long maintain itself, even if the insects had, from some unknown cause, subsequently disappeared.

After the preceding paragraphs were written, it occurred to me, that if this reasoning were correct, New Zealand plants ought to be also deficient in scented flowers ; because it is a part of the same theory, that the odours of flowers have, like their colours, been developed to attract the insects required to aid in their fertilization. I therefore at once applied to my friend Dr. Hooker, as the highest authority on New Zealand botany ; simply asking whether there was any such observed deficiency. His reply was:—

" New Zealand plants are remarkably scentless, both in regard to
the rarity of scented flowers, of leaves with immersed glands
containing essential oils, and of glandular hairs." There are a
few exceptional cases, but these seem even more rare than might
be expected, so that the confirmation of theory is very complete.
The circumstance that aromatic leaves are also very scarce, suggests the idea that these, too, serve as an attraction to insects.
Aromatic plants abound most in arid countries, and on Alpine
heights; both localities where winged insects are comparatively
scarce, and where it may be necessary to attract them in every
possible way.   Dr. Hooker also informs, me that since his *Introduction to the New Zealand Flora* was written, many plants with
handsome flowers have been discovered, especially among the
Ranunculi, shrubby Veronicas, and herbaceous Compositæ.   The
two former, however, are genera of wide range, which may have
originated in New Zealand by the introduction of plants with
handsome flowers, which the few indigenous insects would be
attracted by, and thus prevent the loss of their gay corollas; so
that these discoveries will not much affect the general character
of the flora, and its very curious bearing on the past history of
the islands through the relations of flowers and insects.

   In judging of the relation here supposed to exist, it must be
remembered, that if the New Zealand insects have been introduced from the surrounding countries by chance immigrations at
distant intervals, then, as we go back into the past the insect
fauna will become poorer and poorer, and still more inadequate
than at present to lead to the development of attractive flowers
and odours.   This quite harmonizes with the fact of the ancient
indigenous flora being so remarkably scentless and inconspicuous, while a few of the more recently introduced genera of
plants have retained their floral attractions.

*Concluding Remarks on the Early History of the Australian
Region.*

   We have already discussed in some detail, the various relations
of the Australian sub-regions to the surrounding Regions, and the
geographical changes that appear to have taken place.   A very

few observations will therefore suffice, on the supposed early history of the Australian region as a whole.

It was probably far back in the Secondary period, that some portion of the Australian region was in actual connection with the northern continent, and became stocked with ancestral forms of Marsupials; but from that time till now there seems to have been no further land connection, and the Australian lands have thenceforward gone on developing the Marsupial and Monotremate types, into the various living and extinct races we now find there. During some portion of the Tertiary epoch Australia probably comprised much of its existing area, together with Papua and the Solomon Islands, and perhaps extended as far east as the Fiji Islands; while it might also have had a considerable extension to the south and west. Some light has recently been thrown on this subject by Professor McCoy's researches on the Palæontology of Victoria. He finds abundant marine fossils of Eocene and Miocene age, many of which are strikingly similar to those of Europe at the same period. Among these are Cetaceans of the genus *Squalodon;* European species of Plagiostomous fishes; mollusca and corals closely resembling those of Europe and North America of the same age,—such as numerous Volutes closely allied to those of the Eocene beds of the Isle of Wight, and the genus *Dentalium* in great abundance, almost or quite identical with European tertiary species. Along with these, are found some living species, but always such as now live farther north in tropical seas. The Cretaceous and Mesozoic marine fossils are equally close to those of Europe.

The whole of these remains demonstrate that, as in the northern so in the southern hemisphere, a much warmer climate prevailed in the Eocene and Miocene periods than at the present time. This is a most important result, and one which strongly supports Mr. Belt's view, before referred to, that the warmer climates in past geological epochs, and especially that of the Miocene as compared with our own, was caused by a diminution of the obliquity of the ecliptic, leading to a much greater uniformity of the seasons for a considerable distance from the equator, and greatly reducing the polar area within which the sun would ever disappear during an entire rotation of the earth. During such a period, tropical forms of marine animals would have been able to spread north and south, into what are now cool latitudes; and identical genera, and even species, might then have ranged along the southern shores of the old Palæarctic continent, from Britain to the Bay of Bengal, and southward along the Malayan coasts to Australia.

Numerous Miocene plant-beds have also been found in Victoria, containing abundance of Dicotyledonous leaves, which are said generally to resemble those of the Asiatic flora, and of the Miocene plant-beds of the Rhine. It is to be hoped these beds will be more closely examined for remains of insects, land-shells, and vertebrates, and that the plants will be carefully preserved and critically studied; for here probably lies hidden the key, that will solve much of the mystery that attaches to the past history of the Australian fauna.

TABLES OF DISTRIBUTION.

In drawing up these tables, showing the distribution of the various classes of animals in the Australian region, the following sources of information have been relied on, in addition to the general treatises, monographs, and catalogues used in compiling the 4th Part of this work.

Mammalia.—Gould, Mammals of Australia; Waterhouse on Marsupials; Dr. J. E. Gray's List of Mammalia of New Guinea; Müller, Temminck and Schlegel on Mammals of the Moluccas; papers by Dr. Gray; and personal observations by the Author.

Birds.—Gould's Birds of Australia; Buller's Birds of New Zealand; G. R. Gray's Lists of Birds of Moluccas, &c.; Hartlaub and Finsch on Birds of Pacific Islands; Sclater on Birds of Sandwich Islands; papers by Haast, Hutton, Meyer, Salvin. Schlegel, Sclater, Travers, Lord Walden and the Author.

Reptiles.—Krefft, Catalogue of Snakes; Gunther, List of Lizards in *Voyage of Erebus and Terror* (1875); and numerous papers.

TABLE I.

FAMILIES OF ANIMALS INHABITING THE AUSTRALIAN REGION

EXPLANATION.

Names in *italics* show families which are peculiar to the region.
Names inclosed thus (......) show families which only just enter the region, and not considered properly to belong to it.
Numbers correspond to the series of numbers to the families in Part IV.

Order and Family.	Austro-Malaya.	Australia.	Polynesia.	New Zealand.	Range beyond the Region.
		Sub-regions.			
MAMMALIA.					
PRIMATES.					
3. Cynopithecidæ	—				Oriental and Ethiopian
CHIROPTERA.					
9. Pteropidæ ...	—	—	—	—	Oriental and Ethiopian
11. Rhinolophidæ	—	—			The Eastern Hemisphere
12. Vespertilionidæ	—	—	—	—	Cosmopolite
13. Noctilionidæ...				—	All tropical regions
CARNIVORA.					
25. (Viverridæ) ...	—				Oriental
33. Otariidæ... ...		—		—	N. and S. temperate zones
35. Phocidæ... ...		—		—	N. and S. temperate zones
CETACEA.					
36 to 41.					Oceanic
SIRENIA.					
42. Manatidæ ...	—				Ethiopian, Oriental
UNGULATA.					
47. Suidæ	—				All other regions but Nearctic
50. (Cervidæ) ...	—				All other regions but Ethiopian
52. (Bovidæ) ...	—				All other regions but Neotropical
RODENTIA.					
55. Muridæ	—	—			All other regions
61. (Sciuridæ) ...	—				All other regions
MARSUPIALIA.					
77. *Dasyuridæ* ...	—	—			
78. *Myrmecobiidæ*		—			
79. *Peramelidæ* ...	—	—			
80. *Macropodidæ*...	—	—			

Order and Family.	Austro-Malaya.	Australia.	Polynesia.	New Zealand.	Range beyond the Region.
81. *Phalangistidæ*	—	—			
82. *Phascolomyidæ*		—			
MONOTREMATA					
83. *Ornithorhynchidæ*...		—			
84. *Echidnidæ* ...					
BIRDS.					
PASSERES.					
1. Turdidæ... ...	—	—	—		Cosmopolite
2. Sylviidæ... ...	—	—	—	—	Cosmopolite
3. Timaliidæ ...	—	—		—	Oriental family
5. Cinclidæ ...	—				
8. Certhiidæ ...	—	—			
9. Sittidæ	—	—		—	
10. Paridæ	—			—	
13. Pycnonotidæ...	—				Oriental family
14. Oriolidæ... ...	—	—			Oriental and Ethiopian
15. Campephagidæ	—	—	—	—	Oriental and Ethiopian
16. Dicruridæ ...	—	—			Oriental and Ethiopian
17. Muscicapidæ...	—	—	—	—	The Old World
18. *Pachycephalidæ*	—	—	—		Almost peculiar to region
19. Laniidæ	—	—			The Old World
20. Corvidæ	—	—			Cosmopolite
21. *Paradiseidæ* ...	—	—			
22. *Meliphagidæ*...	—	—	—	—	
23. Nectariniidæ	—	—			Oriental and Ethiopian
24. Dicæidæ ...	—	—	—	—	Oriental and Ethiopian
25. *Drepanididæ*...			—		
30. Hirundinidæ...	—	—	—	—	Cosmopolite
34. Ploceidæ ...	—	—	—		Oriental, Ethiopian
35. Sturnidæ ...	—	—	—	—	The Old World
36. Artamidæ ...	—	—	—		Oriental
37. Alaudidæ ...	—	—			The Old World and N. America
38. Motacillidæ ...	—	—		—	The Old World
47. Pittidæ	—	—			Oriental, Ethiopian
49. *Menuridæ* ...		—			Peculiar to Australia
50. *Atrichiidæ* ...		—			Peculiar to Australia
PICARIÆ.					
51. Picidæ	—				All other regions
58. Cuculidæ ...	—	—	—	—	Cosmopolite
62. Coraciidæ ...	—	—			Oriental and Ethiopian
63. Meropidæ ...	—	—			Oriental and Ethiopian
67. Alcedinidæ ...	—	—	—	—	Cosmopolite
68. Bucerotidæ ...	—	—			Oriental and Ethiopian
71. Podargidæ ...	—	—			Oriental
73. Caprimulgidæ	—	—	—		Cosmopolite
74. Cypselidæ ...	—	—	—		Cosmopolite

Order and Family.	Sub-regions.				Range beyond the Region.
	Austro-Malaya.	Austra-lia.	Polyne-sia.	New Zealand.	
PSITTACI.					
76. *Cacatuidæ* ...	—	—			Philippine Islands
77. *Platycercidæ*	—	—	—	—	
78. Palæornithidæ	—				Oriental
79. *Trichoglossidæ*	—	—	—		
82. *Nestoridæ* ...	—			—	
83. *Stringopidæ*...				—	
COLUMBÆ.					
84. Columbidæ ...	—	—	—	—	Cosmopolite
84a. *Didunculidæ*				—	
GALLINÆ.					
87. Tetraonidæ ...	—	—	—	—	Old World and N. America
88. (Phasianidæ)	—				Oriental
89. Turnicidæ ..	—	—	—	—	The Old World
90. *Megapodiidæ*	—	—	—		
ACCIPITRES.					
96. Falconidæ ...	—	—	—	—	Cosmopolite
97. Pandionidæ ..	—	—	—	—	Cosmopolite
98. Strigidæ ...	—	—	—	—	Cosmopolite
GRALLÆ.					
99. Rallidæ ...	—	—	—	—	Cosmopolite
100. Scolopacidæ...	—	—	—	—	Cosmopolite
103. Parridæ ...	—	—			Tropical
104. Glareolidæ ...	—	—			The Eastern Hemisphere
105. Charadriidæ	—	—	—	—	Cosmopolite
106. Otididæ ...		—			The Eastern Hemisphere
107. Gruidæ ...		—			The Eastern Hemisphere
112. *Rhinochetidæ*				—	
113. Ardeidæ ...	—	—	—	—	Cosmopolite
114. Plataleidæ ...	—	—			Almost cosmopolite
115. Ciconiidæ ...	—	—			Widely distributed
ANSERES.					
118. Anatidæ ...	—	—	—	—	Cosmopolite
119. Laridæ... ...	—	—	—	—	Cosmopolite
120. Procellariidæ	—	—	—	—	Cosmopolite
121. Pelecanidæ ...	—	—	—	—	Cosmopolite
122. Spheniscidæ		—		—	S. temperate regions
124. Podicipidæ ..	—	—	—	—	Cosmopolite
STRUTHIONES.					
127. *Casuariidæ*...	—	—			
128. *Apterygidæ*...				—	
129. *Dinornithidæ*				—	Extinct
130. *Palapteryqidæ*				—	Extinct

Order and Family.	Austro-Malaya.	Australia.	Polynesia.	New Zealand.	Range beyond the Region.
REPTILIA.					
OPHIDIA.					
1. Typhlopidæ ...	—	—			All regions but Nearctic
2. Tortricidæ ..	—				Oriental, S. America, California
3. Xenopeltidæ ...	—				Oriental
5. Calamariidæ ...	—	—			All warm countries
7. Colubridæ ...	—	—			Almost cosmopolite
8. Homalopsidæ	—	—			Oriental, and all other regions
11. Dendrophidæ	—	—			Oriental, Ethiopian, Neotropical
12. Dryiophidæ ...	—				Oriental, Ethiopian, Neotropical
13. Dipsadidæ ...	—	—			Oriental, Ethiopian, Neotropical
15. Lycodontidæ...	—				Ethiopian and Oriental
16. Amblycepha- lidæ) ...		—			Oriental, Neotropical
17. Pythonidæ ...	—	—	—		Tropical regions, California
19. Acrochordidæ	—				Oriental
20. Elapidæ	—	—	—		Tropical regions, Japan, S. Carolina
23. Hydrophidæ ...	—	—	—	—	Oriental, Madagascar, Panama
LACERTILIA.					
30. Varanidæ ...	—	—			Oriental, Africa
33. Lacertidæ ...		—			The Eastern Hemisphere
41. Gymnopthal- midæ ...	—	—	—		Neotropical, Ethiopian, Palæarctic
42. *Pygopodidæ* ...		—			
43. *Aprasiadæ* ...		—			
44. *Lialidæ*		—			
45. Scincidæ ...	—	—	—	—	Almost cosmopolite
48. Acontiadæ ...	—				Ethiopian, Oriental
49. Geckotidæ ...	—	—	—	—	Almost cosmopolite
50. Iguanidæ ...		—			N. and S. America
51. Agamidæ ...	—	—	—		The Eastern Hemisphere
RHYNCOCEPHALINA					
53. *Rhyncocephalidæ*				—	
CROCODILIA.					
54. Gavialidæ ...	—				Oriental
55. Crocodilidæ ...	—	—			Tropical regions
CHELONIA.					
57. Testudinidæ ...	—				All other regions
58. Chelydidæ ...	—	—			Ethiopian, Neotropical
60. Cheloniidæ ...	—	—	—	—	Marine
AMPHIBIA.					
ANOURA.					
7. Phryniscidæ ...		—			Ethiopian, Malayan, Neotropical
9. Bufonidæ ...	—				All other regions
10. *Xenorhinidæ*...	—				
11. Engystomidæ..	—				All regions but Palæarctic
12. Bombinatoridæ				—	Neotropical, Palæarctic

Order and Family.	Austro-Malaya.	Austra-lia.	Polyne-sia.	New Zealand.	Range beyond the Region.
14. Alytidæ	—	—			All regions but Oriental
15. Pelodryadæ ...	—	—			Neotropical
16. Hylidæ	—	—			All regions but Ethiopian
17. Polypedatidæ	—	—	—		All the regions
18. Ranidæ	—	—			Almost cosmopolite
19. Discoglossidæ	—	—			All regions but Nearctic
FISHES (FRESH-WATER).					
Acanthopterygii.					
11. Trachinidæ ...		—			Patagonia (? marine)
35. Labyrinthici ...	—				Oriental, S. Africa
37. Atherinidæ ...	—				Europe, America
38. Mugillidæ ...	—	—		—	Ethiopian, Neotropical
Anacanthini.					
53. *Gadopsidæ* ...		—			
Physostomi.					
59. Siluridæ... ...	—	—	—	—	All warm regions
61. Haplochitonidæ		—			Temperate S. America
65. Salmonidæ ...				—	Palæarctic, Nearctic
67. Galaxidæ ...		—		—	Temperate S. America
78. Ostegolossidæ		—			All tropical regions
85. (Symbranchidæ)	—				Oriental, Neotropical
Dipnoi.					
92. Sirenoidei ...		—			Ethiopian, Neotropical
INSECTS. LEPI-DOPTERA (PART).					
Durini (Butter-flies).					
1. Danaidæ ...	—	—	—	—	All warm regions, and to Canada
2. Satyridæ ...	—	—	—	—	Cosmopolite
3. Elymniidæ ...	—				Oriental, Ethiopian
4. Morphidæ ...	—		—		Oriental, Neotropical
6. Acræidæ... ...	—	—			All tropical regions
8. Nymphalidæ...	—	—	—	—	Cosmopolite
9. Libytheidæ ...	—				All the other regions
10. Nemeobeidæ..	—				All other regions but Nearctic
13. Lycænidæ ...	—	—	—	—	Cosmopolite
14. Pieridæ	—	—	—	—	Cosmopolite
15. Papilionidæ ...	—	—	—	—	Cosmopolite
16. Hesperidæ ...	—	—	—	—	Cosmopolite
Sphingidea.					
17. Zygænidæ ...	—	—	—	—	Cosmopolite
18. Castniidæ ...	—	—	—	—	Neotropical
19. Agaristidæ ...	—	—	—	—	Oriental, Ethiopian
20. Uraniidæ ...	—	—	—	—	All tropical regions
23. Sphingidæ ...	—	—	—	—	Cosmopolite

TABLE II.

GENERA OF TERRESTRIAL MAMMALIA AND BIRDS INHABITING THE
AUSTRALIAN REGION.

EXPLANATION.

Names in *italics* show genera peculiar to the region.
Names enclosed thus (......) show genera which just enter the region, but are not con-
sidered properly to belong to it.
Genera truly belonging to the region are numbered consecutively.

MAMMALIA.

Order, Family, and Genus.	No. of Species.	Range within the Region.	Range beyond the Region.
PRIMATES.			
CYNOPITHECIDÆ.			
(Macacus	1	Lombok to Timor)	Oriental genus
1. Cynopithecus ...	1	Celebes and Batchian	Philippines ?
LEMURIDÆ.			
(Tarsius	1	Celebes)	Indo-Malayan genus
CHIROPTERA.			
PTEROPIDÆ.			
2. Pteropus	15	The whole reg except New Zeal.	Tropics of E. Hemisp.
3. Xantharpyia ...	1	Moluccas and Timor	Oriental, S. Palæarctic
4. Cynopterus ...	1	Morty Island	Oriental
5. Macroglossus ...	1	Celebes, Moluccas, Timor	Indo-Malaya
6. Harpyia	1	Celebes and Moluccas	Philippines
7. *Hypoderma* ...	1	Celebes, Moluccas, and Timor	
8. *Notopteris* ...	1	Fiji Islands	
RHINOLOPHIDÆ.			
9. Rhinolophus ...	7	Moluccas, Timor, Australia	Warmer pts. of E. Hemis.
10. Hipposideros ...	5	Moluccas and Aru Islands	Oriental
11. Phyllorhina ...	2	Moluccas and Timor	Indo-Malaya
12. Asellia	1	Amboyna	Indo-Malaya
13. Megaderma ...	1	Ternate	Oriental, Ethiopian
VESPERTILIONIDÆ.			
14. Scotophilus ...	8	Moluccas, Timor, Australia	Oriental
15. Vespertilio ...	2	Australia	Cosmopolite
16. Miniopteris ...	3	Moluccas, Timor, and Australia	Indo-Malaya, S. Africa
17. Taphozous ...	2	Celebes, Moluccas, N. Australia	Orien.,Ethiop., Neotrop.
18. Plecotus	1	Timor	N. India, S. Palæarctic
19. Nyctophilus ...	5	Australia and Tasmania	India

Order, Family, and Genus.	No. of Species.	Range within the Region.	Range beyond the Region
NOCTILIONIDÆ.			
20. Molossus	1	Australia	Neotrop., Ethiop., S. P
21. *Mystacina* ...	1	New Zealand	
INSECTIVORA. SORICIDÆ.			
22. Sorex	2	Moluccas and Timor	The E.Hemis.& N.Am
CARNIVORA. VIVERRIDÆ.			
(Viverra	1	Celebes and Moluccas)	Oriental genus
(Paradoxurus ...	1	Timor, Ke Islands, ? introduced)	Oriental genus
OTARIIDÆ.			
23. Arctocephalus...	1	S. Australia, New Zealand	S. Temperate shores
24. Zalophus	1	Australia	North Pacific
PHOCIDÆ.			
25. Stenorhynchus	1	New Zealand	Antarctic shores
SIRENIA. MANATIDÆ.			
26. Halicore	1	N. Australia	Oriental Ethiopian
UNGULATA. SUIDÆ.			
27. Sus	4	Celebes to New Guinea	Palæarctic, Oriental
28. *Babirusa*	1	Celebes, Bouru	
CERVIDÆ.			
(Cervus	2	Celebes, Moluccas, Timor)	Oriental genus
BOVIDÆ.			
29. *Anoa*	1	Celebes	
RODENTIA. SCIURIDÆ.			
(Sciurus	5	Celebes)	All the other regions
MURIDÆ.			
30. Mus	13	Australia, Celebes	The Western Hemisphe
31. *Pseudomys* ..	1	Australia	
32. *Hapalotis*... ...	13	Australia	
33. *Hydromys* ...	5	Australia and Tasmania	
34. *Acanthomys* ...	1	N. Australia	
35. *Echiothrix* ...	1	Australia	
MARSUPIALIA. DASYURIDÆ.			
36. *Phascogale* ...	3	New Guinea and Australia	

Order, Family, and Genus.	No. of Species.	Range within the Region.	Range beyond the Region.
37. *Antechinomys* ...	1	S. Australia (interior)	
38. *Antechinus* ...	12	Aru Ids. Australia and Tasmania	
39. *Chætocercus* ...	1	S. Australia	
40. *Dactylopsila* ...	1	Aru Islands and N. Australia	
41. *Podabrus*	5	Australia and Tasmania	
42. *Myoictis*	1	Aru Islands	
43. *Sarcophilus* ...	1	Tasmania	
44. *Dasyurus*	4	Australia	
45. *Thylacinus* ...	1	Tasmania	
MYRMECOBIIDÆ.			
46. *Myrmecobius* ...	1	S. and W. Australia	
PERAMELIDÆ.			
47. *Perameles* ...	8	N. Guinea, Aru Ids., Australia, and Tasmania	
48. *Peragalea* ...	1	W. Australia	
49. *Chœropus*	1	S. E. and W. Australia	
MACROPODIDÆ.			
50. *Macropus*	4	Australia and Tasmania	
51. *Osphranter* ...	5	All Australia	
52. *Halmaturus* ...	18	Australia and Tasmania	
53. *Petrogale*	7	All Australia	
54. *Dendrolagus* ...	2	New Guinea	
55. *Dorcopsis* ...	2	Aru, Mysol, and N. Guinea	
56. *Onychogalea* ...	3	Central Australia	
57. *Lagorchestes* ...	5	N., W., and S. Australia	
58. *Bettongia*	6	W., S., and E. Australia and Tasmania	
59. *Hypsiprymnus*	4	W. and E. Australia & Tasmania	
PHALANIGISTIDÆ.			
60. *Phascolarctos* ...	1	E. Australia	
61. *Phalangista* ...	5	E., S., and W. Australia and Tasmania	
62. *Cuscus*	8	Celebes to N. Guinea, Timor & N. Australia	
63. *Petaurista* ...	1	E. Australia	
64. *Belideus*	5	S., E., & N. Austral., N. Guinea, and Moluccas	
65. *Acrobata*	1	S. and E. Australia	
66. *Dromicia*	5	W. & E. Australia & Tasmania	
67. *Tarsipes*	1	W. Australia	
PHASCOLOMYIDÆ.			
68. *Phascolomys* ...	1	S. E. Australia and Tasmania	
MONOTREMATA.			
ORNITHORHYNCHIDÆ.			
69. *Ornithorhynchus*	1	S. and E. Australia & Tasmania	

Order, Family, and Genus.	No. of Species.	Range within the Region.	Range beyond the Region
ECHIDNIDÆ.			
70. *Echidna*	2	S. & E. Australia, & Tasmania	
		BIRDS.	
PASSERES.			
TURDIDÆ.			
1. Turdus	6	Timor, Austral., New Caledonia, Norfolk Island, Lord Howe's and Samoan Islands	Cosmopolite
2. Oreocincla ...	1	S. E. Australia and Tasmania	Palæarctic, Oriental
3. Geocichla... ...	4	Celebes, Lombok, Timor, Austral.	Oriental
(Monticola ...	1	Gilolo, Celebes)	Palæarctic and Orienta
(Zoothera... ...	1	Lombok)	Oriental genus
SYLVIIDÆ.			
4. Cisticola	7	Celebes, Bouru, Timor, Australia	Palæarctic, Oriental
5. Sphenæacus ...	4	Australia, N. Zealand, Chatham Islands	Ethiopian
6. Megalurus ..	1	Timor	Oriental
7. *Poodytes*	2	Australia	
8. *Amytis*	3	Australia	
9. *Sphenura*... ...	4	Australia	
10. *Stipiturus* ...	1	Australia, Tasmania	
11. *Malurus*	16	Australia, Tasmania, & N.Guinea	
12. *Hylacola*	3	Australia	
13. *Calamanthus* ...	2	Australia and Tasmania	
14. Acrocephalus ..	7	Celebes, Moluccas, Australia, Caroline Islands	Palæarc., Orien., Ethio
15. *Tatare*	2	Samoan to Marquesas Islands	
16. Hypolais	1	Moluccas	Palæarc., Orien., Ethio
17. *Sericornis* ...	7	Australia and Tasmania	
18. *Acanthiza* ...	14	Austral., Tasmania, N.Caledonia	
19. Gerygone	24	The whole region, excl. Moluccas	Philippines
20. *Drymodes*... ...	2	Australia	
21. Oreicola	4	Lombok to Timor	Burmah ?
(Pratincola ...	1	Celebes to Timor)	Oriental, Palæarctic
22. *Epthianura* ...	3	Australia	
23. Petroica	18	Papua to Samoan Ids., Australia	
24. *Myiomoira* ...	3	N. Zealand	
25. *Lamprolia* ...	1	Fiji Islands	
26. Miro...	3	New Zealand	
27. *Cinclorhamphus*	2	Australia	
28. *Origma*	1	Australia	
29. *Orthonyx*	5	N. Guinea, Austral., New Zeald.	
TIMALIIDÆ.			
30. Pomatorhinus...	5	N. Guinea and Australia	Oriental
31. *Cinclosoma* ...	4	Australia and Tasmania	
32. *Turnagra* ...	3	New Zealand	
33. Psophodes... ...	2	S. E. and W. Australia	
34. Alcippe	3	New Guinea	Oriental
(Trichastoma ...	1	Celebes)	Oriental genus

Order, Family, and Genus.	No. of Species.	Range within the Region.	Range beyond the Region.
35. Drymocataphus	1	Timor	Oriental
36. *Struthidea* ...	1	N. and E. Australia	
CINCLIDÆ.			
37. Eupetes	2	New Guinea	Malayan
CERTHIIDÆ.			
38. *Climacteris* ...	8	Australia and N. Guinea	
SITTIDÆ.			
39. *Sittella*	5	Australia and N. Guinea	
40. *Acanthisitta* ...	1	New Zealand	
41. *Xenicus*	3	New Zealand	
PARIDÆ.			
42. *Certhiparus* ...	2	New Zealand	
43. *Sphenostoma* ...	2	E. and S. Australia	
PYCNONOTIDÆ.			
44. Criniger	5	Moluccas, and small islands E. of Celebes	Oriental
ORIOLIDÆ.			
45. *Sphecotheres* ...	3	Timor and Australia	
46. Oriolus	3	Celebes, Sulla Ids., Lombok and Flores	Oriental, Ethiopian
47. *Mimeta*	10	Moluccas, N. Guinea, Timor, & Australia	
CAMPEPHAGIDÆ.			
(Pericrocotus ...	1	Lombok)	Oriental genus
48. Graculus ...	20	Celebes to New Hebrides and N. Zealand	Oriental
49. *Artamides* ...	1	Celebes	
50. *Pteropodocys* ...	1	Australia	
51. Campephaga ...	12	Celebes to Timor & New Guinea	Oriental, Ethiopian
52. Lalage	15	Celebes to Australia & Samoan Ids.	Malayan
53. *Symmorphus* ...	1	E. Australia and Norfolk Id.	
DICRURIDÆ.			
54. Dicrurus	11	Celebes to N. Ireland & Austral.	Oriental, Ethiopian
55. *Chœtorhynchus*	1	New Guinea	
MUSCICAPIDÆ.			
56. *Peltops*	1	Papuan Islands	
57. *Monarcha* ...	30	The whole region (excl. Celebes and N. Zealand)	
58. *Leucophantes* ...	1	N. Guinea	
(Butalis	1	Moluccas and Celebes)	Palæarc., Orien., Ethiop.
59. *Micræca*	6	Timor, N. Guinea, Australia	
60. Cyornis	2	Celebes and Timor	Oriental
61. Siphia	1	Timor	Oriental
62. *Seisura*	5	Moluccas to N. Ireland, Austral.	

Order, Family, and Genus.	No. of Species.	Range within the Region.	Range beyond the Region
63. *Myiagra*... ...	15	Moluccas to Samoan Ids. Austral.	
(Hypothymis	2	Celebes)	Oriental
64. *Machærirhynchus*	4	Papuan Ids. and N. Australia	
65. Rhipidura ...	32	The region to Samoan Ids. and N. Zealand	Oriental
(Myialestes ...	1	Celebes)	Oriental genus
(Tchitrea ...	1	Flores)	Orien. & Ethiop. genu
66. *Todopsis*.. ...	5	Papuan Islands	
67. *Chasiempis* ...	2	Sandwich Islands	
PACHYCEPHALIDÆ.			
68. *Oreœca*	1	Temperate Australia	
69. *Falcunculus* ...	2	Temperate Australia	
70. *Pachycephala*	45	Moluccas to Tonga Ids. and Tasmania	
71. Hylocharis ...	2	Celebes and Timor	Oriental
72. *Eopsaltria* ...	10	Australia to New Hebrides	
LANIIDÆ.			
73. *Colluricincla*...	4	Australia and Tasmania	
74. *Rectes*	18	Papuan to Fiji Ids., N. Austral.	
(Lanius	1	Lombok)	Northern Hemisphere
CORVIDÆ.			
75. *Strepera*	4	Australia and Tasmania	
76. *Barita*	3	Australia and Tasmania	
77. *Cracticus* ...	9	Papuan Ids. to Tasmania	
78. *Grallina* ...	1	Australia	
79. *Streptocitta* ...	2	Celebes	
80. *Charitornis* ...	1	Sulla Islands (Celebes group)	
81. *Corvus*	8	The whole region, excl. N. Zeal.	Almost Cosmopolite
82. *Gymnocorvus*...	2	Papuan Islands	
83. *Corcorax* ...	1	Australia	
84. *Lycocorax* ..	3	Moluccas	
PARADISEIDÆ ...			
85. *Paradisea* ...	4	Papuan Islands	
86. *Manucodia* ...	3	Papuan Ids. and N. Australia	
87. *Astrapia* ...	1	New Guinea	
88. *Parotia*	1	New Guinea	
89. *Lophorina* ...	1	New Guinea	
90. *Diphyllodes* ...	3	Papuan Islands	
91. *Xanthomelus*...	1	New Guinea	
92. *Cicinnurus* ...	1	Papuan Islands	
93. *Paradigalla* ...	1	New Guinea	
94. *Semioptera* ...	1	Gilolo and Batchian	
95. *Epimachus* ...	1	New Guinea	
96. *Drepanornis* ...	1	New Guinea	
97. *Seleucides* ..	1	New Guinea	
98. *Ptilorhis* ...	4	New Guinea and N. Australia	
99. *Sericulus* .	1	E. Australia	
100. *Ptilorhynchus*	1	E. Australia	
101. *Chlamydodera*	4	N. and E. Australia	
102. *Æluredus* ...	3	Papuan Islands and E. Australia	
103. *Amblyornis* ...	1	New Guinea	

Order, Family, and Genus.	No. of Species	Range within the Region.	Range beyond t Region.
MELIPHAGIDÆ.			
104. *Myzomela* ...	20	The region ; excl. N. Zealand	
105. *Entomophila*...	4	Papuan Islands and Australia	
106. *Gliciphila* ...	10	Papuan Ids. Timor, Australia, N. Caledonia	
107. *Acanthorhynchus*	2	Australia and Tasmania	
108. *Meliphaga* ...	1	East and S. Australia	
109. *Ptilotis*	43	Lombok and Gilolo to Tasmania and Samoan Ids.	(Baly)
110. *Meliornis* ...	5	Australia and Tasmania	
111. *Prosthemadera*	1	New Zealand	
112. *Anthornis* ...	4	New Zealand and Chatham Ids.	
113. *Anthochœra* ...	10	New Guinea to Tasmania and Samoan Ids., N. Zealand	
114. *Pogonornis* ...	1	New Zealand	
115. *Philemon* ...	18	Lombok to N. Guinea, N. Caledonia, Australia	
116. *Entomiza* ..	2	Australia	
117. *Manorhina* ...	5	Australia and Tasmania	
118. *Melithreptus* ...	8	N. Guinea, Australia, Tasmania	
119. *Euthyrhynchus*	3	N. Guinea	
120. *Melirrhophetes*	2	N. Guinea	
121. *Melidectes* ...	1	N. Guinea	
122. *Melipotes* ..	1	N. Guinea	
123. *Moho*	3	Sandwich Islands	
124. *Chætoptila* ...	1	Sandwich Islands	
NECTARINIIDÆ.			
125. *Cosmetira* ...	1	Papuan Islands	
(Æthopyga)	1	N. Celebes)	Oriental genus
126. Chalcostetha ...	5	Celebes, Moluccas, Papuan Ids.	Malaya
127. Arachnecthra	5	Austro-Malaya and N. Australia	Oriental
(Nectarophila	1	Celebes)	Oriental genus
Anthreptes ...	1	Celebes and Sulla Islands	Malayan genus
128. Arachnothera	1	Papaun Islands, Lombok	Oriental
DICÆIDÆ.			
129. Zosterops ...	28	The region to Fiji Ids. & N. Zeal.	Oriental, Ethiopian
130. Dicæum	12	Celebes to Solomon Ids.& Austral.	Oriental
131. Pachyglossa ?	1	N. Celebes	Himalayas
132. Piprisoma ...	1	Timor	India, Ceylon
133. *Pardalotus* ...	1	Australia and Tasmania, Timor	
134. Prionochilus ...		Papuan Islands	Malaya
DREPANIDIDÆ.			
135. *Drepanis* ...	3	Sandwich Islands	
136. *Hemignathus*...	3	Sandwich Islands	
137. *Loxops*	1	Sandwich Islands	
138. *Psittirostra* ...	1	Sandwich Islands	
HIRUNDINIDÆ.			
139. Hirundo ..	7	The whole region	Cosmopolite
140. Atticora ...	1	Australia	Neotropical

Order, Family, and Genus.	No. of Species.	Range within the Region.	Range beyond the Region
PLOCEIDÆ.			
141. Estrilda... ...	4	Flores, Timor, Australia	Oriental, Ethiopian
142. *Emblema* ...	1	N. W. Australia	
143. Munia	6	Celebes to N. Guinea and N. Australia	Oriental
144. *Donacola* ...	3	Australia	
145. *Poephila* ...	6	Australia	
146. Amadina ...	9	Flores to Tasmania and Samoan Islands	Ethiopian
147. Erythrura ..	7	Moluccas to Caroline and Fiji Islands, Timor, N. Caledonia	Java, Sumatra
STURNIDÆ.			
148. Eulabes	4	Sumbawa, Flores, Papuan and Solomon Islands	Oriental
149. *Basilornis* ...	2	Celebes and Ceram	
(Acridotheres	1	Celebes)	Oriental genus
150. *Creadion* . .	2	N. Zealand	
151. *Heterolocha* ...	1	N. Zealand	
152. *Callœas*	2	N. Zealand	
153. *Aplonis*	8	N. Caledonia to Tonga Islands	
154. Calornis... ...	13	Celebes to Solomon Islands and N. Australia	Malaya
155. *Enodes*	1	Celebes	
156. *Scissirostrum*...	1	Celebes	
ARTAMIDÆ.			
157. Artamus... ...	15	Celebes to Fiji Ids. and Tasmania	Oriental
ALAUDIDÆ.			
158. Mirafra	2	Flores and Australia	Oriental, Ethiopian
MOTACILLIDÆ.			
159. Budytes... ...	11	Moluccas, Timor, Australia	Palc., Ethiopian, Orien
160. Corydalla ...	5	Lombok and Moluccas to N. Zealand	Palæarctic, Oriental
PITTIDÆ.			
161. Pitta	12	Celebes and Lombok to N. Guinea and Australia	Oriental
162. Hydrornis ...	1	Gilolo, Batchian	Himalayas to Java
163. *Melampitta* ...	1	N. Guinea	
MENURIDÆ.			
164. *Menura*	2	E. Australia	
ATRICHIIDÆ.			
165. *Atrichia*... ..	2	W. Australia and Queensland	
PICARIÆ.			
PICIDÆ.			
166. Yungipicus ...	2	Celebes, Lombok, and Flores	Oriental
(Mulleripicus...	1	Celebes)	Oriental genus

Order, Family, and Genus.	No. of Species.	Range within the Region.	Range beyond the Region.
CUCULIDÆ.			
167. *Rhamphococcyx*	1	Celebes	
168. Centropus ...	13	Austro-Malaya and Australia	Oriental, Ethiopian
169. Cuculus	5	Austro-Malaya and Australia	Palc., Orien., Ethiopian
170. *Caliechthrus* ..	1	Papuan Islands	
171. Cacomantis ...	10	Austro-Malaya and Australia	Oriental
172. Chrysococcyx	5	Austro-Malaya to Fiji Islands and N. Zealand	Oriental, Ethiopian
(Hierococcyx...	1	Celebes)	Oriental genus
173. Eudynamis ...	6	The whole region; excl. Sandwich Islands	Oriental
174. *Scythrops* ...	1	Celebes, Moluccas, and Australia	
CORACIIDÆ.			
(Coracias ..	1	Celebes)	Oriental and Ethiopian
175. Eurystomus ...	4	Austro-Malaya and Australia	Oriental and Ethiopian
MEROPIDÆ.			
176. *Meropogon* ...	1	Celebes	
177. Merops.	2	Austro-Malaya and Australia	Palc., Orien., Ethiopian
ALCEDINIDÆ.			
178. Alcedo	4	Celebes to New Ireland	Palc., Orien., Ethiopian
179. *Alcyone*	6	Batchian to Tasmania	Philippines
180. Pelargopsis ...	2	Celebes, Flores	Oriental
181. Ceyx ...	7	Celebes to New Guinea	Oriental
182. *Ceycopsis* ...	1	Celebes	
183. *Syma*	2	Papuan Islands and N. Australia	
184. Halcyon... ...	19	The whole region; excl. Sandwich Islands	Oriental, Ethiopian
185. *Todirhamphus*	3	Central Pacific and Sandwich Ids.	
186. Dacelo	6	Papuan Islands and Australia	
187. *Monachalcyon*	1	Celebes	
188. *Caridonax* ...	1	Lombok and Flores	
189. Tanysiptera ...	14	Batchian to N. Guinea and N. Australia	
190. *Cittura*	2	Celebes and Sanguir Islands	
191. *Melidora* ...	1	New Guinea	
BUCEROTIDÆ.			
192. Hydrocissa ? ...	1	Celebes	Oriental
193. Calao	1	Moluccas to Solomon Islands	Malayan
194. *Cranorrhinus* ?	1	Celebes	Malayan
PODARGIDÆ.			
195. *Podargus* ...	10	Papuan Islands to Tasmania	
196. Batrachostomus	2	Moluccas	Oriental
197. *Ægotheles* ...	5	Papuan Islands to Tasmania	
CAPRIMULGIDÆ.			
198. Caprimulgus ...	4	Lombok to Australia, N. Guinea to Pelew Islands	Palc., Ethiopian, Orien.

I I

Order, Family, and Genus.	No. of Species.	Range within the Region.	Range beyond the Region
199. *Eurostopodus*...	2	Aru Islands and Australia	Oriental genus
(*Lyncornis* ...	1	Celebes)	
CYPSELIDÆ.			
200. Dendrochelidon	2	Celebes to N. Guinea	Oriental
201. *Collocalia* ...	4	Celebes to Pacific Islands	Oriental
202. *Cypselus* ...	1	Australia	Palæ., Orien., Ethiopi
203. *Chætura*... ...	2	Celebes, Australia	Ethio., Orien., Ameri
PSITTACI.			
CACATUIDÆ.			
204. *Cacatua*... ...	17	Celebes and Lombok, to Solomon Islands and Tasmania	Philippines
205. *Calopsitta* ...	1	Australia	
206. *Calyptorhynchus*	8	Australia and Tasmania	
207. *Microglossus* ...	2	Papuan Islands and N. Austral.	
208. *Licmetis*... ...	3	Austr., Solmn. Ids., & N.Guin.?	
209. *Nasiterna* ...	3	Papuan and Solomon Islands	
PLATYCERCIDÆ.			
210. *Platycercus* ...	14	Austral., Tasmania, Norfolk Id.	
211. *Psephotus* ...	6	Australia	
212. *Polytelis*... ...	3	Australia	
213. *Nymphicus* ...	1	Australia and N. Caledonia	
214. *Aprosmictus* ...	6	Moluccas, Timor, Papuan Islands, Australia	
215. *Pyrrhulopsis* ...	3	Tonga to Fiji Islands	
216. *Cyanoramphus*	14	N. Zealand, Norfolk Island, N. Caledonia, Society Islands	
217. *Melopsittacus* ..	1	Australia	
218. *Euphema* ...	7	Australia	
219. *Pezoporus* ...	1	Australia and Tasmania	
220. *Geopsittacus* ...	1	W. Australia	
PALÆORNITHIDÆ.			
221. *Prioniturus* ...	2	Celebes	Philippines
222. *Geoffroyus* ...	5	Borneo to Timor & Solomon Ids.	
223. *Tanygnathus*...	4	Celebes to New Guinea	Philippines
224. *Eclectus*... ...	8	Moluccas and Papuan Islands	
225. *Cyclopsitta* ...	7	Papuan Ids. and N.E. Austral.	Philippines
226. *Loriculus* ...	7	Celebes to Mysol, Flores	Oriental
227. *Trichoglossus*	29	The whole region, excl. Sandwich Islands, and N. Zealand	
228. *Nanodes*... ...	1	Australia and Tasmania	
229. *Charmosyna* ...	1	New Guinea	
230. *Eos*...	9	Sanguir Ids. and Moluccas to Solomon Ids.	
231. *Lorius*	23	Bouru and Gilolo to Solomon Ids	
232. *Coriphilus* ...	4	Samoan to Marquesas Islands	
NESTORIDÆ			
233. *Nestor* ...	5	New Zealand and Norfolk Ids.	
234. *Dasyptilus* ..	1	New Guinea	

Order, Family, and Genus.	No. of Species.	Range within the Region.	Range beyond the Region.
STRINGOPIDÆ.			
235. *Stringops* ...	1	N. Zealand, Chatham Islands ?	
COLUMBÆ.			
COLUMBIDÆ.			
236. Treron	5	Celebes, Bouru, and Ceram, Flores and Timor	Oriental, Ethiopian
237. Ptilopus... ...	50	The whole region ; excl. N. Zealand	Indo-Malaya
238. Carpophaga ...	40	The whole region	Oriental
239. Ianthænas ...	6	Gilolo, Timor, Papuan Ids. to Samoan Islands	Japan, Philippines, Andaman Islands
240. *Leucomelœna*...	1	Australia	
241. *Lopholœmus* ...	1	Australia	
242. Geopelia... ...	5	Lombok to Tasmania	Malaya, China
243. Macropygia ...	6	Austro-Malaya, Australia	Indo-Malaya
244. *Turacœna* ...	3	Celebes, Timor, Solomon Ids.	
245. *Reinwardtœnas*	1	Celebes to New Guinea	
246. Turtur	2	Austro-Malaya	Palæarc., Orien., Ethiop.
247. *Ocyphaps* ...	1	Australia	
248. *Petrophassa* ...	1	N. W. Australia	
249. Chalcophaps ...	4	Austro-Malaya, Australia	Oriental
250. *Trugon*	1	N. Guinea	
251. *Henicophaps* ...	1	Papuan Islands	
252. Phaps	3	Australia and Tasmania	
253. Leucosarcia ...	1	Australia	
254. *Geophaps* ...	2	Australia	
255. *Lophophaps* ...	3	Australia	
256. Calœnas... ...	1	Austro-Malaya	Indo-Malaya
257. *Otidiphaps* ...	1	N. Guinea	
258. Phlogœnas ...	7	Celebes, N. Guinea to Madagascar	Philippine Islands
259. *Goura*	3	Papuan Islands	
DIDUNCULIDÆ.			
260. *Didunculus* ...	1	Samoan Islands	
GALLINÆ.			
TETRAONIDÆ.			
261. *Coturnix* ...	9	Celebes, Timor, Australia, N. Zealand	Palæarc., Orien., Ethiop.
PHASIANIDÆ.			
(Gallus	2	Celebes to Timor)	Oriental genus
TURNICIDÆ.			
262. Turnix	9	Celebes & Moluccas to Tasmania	Palæarc., Orien., Ethiop.
MEGAPODIIDÆ.			
263. *Talegallus* ...	3	Papuan Islands and Australia	
264. *Megacephalon*	1	Celebes	
265. *Lipoa*	1	S. Australia	
266. *Megapodius* ...	12	Celebes to Austral. & Samoan Ids.	Philippines, Nicobar Ids.

Order, Family, and Genus.	No. of Species.	Range within the Region.	Range beyond the Region
ACCIPITRES.			
FALCONIDÆ.			
267. Circus	2	Celebes, S. and E. Austral	Almost Cosmopolite
268. Astur	20	The region, to Fiji Islands	Almost Cosmopolite
269. Accipiter ...	6	The whole region, to Fiji Islands	Almost Cosmopolite
270. *Urospiza* ..	1	Australia	
271. *Uroaëtus* ..	1	Australia and Tasmania	
272. Nisaëtus... ...	1	Australia	S. Palæarc., Ethiopia, Oriental
273. Neopus	1	Celebes and Ternate	Oriental
274. Spizaëtus ...	2	Celebes and N. Guinea	Neotrop.,Ethiop., Orie
275. Circaëtus ...	1	Timor and Flores	Palæarc., Ethiop., Ori
(Spilornis ...	2	Celebes and Sulla Islands)	Oriental genus
276. Butastur ...	1	Celebes to New Guinea	Oriental, N. E. Africa
277. Haliæetus ...	1	The whole region	Cosmop., excl. Neotro region
278. Haliastur ...	2	Australia and N. Caledonia	Oriental
279. Milvus	1	Celebes to Australia	Palæarc., Orien., Ethio
280. *Lophoictinia* ...	1	Australia	
281. *Gypoictinia* ...	1	Australia	
282. Elanus	3	Celebes and Australia	Oriental, Ethiopian
283. *Henicopernis* ...	1	Papuan Islands	
(Pernis	1	Celebes)	Palæarctic, Oriental, a Ethiopian
284. Baza	4	Moluccas and Australia	Oriental
285. *Harpa*	1	N. Zealand and Auckland Ids.	
286. Falco	6	Austro-Malaya and Australia	Almost Cosmopolite
287. *Hieracidea* ...	2	Australia and Tasmania	
288. Cerchneis ...	2	Austro-Malaya and Australia	Almost Cosmopolite
PANDIONIDÆ.			
289. Pandion... ...	1	The whole region	Cosmopolite
290. Polioaëtus ...	1	Celebes and Sandwich Islands	Oriental
STRIGIDÆ.			
291. Athene	21	The whole reg., excl. Pacific Ids.	Palæarc., Orien., Ethio
292. Scops	6	Celebes, Moluccas, N. Zealand	Almost Cosmopolite
(Asio	1	Sandwich Islands)	Almost Cosmopolite, ex Australian region
293. Strix	7	The whole region	Cosmopolite

Peculiar or very Characteristic Genera of Wading and Swimming Birds.

GRALLÆ.			
RALLIDÆ.			
Ocydromus ...	5	New Zealand	
Cabalus	1	Chatham Islands	
Notornis... ...	2	New Zealand, Norfolk and Lord Howe's Islands	
Tribonyx ...	4	Australia and N. Zealand	
Habroptila ...	1	Moluccas	

Order, Family, and Genus.	No. of Species.	Range within the Region.	Range beyond the Region.
Rallina	6	Austro-Malaya	Oriental
Pareudiastes ...	1	Samoan Islands	
SCOLOPACIDÆ.			
Cladorhynchus	1	Australia	
CHARADRIIDÆ.			
Esacus	1	Austro-Malaya, Australia	Oriental
Erythrogonys...	1	Australia	
Thinornis ...	2	New Zealand	
Anarhynchus	1	New Zealand	
Pedionomus ...	1	Australia	
RHINOCHETIDÆ.			
Rhinochetus ...	1	New Caledonia	
ANATIDÆ.			
Nesonetta ...	1	Auckland Islands	
Malacorhynchus	1	Australia	
Hymenolæmus	1	New Zealand	
Biziura	1	Australia	
Anseranas ...	1	Australia	
Cereopsis ...	1	Australia and Tasmania	
PROCELLARIIDÆ.			
Prion	6	New Zealand	Antarctic Seas
SPHENISCIDÆ.			
Eudyptes ...	4	Australia and N. Zealand	Antarctic shores
STRUTHIONES.			
CASUARIIDÆ.			
294. *Dromæus* ...	2	Australia	
295. *Casuaruis* ...	9	Ceram to New Britain, N. Austrl.	
APTERYGIDÆ.			
296. *Apteryx*	4	New Zealand	
DINORNITHIDÆ.		(Extinct)	
297. *Dinornis*... ...	5	N. Zealand	
298. *Mionornis* ...	2	N. Zealand	
PALAPTERYGIDÆ.		(Extinct)	
299. *Palapteryx* ...	2	N. Zealand	
300. *Euryapteryx* ...	2	N. Zealand	

INDEX TO VOL. I.

INDEX TO VOL. I.

NOTE.—In this Index the names in Italics all refer to fossil genera or families mentioned in Part II. The systematic names of genera and families occurring in almost every page of Part III. are not given, as they would unnecessarily swell the Index; but they can be readily referred to by the Class or Order, or by the Geographical Division (Region or Sub-region) under which they occur. They will, however, all be found in the General Index, with a reference to the page (in Vol. II., Part IV.) where a systematic account of their distribution is given.

K K

END OF VOL. I.

LONDON: R. CLAY, SONS, AND TAYLOR, PRINTERS, BREAD STREET HILL.

Printed in the United States
By Bookmasters